EMC for Systems and Installations

EMC for Systems and Installations

Tim Williams
&
Keith Armstrong

Newnes

OXFORD AUCKLAND BOSTON JOHANNESBURG MELBOURNE NEW DELHI

Newnes
An imprint of Butterworth-Heinemann
Linacre House, Jordan Hill, Oxford OX2 8DP
225 Wildwood Avenue, Woburn, MA 01801-2041
A division of Reed Educational and Professional Publishing Ltd

 A member of the Reed Elsevier plc group

First published 2000

British Library Cataloguing in Publication Data
A catalogue record for this book is available from the British Library

Library of Congress Cataloguing in Publication Data
A catalogue record for this book is available from the Library of Congress

ISBN 978 0 7506 4167 8

FOR EVERY TITLE THAT WE PUBLISH, BUTTERWORTH-HEINEMANN
WILL PAY FOR BTCV TO PLANT AND CARE FOR A TREE.

Transferred to digital printing 2007

Contents

Chapter 4

Interference sources, victims and coupling   67

Chapter 5

Earthing and bonding   101

Chapter 6

Cabinets, cubicles and chambers   **133**

Chapter 7

Cabling   **155**

Chapter 8

Filtering   **193**

Chapter 9

Lightning and surge protection   **211**

Chapter 10

In situ testing   **235**

Appendix A

Appendix B

Appendix C

Appendix D

Preface

Electromagnetic compatibility (EMC) is defined as "the ability of a device, equipment or system to function satisfactorily in its electromagnetic environment without introducing intolerable electromagnetic disturbance to anything in that environment".

Historically, EMC has been concerned principally with ensuring the proper operation of collections of electrical and electronic apparatus. Since interference is a function of separation distance, equipment used in close proximity to other equipment had to be "compatible" with its neighbours. Putting together a "system" out of several often disparate items of "apparatus" meant that these items were naturally close to each other, and their EMC was necessary in order for the system to work successfully. Hence the discipline of EMC grew up in those industry sectors where systems integration was the norm – notably in the military, where the majority of electrical and electronics equipment is used on "platforms", i.e. ships, aircraft and land vehicles; in their civilian equivalents, aerospace, rail, automotive and marine transport; and in the process control industry. The consumer, IT and professional equipment sectors largely escaped this discipline, because their individual products could assume a large enough separation distance that EMC could be regarded as a luxury rather than a necessity.

Exceptions to this state of affairs developed in industry sectors whose products created an identifiable threat to third-party radio reception: so for instance products using spark ignition, electric motors, power switching devices, fluorescent luminaires and of course digital devices began to fall under suspicion, and eventually were regulated in their radio frequency emissions, in a more-or-less piecemeal fashion. But system builders were unaware of this, since it affected only the equipment itself, not the systems that were built from it – unless those systems included radio receivers. Commercial systems that faced issues of safety integrity had often to meet requirements for immunity from various phenomena, such as radio frequency fields, electrostatic discharge, and various types of conducted transients. But these were contractual requirements, agreed between the equipment suppliers, and the system designers and operators; they were instigated as a result of operational experience, not because of legislation.

This somewhat *ad hoc* framework for EMC has changed enormously as a result of the EMC Directive. The consequence of the piecemeal process of regulation of electrical and electronic products, was that different countries within Europe took different approaches to that regulation. These discrepancies led to a particular instance of a "technical barrier to trade". The European Commission, having identified this barrier in the mid-80s, put forward a Directive whose intent was to harmonise EMC regulations throughout the EU and therefore encourage the free movement of apparatus, while at the same time protecting the electromagnetic environment.

The EMC Directive was clearly intended from the start to apply to apparatus, that is, individual product items. But its scope was drawn more widely than that: in Article 1.1., apparatus was defined as "all electrical and electronic appliances together with

equipment and installations containing electrical and/or electronic components",
therefore catching not just apparatus as it is commonly understood, but also whole
systems and installations. In the early days of interpretation there was some concern
that large installations such as nuclear power plants or telephone exchanges would have
to carry the CE Mark, but such excess zeal was quickly dampened by the definition of
the "excluded installation" in the UK regulations of 1992, which aligned with the first
round of Commission Guidance published later. This meant that installations
comprised of items put together at a given place to fulfil a specific objective but not
supplied as a single functional unit, were excluded from the scope of the regulations.

Nevertheless, many classes of system do still remain within the Directive's scope,
and this now presents system designers and installers with a set of legal requirements,
and consequent technical requirements, with which they are wholly unfamiliar. This
arises because the system is now regarded "as a whole" and its interaction with its
external environment is regulated. As explained earlier, historically this was hardly
considered, whereas the "internal" EMC was an operational necessity.

A significant issue in this context is the insistence by the European Commission that
products (including those intended for use in systems) should come with "instructions
for use" which have the purpose of ensuring that no EMC problem is encountered when
the products are put into service. With a few rare exceptions, this concept is totally new
to system builders, and to their equipment suppliers as well. Installation instructions
have traditionally only been concerned with functional requirements and knowledge of
EMC issues is not only rare among installation engineers but also among the
applications engineers of equipment suppliers, who are normally responsible for
advising their customers on product use. Although the Commission have made the
superficially reasonable assumption that EMC instructions are capable of being
supplied and implemented, the reality is rather different from this, at least at the time of
writing.

Fortunately, the techniques needed to deal with internal EMC also to some extent
affect the external interactions, so that system builders are able in part to kill two birds
with one stone. But the overlap is by no means exact, and there are a number of areas
where external EMC puts demands on design, procurement and installation of
apparatus which have no real precedent. And even in the realm of internal, operational
EMC, for many designers and installers this has traditionally been dealt with by totally
ad hoc methods: an operational problem, traced to internal interference, would be
solved on the spot by application of a range of "fixes" such as one-off filters or shielded
cable. In by no means all cases have lessons learned from this one problem been applied
in subsequent designs, and equally rare is the abstraction of common principles of
interference control which form the basis of design techniques to be applied across the
board.

This book has been written to address the resulting problem, of designers and
installers being suddenly faced with responsibilities for which they have neither an in-
house solution nor the expertise to handle.

It is presented in two parts: the first is relatively non-technical, and looks at the need
for EMC in the context of systems, followed with a discussion of the somewhat
complex way in which the EMC Directive applies to systems and installations. A
further chapter then covers the management aspects of systems EMC, including control
and test plans, and the important aspect of procurement specifications.

The second part covers the technical aspects of systems EMC, looking at the various
established methods which can be applied to ensure compatibility, and putting these in

the context of the new responsibilities facing system builders. These techniques include earthing and bonding, the use of shielded and unshielded enclosures, choice and routing of cables, filtering and transient suppression. The assumption is made that the system builder is faced with a number of electrical and electronic modules which have to be interconnected in a way that is compatible both within the system and with the external environment, and does not have any say or indeed interest in the way that the modules themselves are designed and built. (That is the subject of a different book [13].) The question of lightning protection sits uneasily with the context of the EMC Directive, but it is of prime concern to the operators of systems and installations, so we have included a short chapter covering it. There is an initial chapter on the fundamental issues of interference sources and coupling paths between source and victim, and a final chapter looks at the question of how you might prove the compatibility of a system through testing. Most larger systems can only be tested in situ, once they have been installed, and such testing has its own problems and solutions, which are inadequately dealt with in most of the existing test standards, since these were written around laboratory testing of apparatus.

This is not an academic book. Where theory is necessary, it is presented with as few trappings as possible. Its intended readership is those engineers who are faced with the practical problem of implementing EMC requirements in their day-to-day work, who need a guide to the techniques that they can use straight away, and who do not have the time for dealing with the complex mathematics that undoubtedly underlies much rigorous electromagnetic theory. At the same time, though, a list of prescriptive actions by itself is less than helpful, without the understanding needed to adapt each technique to a specific set of circumstances. As Dr Jasper Goedbloed has elegantly put it, "insight is considered more valuable than the recipe for tackling a symptom". So we hope that the treatment herein, while acknowledging the value of recipes and indeed providing a few of them in suitably pre-digested form at the back, will be seen as a foundation on which to create your own variations as each case dictates.

Chapter 1

Introduction: the EMC needs of systems

Any sufficiently advanced technology is indistinguishable from magic.

    – Arthur C. Clarke

Control of EMC is an increasingly necessary discipline in these days of encroaching technology. Even where there are no laws or regulations on EMC, control of EMC is still needed for the financial benefit of both suppliers and users. This book describes how EMC applies to systems and installations, and also describes proven best EMC practices in design, assembly, and installation. The correct application of this knowledge and these techniques will generally reduce project timescales, help ensure reliable operation, and reduce commercial risk, and meet regulations such as the EMC Directive where they apply. EMC techniques need to be employed from the very beginning of a project (even at the project specification and tendering stages) to achieve the greatest benefits and lowest risks.

The best EMC practices described here are taken from several IEC and European standards and moderated by many years of experience in their use, solving real-life interference problems, and EMC testing. EMC concerns all electromagnetic phenomena including power-frequency interference, so these best EMC practices are excellent at preventing "ground loop" problems and achieving good analogue and digital signal quality, improving signal integrity and signal-to-noise ratios and actually improving the functioning of all electrical/electronic/programmable apparatus and their systems and installations. Some modern best EMC practices contradict "traditional" techniques. Professional engineers have a duty to be up-to-date and to use modern best practices, and this often means revising or replacing traditional practices as technology advances.

Whether or not the EMC Directive applies, its "Protection Requirements" are nothing more than good engineering practice that all suppliers of systems and installations can follow. On emissions they should follow the relevant harmonised standard to protect radio reception in the world outside their control, unless there is very sensitive apparatus nearby which makes even lower levels of emissions necessary. On immunity they need to make sure that their products will function with adequate performance and reliability in their intended environment. This may not require following a harmonised immunity standard exactly: sometimes they can reduce a test level or skip some immunity phenomena altogether; sometimes they need to increase a test level or cover extra phenomena. However, for both emissions and immunity they still need to do the same EMC work in design and installation as if they were applying the CE mark, for reasons of good engineering and medium- to long-term financial success through the avoidance of commercial risk. Since many customers and users (and not just in the EU) expect, rightly or wrongly, to see a CE mark anyway it is then a simple matter to complete the necessary documentation and apply the mark.

1.1   The definition of electromagnetic compatibility

Before embarking on a discussion of systems EMC it would be as well to define the terms. Electromagnetic compatibility is defined in IEC publication 50, the International Electrotechnical Vocabulary (IEV) [79], as

> "The ability of a device, equipment or system to function satisfactorily in its electromagnetic environment without introducing intolerable electromagnetic disturbance to anything in that environment."

There are several things to note about this definition. The first is that it is about the environment, which is itself defined in the IEV as "the totality of electromagnetic phenomena existing at a given location". Compatibility is defined in terms of the environment in which the apparatus is used. Implicit in this definition is the existence of a boundary between the apparatus and its environment. This is relatively straightforward to conceive for individual items of apparatus. But when we come to consider systems and installations, these both create their own environment within their boundary, and they also exist within a larger environment. So it is necessary at the beginning to make the distinction between EMC *internal* to a system, and EMC *external* to it. This gives us two modified definitions:

> Intra-system (internal) EMC: "The ability of components of a system to function satisfactorily within the electromagnetic environment created by that system, without introducing intolerable electromagnetic disturbance to other components of that system."

> Inter-system (external) EMC: "The ability of a whole system to function satisfactorily within the electromagnetic environment created by external sources, without introducing intolerable electromagnetic disturbance to anything within the external environment."

The second point to note about the IEV's definition is that it covers both

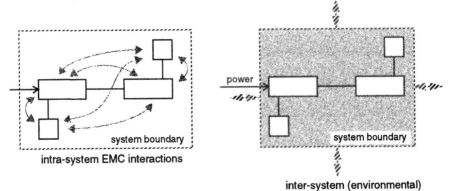

intra-system EMC interactions

inter-system (environmental) EMC interactions

Figure 1.1 Intra- and inter-system EMC

complementary aspects of EMC: control of emissions, and the provision of adequate immunity. Historically, the prevention of interference to other equipment has been treated as largely separate from the correct operation of the equipment itself; but it is now clear that "compatibility" means compatibility in both directions. As we shall see shortly, there are significant differences in the way that these two requirements are defined and met.

The third point is that the definition of EMC makes no mention or classification of what is meant by "disturbance", nor does it place any limits on any of the possible parameters or modes of coupling of such disturbances. It does, at least, make it clear that perfection is not expected. The equipment should not introduce *intolerable* disturbance and it should only need to perform *satisfactorily* in its environment. This implies that it is necessary to know what is a "tolerable" disturbance and to decide what is "satisfactory" performance: more of this later. The IEV does give us a definition of electromagnetic disturbance:

> "Any electromagnetic phenomenon which might degrade the performance of a device, equipment or system, or adversely affect living or inert matter. An electromagnetic disturbance might be electromagnetic noise, an unwanted signal or a change in the propagation medium itself"

but this is not detailed enough for practical use. The generic standards [69],[70], which explicitly define the essential protection requirements of the EMC Directive, do include the somewhat academic statement that

> "Disturbances in the frequency range 0Hz to 400GHz are covered"

even though they actually refer to a narrower and more practical frequency range. The most useful listing of disturbance phenomena is found in IEC 61000-2-5, "Classification of electromagnetic environments" [87], from which Table 1.1 has been abstracted, and which is discussed further in Chapter 4. The right-hand columns labelled "coverage in EMC standards" are an attempt to list the extent to which each phenomenon is included in the various European generic and product standards. Note that military and aerospace standards do have a wider coverage of phenomena.

1.1.1   A description of EMC phenomena

As Table 1.1 shows, electromagnetic disturbances can be classified into four types of phenomena.

1.1.1.1   *Conducted low frequency phenomena*

With respect to these phenomena, EMC can mostly be considered as "compatibility with the power supply network", which to all intents and purposes can be taken to mean the low-voltage AC mains supply. DC supplies, while important for many types of equipment, have been far less studied and are not nearly so well characterised as AC supplies; in any case, they tend to be specific to a particular system, whereas the AC supply is in a manner of speaking "public property", and therefore both acts as a carrier of disturbances and needs to be protected from them. Most of these disturbances appear between phases of the supply, though earth potentials can be more widely distributed.

Harmonics and interharmonics

Non-linear loads draw current at harmonics of the supply frequency (significant up to 2–3kHz), which when flowing in the supply network impedance cause harmonic voltage distortion. In the low voltage public supply, the main threat is from large numbers of products with electronic power supplies, notably personal computers and TV sets. Large industrial loads such as variable speed drives can also cause harmonic distortion in low, medium or high voltage supplies. Interharmonics are due to other loads which are modulated in the same frequency range but not related to the supply frequency.

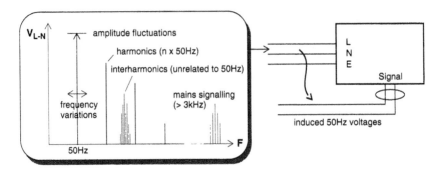

Figure 1.2 Conducted low frequency phenomena

Table 1.1 Disturbance phenomena

Electromagnetic disturbance phenomena	Coverage in EMC standards	
	Emissions	Immunity
Conducted low frequency phenomena (< 9kHz)		
Harmonics and interharmonics	Increasing	None
Signalling voltages	Existing	None
Voltage fluctuations, dips and interruptions	Increasing	Increasing
Voltage unbalance	None	None
Power frequency variations	None	None
Induced low frequency voltages	None	Rare
DC in AC networks	None	None
Radiated low frequency phenomena (<9kHz)		
Magnetic fields	Rare	Rare
Electric fields	None	None
Conducted high frequency phenomena (>9kHz)		
Induced CW voltages or currents	Widespread	Increasing
Unidirectional and oscillatory transients	None*	Widespread
Radiated high frequency phenomena (>9kHz)		
Magnetic fields	Rare	None
Electric fields	None	None
CW electromagnetic fields	Widespread	Widespread
Transient electromagnetic fields	None	None
Electrostatic discharge	N/A	Widespread

excepting the discontinuous disturbance provisions of EN 55014

Signalling voltages

A secondary use of power networks is for information transfer via "mains signalling". A specific standard (EN 50065–1) [65] exists for such systems, which regulates the signalling voltages applied to the network in the frequency range from 3kHz to 148.5kHz (the bottom of the long wave broadcast band). Other users of the mains supply should expect these voltages to be present and be able to ignore them.

Power system voltage and frequency variations

Rapid fluctuations at low level in the supply voltage, occurring from once a minute to 25 per second, are known as "flicker" since they cause this visual effect in electric lighting, and are caused usually by fluctuating industrial loads such as arc furnaces, motor switching or welding equipment. Voltage dips and short interruptions also occur, due to fault clearance in the power network. Voltage unbalance is a feature of three phase supplies caused by large asymmetrical loads. Power frequency stability is normally better than 0.1Hz, but may depart from this by up to 3% during network disturbances. The general quality of the power supply in Europe, covering all these features and others, is detailed in the European standard EN 50160 [71], which gives the acceptable parameters for power supply quality.

Induced LF voltages

Power supply cables may induce interference voltages at the power frequency and its harmonics (possibly up to 20kHz) in adjacent signal cables. The effect is usually most significant in audio circuits such as are found in studio or telecomm applications. Cable layout and effective coupling length will affect the magnitude of the induced voltages, which may also appear in differential mode on some signal cables. Under power system fault conditions, the induced voltages can increase by several orders of magnitude.

1.1.1.2   Radiated LF phenomena

Local magnetic fields exist around components of the power network as a result of current flow. Overhead and buried lines will exhibit magnetic fields, typically up to 40μT depending on proximity, and the nature and configuration of the line. Fault conditions may significantly exceed this figure; for instance, a common problem in offices is VDU screen wobble due to earth faults in the building supply wiring, which allow return currents (and hence high magnetic fields) to flow outside of the normal wiring routes. Power transformers will have stray fields around them, as will many operating appliances, especially those containing motors. VDUs and TV sets will suffer perceptible effects on-screen at a threshold of 0.5–2μT, and themselves generate magnetic fields at the screen scan frequencies.

High-level electric fields of 10kV/m or more occur underneath overhead lines and in substations, but the magnitude is reduced by a factor of 10–20 within buildings due to the shielding effect of the structure (no such effect occurs for magnetic fields). Electric fields due to internal building wiring or operating appliances are generally low unless there is a fault in the wiring, typically due to a broken earth connection. Again, VDUs and TV sets may generate electric fields perpendicular to the screen if no design precautions are taken.

1.1.1.3   Conducted HF phenomena

High frequency disturbances can be either transient or continuous. They are denoted as "conducted" if the principal coupling occurs in either differential mode or common

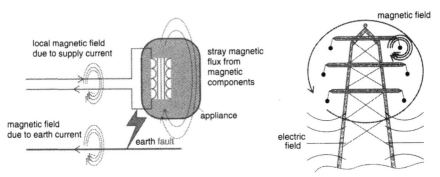

Figure 1.3 Radiated LF phenomena

mode[†] via the cables, either power or signal, which are connected to the affected or disturbing equipment. For evaluation purposes in the EMC test standards the distinction between conducted and radiated is normally made on the basis of frequency, but in real life both conducted and radiated coupling can occur over a wide and overlapping frequency range.

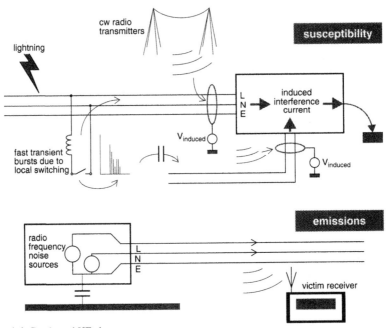

Figure 1.4 Conducted HF phenomena

Induced continuous wave voltages or currents

The fields from nearby radio transmitters induce currents on any conductor exposed to them. This current when appearing at the interface to the affected equipment then induces a voltage at that interface whose amplitude depends on the common mode source and port impedances. The amplitude of the induced current depends on the

† The distinction between differential and common mode is treated in depth in Chapter 4, section 4.2.4

conductor length, its separation from the ground reference, loops formed by stray capacitance and any resonance effects. Emissions from the equipment are coupled out onto conductors and radiated from them, and hence into the aerials of victim receivers, by the reverse process.

Transients

These are classified into a number of subsets depending on their source and nature. Surges can be classified as oscillatory or unidirectional and have relatively high energy, i.e. are capable of inducing damage in coupled equipment. Their risetime and duration is generally in the order of microseconds to milliseconds, and their spectral distribution is limited to the lower radio frequencies. Typical sources are lightning, power system fault clearance or capacitor switching, and collapse of stored energy in inductive loads such as large motors.

Fast transient bursts are due to switching of local inductive loads, normally by electromechanical means (switches or relays). The resultant arcing causes a burst of short pulses each of a few nanoseconds duration, with little energy, but capable of causing severe upset when they couple into electronic circuits.

Coupling in each of these cases can be either differential mode between conductors or common mode with respect to a ground reference. Emissions of fast transient bursts by equipment which includes switching functions, though widespread, are generally not considered worthy of limitation, although the radio frequency content of automatically-produced, repetitive burst-type noise is regulated in the provisions for discontinuous disturbance of EN 55 014 ([67], clause 4.2 etc.).

1.1.1.4   Radiated HF phenomena

Continuous radiated interference

Due to nearby radio transmitters and other radio-frequency generating equipment (RF heating being the most common), this will impinge upon the affected equipment in different ways depending on the source impedance of the field. Three cases can be distinguished:

- magnetic field interference, low source impedance;
- electric field interference, high source impedance;
- electromagnetic wave interference, plane wave (free space) impedance.

The first two cases occur in the near field of the transmitter while the last occurs in the far field. The distinction between the two is treated in section 4.2.3 and is a function of wavelength. At 30MHz, the wavelength is 10m and the near field/far field transition occurs at around 1.6m. As a very broad assumption, the separation distance between source and victim may be taken to be of this order of magnitude and so the 30MHz breakpoint is often used to divide field-related phenomena into electromagnetic (above 30MHz) and electric or magnetic (below 30MHz). Conveniently, 27MHz is the lowest frequency at which portable high power transmitters (for CB radio) are likely to be encountered in typical environments. Below this frequency, the spectrum is mostly allocated to broadcasting and fixed communications links, using fixed transmitters; exceptions occur for marine and military purposes.

Transient radiated field interference

This tends to arise from lightning strikes and from high power switching events in substations or near other parts of the electrical supply infrastructure. For most types of equipment it is less relevant than conducted transients. Equipment in close proximity to sources of fast transient bursts may see such bursts coupled by radiation but such proximity is unusual by comparison to the more typical conducted coupling path.

Electrostatic discharge

This is a specific phenomenon which is hard to classify in the same way as others treated here. It occurs as the result of equalization of charge between two objects carrying different levels of charge, one of which is often a person. The ESD victim sees first the approaching electric field associated with the charge and then, when air breakdown occurs, there is a transient current with associated magnetic and electric fields. Usually the most disturbing effect of the discharge event is the very high rate of rise of current or field. The duration of the event is no more than a few nanoseconds but the current di/dt may approach 10^9 or 10^{10} amps per second, and the field rate of change may be greater than 1kV/m per nanosecond or 10 A/m per nanosecond. The ESD threat in a given environment is greatly affected by the nature of the insulating materials contained in the environment – that is, whether they are effective at supporting and generating electrostatic charge – and by the ambient relative humidity, since high relative humidity allows charge levels to dissipate more readily.

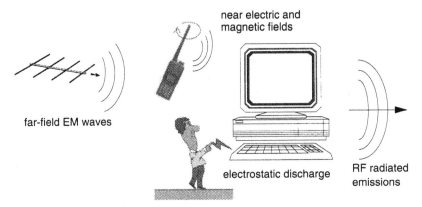

near electric and magnetic fields

far-field EM waves

electrostatic discharge

RF radiated emissions

Figure 1.5 Radiated HF phenomena

1.2   The need for EMC

As we have seen, EMC has the complementary aspects of emissions and immunity. But the roots of these two aspects are different, as is discussed in the following sections.

1.2.1   Control of emissions

In most countries of the world, the radio spectrum is heavily used for many kinds of traffic. Broadcasting and telecommunications are the most obvious uses, but telemetry, radar, radionavigation and space research are some other purposes. Spectrum users pay a licence for the privilege of being allowed to transmit and receive and in return they expect this privilege to be unaffected by interfering sources. Services which rely on radio communication for the economics of their operation are normally able to put a price on its continued reliability. In addition, safety related services demand assured reliability of spectrum access.

For these reasons governments have found it necessary to regulate the spread of types of apparatus which, though not licenced or intended as radio transmitters ("unintentional radiators" in US jargon) have the potential to disrupt such services. Historically, such apparatus has been dominated by broadband sources, notably motor-driven equipment, fluorescent lights and pulsed ignition (from petrol engines). For decades emission limits have been placed on these kinds of equipment and have been enforced by many countries. This has been a well established aspect of national and international EMC control.

Electronic equipment also has the potential to generate radio frequency interference as a by-product of its operation, unless it is specifically designed to avoid doing so. The main culprits are digital electronics incorporating microprocessors, and power switching circuits using fast electronic switches – MOSFETs, IGBTs, transistors, triacs and so on. Even though they are not intended deliberately to transmit RF energy, sufficient is produced in sensitive parts of the spectrum to make it necessary to apply the same types of legislative control as has been applied to other types of unintended source in the past.

With the expansion of the concept of EMC to include compatibility with the supply network, control of other types of emissions are becoming necessary. These are principally power frequency harmonic currents, and flicker (section 1.1.1.1). Because the mains power supply at a given node may be shared among several users, and has a finite source impedance, any load current disturbances attributed to one user will cause voltage distortion which is presented to all other users at that node. The amplitude of such distortion has to be limited, and since the source impedance cannot economically be made arbitrarily low, this implies some limit on the allowable current disturbances.

1.2.2   Control of immunity

The immunity aspect of EMC is a rather more complex and contentious issue. It has several strands which need to be separated and examined:

- the need for continued safe operation of safety related systems in the face of external interference
- the idea of an economic tradeoff in apportioning requirements as between emissions and immunity
- the question of fitness for purpose of apparatus intended for a given environment
- the desire to establish a baseline for product quality, and hence remove the consumer's option to pay less for less reliable products.

Safety is addressed separately in section 1.2.3 shortly. The other issues are discussed here.

1.2.2.1   The "virtual tradeoff" between emissions and immunity

There is a common misconception that there is some relationship between emission and immunity requirements, so that in order fairly to distribute the compatibility burden, the requirements for emissions limits and immunity limits must be shared. This idea totally ignores the reality that emissions and immunity limits apply to two quite different phenomena. RF emissions limits are placed on unintentional emitters in order to protect radio reception, they are specified in values of microvolts per metre or millivolts, and one cannot expect a radio receiver to be immune from the very signals it is expected to receive. On the other hand, immunity requirements are necessary because of the inherent nature of the operation of radio *transmitters* and other man-made and natural phenomena, and are specified in values of volts per metre or kilovolts (a million times greater). In other words, these two aspects of EMC address quite different issues and cannot be traded off one against the other.

One area where there could be a sharing of the burden is in the phenomenon of transients due to switching operations. In this case, control of the amplitudes of emitted transients could, with time and with the eventual turnover of the installed base of equipment, lead to a reduced immunity requirement against such transients; but there are no serious moves in the standardization committees to address the problem of regulating such transients.

1.2.2.2   Fitness for purpose

As explained earlier, EMC is an environmental issue. If a particular product is intended to be used in a given environment, the user ought reasonably to be able to assume that it will cope with any disturbances that might be expected in that environment. If it has a particular function, then that function should be unaffected by typical levels of interference. In other words, a product should be fit for its intended purpose, and of course this purpose includes being used in a particular electromagnetic environment.

Clearly there must be a match between the actual environment of use and the intended environment. It would be unreasonable, for instance, to expect components of a fire alarm system that had been designed for a maximum RF field strength of 10V/m to work correctly when exposed to 50V/m because they had been placed within a few inches of a communications base station transmitter. But if it is in fact reasonable to expect such transmitters to be located in close proximity to such components, then what is wrong is the characterization of the environment at 10V/m. For this reason EMC standards for various types of product are written around an environment classification, as are the generic EMC standards [69], [70]. In fact, because there is a strong statistical component to the make-up of any real environment – sources of a given type, strength and frequency are only present at a particular location for a proportion of the time, often a small proportion – it is accepted that occasional interference problems will occur which have to be dealt with on an *ad hoc* basis. This does not invalidate the expectation of fitness for purpose in the general case.

A further difficulty lies in defining what is adequate "fitness" – in other words, if the product suffers a degradation in performance due to interference, to what degree is this acceptable. The question of performance criteria is fundamental to immunity compliance and will appear at various points in this book – see Appendix B in particular.

1.2.2.3   Mandatory product quality

The effect of regulating the electromagnetic immunity of products is to set a baseline for product quality in EMC terms. If a product has to meet certain minimum standards for immunity before it can be marketed, this effectively removes the consumer's right to trade off price against quality – even though a manufacturer can specify his own performance criteria in many cases. As the EMC Directive has taken hold, there have been many cases of customers discovering that the price of something that they have been specifying for years has gone up sharply, or worse, that the product is suddenly unavailable. On inquiring, they are told that the supplier can no longer supply that product, or has had to redesign it, "because of the new European regulations". The natural reaction on hearing this news is to rail against the idiocy of the European regulations, and to ask the supplier if he will accept a waiver against the regulations in this case. If the supplier wishes to continue trading legally, he has regretfully to decline.

The fact that more often than not, the reason the supplier can't supply as before is because the product was never designed for EMC and failed the newly-required tests miserably, is of no concern to the customer. It always worked in his application, and he sees no reason why the imposition of a new law should disadvantage him.

This is a political question rather than a technical one. There has certainly been heated debate over the past few years as to whether the European Union has a right to insist on minimum levels of product quality in its marketplace; but the weight of opinion so far seems to be that it has, or at least that those who would argue the opposite case are unable to muster much support. There is of course a valid argument – made throughout this book – that good EMC design and installation practices will save many problems and costs in commissioning and operational reliability. The question is whether these costs should be made transparent to the customer.

1.2.3   Safety aspects

For any system which could pose a hazard to people or property if it was incorrectly controlled, safety questions must invariably be addressed, and these questions must include the issue of safety in the face of external interference.

Because the inherent susceptibility to electromagnetic disturbances of electronic control devices has been understood for many years, it is usual for true safety-critical functions to be controlled only by electro-mechanical means. For instance, the emergency stop on a large machine will be a simple switch which physically disconnects the power to the machine when it is activated. On more complex systems, relay logic might be used for the same function. The great merit of such an approach is that electromechanical devices are immune to normal electromagnetic disturbance sources such as radio transmitters, welders and switch arcing; such disturbances simply don't have enough energy to cause a change of state.

However, the more complex control systems become, the harder it becomes to implement safety critical functions only in such simple form, and the question of the immunity of the electronic control devices has to be addressed.

Documents linking EMC and safety

IEC publication 61000-1-2 on "Methodology for the achievement of functional safety of electrical and electronic equipment" [84], in draft at the time of writing, offers guidance on the safety of equipment which is exposed to electromagnetic disturbances. It points out that as well as testing, an understanding of the relevant electromagnetic environment and its relationship to safety requirements is needed, and there should be a "dependability analysis" to identify the hazards which can cause safety risks due to electromagnetic disturbances. It stresses that the levels of electromagnetic disturbances indicated in various standards must be considered very cautiously as regards safety implications, in particular:

- the disturbance levels vary according to a statistical distribution and the given values can be significantly exceeded in some circumstances, i.e. infrequently or on particular sites

- the standard levels and performance criteria are generally related to functional requirements and not safety

- equipment immunity characteristics may worsen with age.

With regard to immunity performance criteria, degradation or loss of function can have safety implications which must be analysed; fail-to-safe behaviour in the face of interference might be acceptable. Equipment failure or degradation may not be equivalent to system failure or degradation, for instance if redundancy is used, but this means that parallel channels should not be affected in the same way by a disturbance.

The document stresses that higher than "functional" immunity test levels, and/or additional EMC testing, may be needed to cover safety implications. This is especially true if the only available immunity standard is a generic standard that does not anticipate any special operating conditions. With respect to safety requirements, functional immunity levels should be increased by some appropriate margin. It may also be that a type of disturbance that was neglected in the normal standards could have a safety implication and should be included in the test regime. Due to the complexities of system interaction, a system should be tested wherever possible at the highest degree of integration.

The Institution of Electrical Engineers (IEE) has a working group on EMC and Functional Safety, which is producing a report, due for publication in late 1999.

The concept of multiple immunity test series, one for system parts not relevant for safety and one for safety-related parts with more severe requirements, is becoming common in some industries. This approach is typically taken by specifiers of safety-critical systems such as aircraft, cars or military systems. To take an automotive example, the engine management and anti-lock braking systems must have an assured immunity at the highest possible level, but the operation of electric windows and sunroof can have a relaxed immunity, and the immunity of the dashboard clock is purely a matter of convenience. In the context of safety certification of civil airliners, two levels are defined: a "critical" function is one whose failure would prevent the continued safe flight and landing of the aircraft, while an "essential" function is one whose failure would reduce the capability of the aircraft or the ability of the crew to cope with adverse operating conditions. A critical function must be maintained in the presence of a "severe" level of external RF interference while an essential function is only required to function correctly within a "normal" environment of RF interference.

Another example is provided by the Navy's specification, as given in DEF STAN 59-41 part 1 [116]:

> For Naval applications an additional type of equipment to those defined in Part 3 of this Standard has been added. This is to cover equipment that is not essential to the float, move or fight function of the ship or submarine. Equipment in this category includes those for the function of the galley, laundry and crew entertainment.

This flexibility of approach allows for a reasonable matching of protection measures against the electromagnetic threat – provided, of course, that the definition of the functions and of the environment is done adequately. For instance, the critical functions need to consider foreseeable environmental extremes (lightning, proximity to high-power radars), misuse (ignoring the "turn off your cell-phone" sign) and faults (a blown fuse).

Chapter 2

The EMC Directive's requirements for systems and installations

... Finally, there is something of a problem concerning systems comprising assemblies of apparatus. In some cases, the EMC of a system can differ from the EMC of its constituent parts, and a system which is marketed as such requires separate certification under the Directive. To a large extent, the problems can be resolved by ensuring that the standards contain adequate practical test methods for systems, but some systems can only be tested "in situ" – ie long after the marketing stage. A similar problem applies to "one off" items of large capital equipment. We are trying to obtain clarification of the legal implications and will report back.

– John Ketchell, DTI, 28th February 1989, in a circular to interested parties regarding progress on the negotiations prior to introducing the EMC Directive

The application of the EMC Directive to installations is, from the experience obtained over the last four years, a very controversial issue.

– EC Guidelines on the application of the EMC Directive, July 1997, section 6.5.2.1

2.1   Introduction to the EMC Directive

Of the various aims of the creation of the Single European Market, the free movement of goods between European states† is fundamental. Before European integration, countries would impose standards and obligations on the manufacture of goods in the interests of quality, safety, consumer protection and so forth. Because of detailed differences in procedures and requirements, these acted as technical barriers to trade, fragmenting the European market and increasing costs because manufacturers might have to modify their products for different national markets.

Initially the European Commission tried to remove these barriers by proposing Directives which gave the detailed requirements that products had to satisfy before they could be freely marketed throughout the Community, but this proved difficult because of the detailed nature of each Directive and the need for unanimity before it could be adopted. In 1985 the Council of Ministers adopted a resolution setting out a "New Approach to Technical Harmonisation and Standards".

Under the "new approach", directives are limited to setting out the essential requirements which must be satisfied before products may be marketed anywhere within the EU. The technical detail is provided by standards drawn up by the European standards bodies CEN, CENELEC and ETSI. Compliance with these standards is presumed to demonstrate conformity with the essential requirements of each Directive. All products covered by each Directive must meet its essential requirements, but all products which do comply, and are labelled as such (the "CE Marking"), may be

† Appendix D lists the EU and EEA Member States.

circulated freely within the Community; no member state can refuse them entry on technical grounds.

Contents

A new approach Directive contains the following elements:

- the *scope* of the Directive
- a statement of the *essential requirements*
- the *methods of satisfying* the essential requirements
- how *evidence of conformity* will be provided
- what *transitional arrangements* may be allowed
- a statement confirming entitlement to *free circulation*
- a *safeguard procedure*, to allow Member States to require a product to be withdrawn from the market if it does not satisfy the essential requirements.

It is the responsibility of the European Commission to put forward to the Council of Ministers proposals for new Directives. Directorate-General III (DGIII) of the Commission has the overall responsibility for the EMC Directive. The actual decision on whether or not to adopt a proposed Directive is taken by the Council of Ministers. At the time of adoption of the EMC Directive this required a qualified majority of 54 out of 76 votes (the UK, France, Germany and Italy each have ten votes; Spain has eight votes; Belgium, Greece, The Netherlands and Portugal each have five votes; Denmark and Ireland have three votes and Luxembourg has two votes). Texts of Directives proposed or adopted are published in the *Official Journal of the European Communities*. Consultation on draft Directives is typically carried out through European representative bodies and in working parties of governmental experts; this was the case for the EMC Directive.

2.1.1   Scope and purpose

The EMC Directive, 89/336/EEC [119], applies to apparatus which is liable to cause electromagnetic disturbance or which is itself liable to be affected by such disturbance. "Apparatus" is defined as all electrical and electronic appliances, equipment *and installations*, and it is the extension of this definition of apparatus to include installations that forms the subject of this book. Essentially, anything which is powered by electricity is covered, regardless of its power source.

The Directive finally came into effect on 1st January 1996, with transitional arrangements before this date.

2.1.1.1   *Essential requirements*

The essential requirements (also known as the "protection requirements") of the Directive (Article 4) are that the apparatus shall be so constructed that:

- the electromagnetic disturbance it generates does not exceed a level allowing radio and telecommunications equipment and other apparatus to operate as intended;
- the apparatus has an adequate level of intrinsic immunity to electromagnetic disturbance to enable it to operate as intended.

The intention is to protect the operation not only of radio and telecommunications equipment but any equipment which might be susceptible to EM disturbances, such as information technology or control equipment. At the same time, all equipment must be able to function correctly in whatever environment it might reasonably be expected to occupy. Notwithstanding these requirements, any member state has the right to apply special measures with regard to the taking into service of apparatus, to overcome existing or predicted EMC problems at a specific site or to protect the public telecommunications and safety services.

2.1.1.2   Sale and use of products

The Directive applies to all apparatus that is placed on the market or taken into service. The definitions of these two conditions, especially as they apply to systems and installations, do not appear within the text of the Directive but have been the subject of the Commission's guidance as well as interpretation by national authorities.

Placed on the market

The "market" means the market in any or all of the EU member states; products which comply within one state are automatically deemed to comply within all others. "Placing on the market" means the *first* making available of the product within the EU, so that the Directive covers only newly manufactured products within the EU, but both new and used products imported from a third country. Products sold second-hand within the EU are outside its scope. Where a product passes through a chain of distribution before reaching the final user, it is the passing of the product from the manufacturer into the distribution chain which constitutes placing on the market. If the product is manufactured in or imported into the EU for subsequent export to a third country, it has not been placed on the market.

The Directive applies to each individual item of a product type regardless of when it was designed, and whether it is one-off or high volume. Thus items from a product line that was launched at any time before 1996 must comply with the provisions of the Directive after 1st January 1996. Put another way, there is no "grandfather clause" which exempts designs that were current before the Directive took effect. However, products already *in use* in the EU before 1st January 1996 do not have to comply retrospectively.

If the manufacturer resides outside the EU, then the responsibility for maintaining the declaration of conformity with the Directive rests with the person placing the product on the market for the first time within the EU, i.e. the manufacturer's authorized representative or the importer. Any person who produces a new finished product from already existing finished products, such as a system builder, is considered to be the manufacturer of the new finished product.

Taken into service

"Taking into service" means the first *use* of a product in the EU by a user. If the product is used without being placed on the market, if for example the manufacturer is also the end user, then the essential requirements of the Directive still apply. This means that sanctions are still available in each member state to prevent the product from being used if it does not comply with the essential requirements or if it causes an interference problem. On the other hand, it should not need to go through the conformity assessment procedures to demonstrate compliance (article 10, which describes these procedures, makes no mention of taking into service). Thus for instance an item of special test gear built by a technician for use within a company's on-site test and installation procedures must still be designed and used so as not to cause or suffer from interference, but should not need to follow the procedure for applying the CE mark.

There is a special concern with the question of taking into service of installations, discussed shortly in section 2.2.

2.1.1.3   Exceptions

There are a few specific exceptions from the scope of the Directive, but these are not such as to offer much cause for relief. Self-built amateur radio apparatus (but not CB equipment) is specifically excluded. In the UK regulations, apparatus for use in a sealed electromagnetic environment is also excluded.

Military equipment (munitions of war) is excluded as a result of an exclusion clause in the Treaty of Rome, but equipment which has a dual military/civil use will be covered when it is placed on the civilian market. Education and training equipment, according to the UK regulations, need not meet the essential requirements – since its whole purpose is deliberately to emit or be susceptible to interference – provided that its user ensures that it does not cause interference outside its immediate environment, and provided that it is accompanied by a warning that its use outside the classroom or lab invalidates its EMC conformity. Immunity requirements are waived.

Other exclusions are for those types of apparatus which are subject to EMC requirements in other directives. These include automotive equipment, medical devices (with a transition period ending on 14th June 1998), and marine equipment (with a transition period ending on 31st December 1998). Immunity requirements for electricity meters and non-automatic weighing instruments are covered in their own specific directives.

2.1.1.4   Benign apparatus

An important exclusion from the scope of the Directive is that class of products which have become known as "electromagnetically benign" apparatus. This is defined in the UK regulations as apparatus "the inherent qualities of which are such that neither is it liable to cause, nor is its performance liable to be degraded by, electromagnetic disturbance". Such products are excluded from the CE Marking and essential requirements of the EMC Directive. The EC Guidelines (see 2.2.1) include the following expanded definition of what they prefer to call "passive-EM" equipment:

> Equipment is considered a passive-EM equipment if, when used as intended (without internal protection measures such as filtering or shielding) and without any user intervention, it does not create or produce any switching or oscillation of current or voltage and is not affected by electromagnetic disturbances.

The guidelines offer the following examples:

- cables and cabling systems, cable accessories;
- equipment containing only resistive loads without any automatic switching device; e.g. simple domestic heaters with no controls, thermostat, or fan;
- batteries and accumulators.

At first sight this is good news, especially for producers of such items: one less set of regulations to take into account. But it does not necessarily mean that no CE mark is required at all, since another directive (see 2.1.2) may well apply; for instance the Low Voltage Directive will apply to many products that fall under the second example above. For the system builder this means that it is impossible to rely on the presence of the CE mark to prove EMC performance (see also 2.1.3 shortly). Many system parts will fall into the category of benign apparatus – the interconnecting cabling and cabinets or other enclosures are prime examples – yet their specification can have a profound effect on the EMC of the system as a whole. It would be very attractive to many system builders if the whole question of EMC could be sidestepped by specifying *all* components to be CE marked. Sadly for the cause of simplicity, this is neither reasonable nor sensible (see section 2.3.3 later).

2.1.2   Other CE marking directives

There are two other directives, compliance with which is signified by the CE Mark, which are relevant to many system builders. These are the Low Voltage Directive (LVD, 73/23/EEC, modified by the CE Marking Directive 93/68/EEC) and the Machinery Safety Directive (MSD, 89/392/EEC). These are briefly described below.

The directives mentioned so far tend to be catch-all or "horizontal" types, and there are a number of more specific "CE marking" directives which could also be relevant to system builders and installers. These include directives for: Medical Devices; Telecommunications Terminal Equipment; Personal Protective Equipment; Simple Pressure Vessels; Explosive Atmospheres; Lifts (elevators); Recreational Craft, Marine Equipment, etc. Where some of these directives apply, they sometimes remove the need to meet the EMC, LVD, and/or Machinery directives.

There are also a number of directives which do not require CE marking, such as the "six-pack" of Health and Safety at Work Directives. Although Health and Safety at Work directives apply mainly to employers, they will also be of concern to systems builders and installers because their "products" must be able to be comply with those directives when used. The General Product Safety and Product Liability directives will also be of concern to anyone engaged in trade.

2.1.2.1   *Machinery Safety Directive (MSD)*

As its name suggests, this Directive applies to the safety of machinery, broadly defined as an assembly of linked parts, at least one of which moves, powered by a source of energy other than human effort. Its main concern is that hazards presented by the machinery to its operators are taken into account in its design and construction. Included within these potential hazards are those which may result from incorrect operation of the machinery due to external interference, and hazards that may result due to emission of radiation. These are evidently related to the immunity of the electronic control equipment that may form part of the machine, and to the control of high-power electromagnetic source emissions, such as from RF induction heaters.

Any such electrically-operated machinery is also subject to the EMC Directive. In this case the objective is not safety but correct operation of the equipment in its expected environment, and electromagnetic protection of other apparatus in that environment. The aims of the two directives are therefore complementary and each must be applied in full.

For machinery which falls under the description of an electrical system or installation (see section 2.2) then the provisions discussed under that section should be followed as far as compliance with the EMC Directive is concerned. In some cases, the requirements of the two directives are subtly different; for instance, whereas there is the provision for a "Declaration of Incorporation" for a sub-assembly under the MSD, no such possibility exists for the EMC Directive. Also, the Technical File for the MSD, which is mandatory and contains the manufacturer's own safety assessment for the product, is a different animal from the Technical *Construction* File for the EMC Directive, which is just one option available to the manufacturer and which, if taken, must contain an assessment from a competent body.

Note that the EMC Directive does not cover safety issues at all, so where the functional safety of a machine (or any other apparatus) may be affected by the effects of electromagnetic disturbances, these must be considered as part of the hazards and risk assessment carried out by the manufacturer as required by the MSD. (Refer to section 1.2.3 for more on EMC and safety). The EMC work required to reduce safety hazards and risks to acceptable levels may require significantly more effort than merely meeting harmonized EMC standards.

2.1.2.2   Low Voltage Directive (LVD)

This directive applies to the safety of all electrical equipment operating between the voltages of 50–1000V AC or 75–1500V DC. It does not have any direct relationship to the EMC Directive. But the CE Mark which is applied to a product for EMC will also usually mean that the apparatus complies with the LVD, assuming that it operates within the voltage limits cited above.

The LVD applies to all electrical equipment in its scope, but where the equipment is a machine a risk analysis (using EN 1050 as the recommended method) should be carried out. Where safety risks are mainly electrical, only the LVD is applied. Where the machine has significant mechanical safety risks, the MSD is applied as well as the LVD. The EC's 1997 "Guidelines on the Application of Council Directive 73/23/EEC" state:

> When the results of the risk assessment by the manufacturer show that the risks are mainly of electrical origin, the machinery equipment will be covered exclusively by the "Low Voltage" Directive, which anyhow deals with all the safety aspects, including mechanical safety.

And also:

> Apart from machinery covered by Article 1(5) (of the MSD), all machinery having an electrical supply and designed to operate between 50 and 1000 V in AC or 75 and 1500 V in DC is covered by both the "Machinery" Directive and the "Low Voltage" Directive, applying in a complementary way.

The essential requirement of both is that the equipment should be safe. As with the MSD, the LVD requires that a hazard analysis is carried out and the result kept in a Technical File – as before, not the same as the EMC Technical Construction File.

It would not be atypical, though, for a manufacturer to keep one set of documentation covering compliance with all the required directives. The Technical File for the LVD and/or MSD would form one part, and the rationale for compliance with the EMC Directive would form another – whether this was simply test results against harmonized standards to support the self-certification route (section 2.3.1) or a full Technical Construction File with competent body certificate (section 2.3.2).

2.1.3   The meaning of the CE Mark

There is a great deal of confusion about the meaning of the CE Mark, particularly amongst people who are only peripherally involved with its implementation.

The CE Mark is a **passport**.

It is *not* a mark of **approval**.

The purpose of the CE Mark is to allow free passage of equipment bearing it, within the European Economic Area – nothing more, nothing less. Assuming that the presence of the CE Mark signifies some guaranteed performance level is equivalent to thinking that because Messrs Williams and Armstrong have EC passports, they must therefore be highly qualified EMC consultants. While both may be true, they are unrelated.

The directives which require CE Marking, by and large allow self-declaration as a route to applying it. This means that it is the manufacturer's own responsibility to declare conformance with a particular directive's requirements. There is no third party oversight of his justification for doing so – except the very imperfect and uneven mechanism of enforcement after the fact in each of the Member States – and therefore *no third party approval is implied by the presence of the mark*. This is in direct contrast to the meaning of other marks, usually connected with safety (such as the BSI Kitemark, or the UL, VDE, and SEMKO/DEMKO/NEMKO symbols) which are granted once a product has survived independent testing by the relevant laboratory. There *is* now an independent international EMC mark, offered by a consortium of test laboratories, but it has yet to establish itself as desirable to customers.

Legally, the presence of the CE Mark signifies that the manufacturer believes that his product complies with the requirements of all relevant directives, which are generally viewed as minimum requirements for entry to the European market. It also implies that the manufacturer has made an appropriate declaration of conformity (for the form of a sample declaration, see Figure 3.3 on page 58). It is not an indicator of product quality and, especially relevant to system integrators, it does not on its own prove that a particular product, if incorporated into a system or installation, will automatically allow that system or installation itself to comply with any relevant directive. We look at this question in more detail in section 2.3.3 and in Chapter 3. But meanwhile the situation has arisen that many less-informed customers are demanding the CE Mark as a condition of procurement even on items for which it is unnecessary. Therefore suppliers of some classes of products, particularly benign or apparently benign equipment as discussed earlier, are attempting to find ways to apply the CE Mark when the legislation does not compel them to, just to satisfy customer pressure. In any business, after all, whatever the customer wants is right.

The confusion surrounding the introduction of the CE Mark has been partly due to lack of education and partly because of the differing requirements of the various directives. This has been well illustrated by an article in BSI News [23]:

> First, its blanket nature means it cannot be used for product differentiation and secondly, the fact that it means different things for different products is likely to cause confusion both for the consumer and the installer. The inconsistency across the directives of the independent testing requirement makes it very difficult for purchasers to assess on what basis certification has taken place.

Or, to put it another way [54],

> Safety marks came into being because the legal system did not reliably give consumers confidence in a product's safety. The underlying basis for applying the CE mark is no more reliable...

2.2   Applicability to systems and installations

As implied by the quotations at the head of this chapter, the application of the EMC Directive to systems and installations has been especially vexatious. This section expands on some of the issues.

2.2.1   The 1997 guidelines

In October 1993 the Commission issued a slender booklet titled *Guidelines on the application of Council Directive 89/336/EEC*. This was its first attempt to clarify many of the issues that had arisen in interpreting the Directive during its first year of operation. After four more years and a number of half-leaked revisions, an updated and greatly expanded version was published in July 1997 [121]. Distribution of these guidelines has been somewhat haphazard, with many people who were not in the centre of the EMC community being unaware of their existence. The July 97 version was released only in electronic form, with the printed document becoming available only some months later. More seriously, almost as soon as they had been published in the UK the guidelines were the subject of serious concern both from the Trading Standards organizations (who are the enforcement agents in the UK) and from the Department of Trade and Industry's own lawyers, who identified a number of inconsistencies between the guidelines and the UK's legislation. This has led the DTI to be very cautious in recommending the new guidelines, with a clear statement advising that in cases of conflict, the UK legislation takes precedence. To be fair, the guidelines themselves emphasize this same point in their preamble, but it does leave commentators in a difficult position when it comes to attempting to offer a definitive position on a point of application of the Directive. In addition, the language is at times obscure and not conducive to a clear understanding of the concepts being expressed.[†]

Nevertheless, the guidelines have an extensive chapter devoted to the problem of systems and installations and it is worth quoting this in full so that any comment that this book makes can be seen in the context of the official recommendations. The text of this chapter, or at least, that part of it referring explicitly to systems and installations, is therefore reproduced verbatim below.

---

† The EC is not the only body to suffer from guidance proliferation. An article in *Compliance Engineering* magazine (March/April 1997) revealed that the American Food & Drug Administration has published more than a thousand guidance documents; but they will all include the warning, "Although this guidance document does not create or confer any rights for or on any person and does not operate to bind FDA or the public, it does represent the agency's current thinking on the topic". As the article pointed out, this means "heads we win, tails you lose".

**Extract from "Guidelines on the application of Council Directive
89/336/EEC", European Commission, July 1997**

6.   **Application of the Directive to components, finished products, systems and installations.**

6.1.   General.

In order to make the EMC Directive easier to understand, particularly its scope and the conformity
assessment procedures laid down in it, it is necessary to explain or clarify some terms used in the
Directive by taking account of practice in this sector, particularly for:

> components;
> finished products
> systems;
> installations.

NOTE 5 - The contents of these guidelines define the "status" for different types of apparatus as
regards as the application of the EMC Directive. These guidelines do not prejudice the
application of EMC requirements for those apparatus excluded from the Directive according to this document, when established through a contractual framework between suppliers, subcontractors, etc.

....
(Sections 6.2 and 6.3 relating to components and finished products omitted)

6.4   Application of the Directive to systems

6.4.1   A common understanding of "Systems"

In normal usage, the word "system" is sometimes used for an optional combination of several apparatus to perform a specific task where the end-user is the person who decides which apparatus are
used to construct this so-called "system", and where the apparatus were not intended to be placed
together on the market as a single functional unit.

A computer "system" consisting of a CPU, keyboard, printer, monitor, etc. is a good example. Each
one of those parts is an apparatus placed on the market independently from the others and complying in full with the EMC Directive. *They are all CE marked*. They can be interconnected by a person not technically proficient in EMC matters. In accordance to chapter 10 they are supplied with
clear instructions for interconnection, integration, use and maintenance (when applicable), as well
as limitations for connection and use. Following those instructions, in particular those related to
cabling, in the manner intended by the manufacturer(s) of the constituent parts incorporated into the
system justifies the assumption that the system is electromagnetically compatible.

The manufacturer of each constituent piece of apparatus in the system has already fully applied the
Directive, and particularly taken into account the expected electromagnetic environment and the
intended use.

Clearly, for such a so-called "system", the EMC Directive has already produced its effect. As the
parts are **not placed on the market as one functional unit**, further measures that might be needed
are outside the application of the EMC Directive. This kind of "system" neither needs an additional
CE marking nor an additional EC declaration of conformity for the "system" as a whole.

If the EMC environment in which the "system" is used is different from that intended by the manufacturer(s) of the apparatus incorporated into the "system", the "system" may be subject to EMC
problems. The user, the assembler or the installer must therefore overcome these specific unforeseen EMC problems, or else, purchase other apparatus suitable for that environment. But here such
initiatives fall outside the scope of the Directive.

6.4.2   "Systems" within the EMC Directive

For the purposes of the EMC Directive, a system is defined as a combination of several equipment, finished products, and/or components (hereinafter called "parts") combined, designed and/or put together by the *same person (system manufacturer) intended to be placed on the market* for distribution as *a single functional unit for an end user and intended to be installed and operated together to perform a specific* task.

The system as a whole is a final apparatus; within the meaning of the EMC Directive it is an apparatus; it can enjoy free movement within the EEA. It must therefore be designed and put together so as to comply with the essential requirements of the EMC Directive. This compliance should include any reasonably foreseeable situation, in any intended electromagnetic environment and in any of its configurations.

A combination of "parts" may only be considered as a system if the manufacturer lists all "parts" in the instructions for use and declares for the attention of the installer and/or end user that **this combination forms a system**. The system manufacturer assumes responsibility for the compliance of the system as a whole with the Directive, and must therefore, in accordance with chapter 10, provide clear instructions for assembly, interconnection, integration, installation, use and maintenance (where applicable), as well as limitations for connection and use. Given that the assembler, installer and/or end-user have only to follow these instructions, they may assume that they install and operate the system in conformity with the relevant provisions.

An apparatus, that could also be called a system, composed of other apparatus and/or components (whether or not they are CE marked) **and** which is a single commercial unit, **must comply** fully with the EMC Directive. An illustrative example is a computer CPU, composed of a power supply, CD ROM, mother board and disk drive supplied in **an enclosure**. This "system" is regarded as an apparatus and therefore subject to the EMC Directive.

Several cases of systems in the sense of the Directive ought to be considered:

6.4.2.1   Systems assembled from only CE marked apparatus.

As a good example we can take again the computer system consisting of a CPU, keyboard, printer, monitor, etc., as described in chapter 6.4.1. The difference between that chapter, and the case described here, is that in this instance, the above mentioned parts are *put together by the same person* (the system manufacturer) and placed on the market as a single functional unit, and that *this person assumes responsibility for the compliance of the system as a whole with the Directive*. Since the manufacturer(s) of each part of the system has/have already fully applied the Directive, and particularly taken into account the expected electromagnetic environment and the intended use, there are additional requirements for the system manufacturer to apply to comply with the EMC Directive:

The EC declaration of conformity, as well as the instructions for use must refer to the system as a whole. It must be clear (e.g. by enclosing a list of all parts) which is/are the combination(s) that form(s) the system placed on the market for distribution and/or use. The manufacturer assumes responsibility for compliance with the Directive, in particular with the protection requirements in all expected electromagnetic environments, and must therefore, in accordance with chapter 10, provide clear instructions for assembly/installation/operation/maintenance in the instructions for use. The system as a whole does not need to bear the CE marking (all this applies even if it is offered on the market as a single functional unit, as long as each part bears the CE marking).

If the electromagnetic environment in which the system is used is different from that intended by the manufacturer(s) of the apparatus intended to be incorporated into the system, the system may be subject to unforeseen EMC problems. The user, the assembler or the installer must therefore overcome these specific unforeseen EMC problems e.g. by following the procedure within 6.4.3, or else, purchase another system suitable for that environment.

Additional comment: Manufacturers of systems described above should be aware that combining two or more CE marked subassemblies may not automatically produce a system which meets the requirements of the relevant standard. E.g.: a combination of CE marked PLC's (Programmable Logic Controllers) and motor drives within a machine tool put together to be placed on the market as a system may fail the requirements, whereas a HI-FI system composed of a separately CE marked amplifier, tuner, CD player and cassette deck, wired up correctly is quite likely to maintain its compliance

6.4.2.2 Systems assembled from apparatus including some not CE marked.

Constituent parts considered in this section are:

CE marked apparatus, finished products and components with a direct function, which fully comply with the Directive.

Non-CE marked apparatus, finished products or components intended exclusively for an industrial assembly operation for incorporation in other "apparatus".

Systems discussed in this section are composed of non-CE marked apparatus, finished products or components and may also include CE marked apparatus. They must only be combined into a system (intended to be placed on the market in view of its free movement as a single functional unit) by a professional person.

As a professional person, he is supposed to understand the EMC related technical implications of the parts when combined into a system and make the right judgements so as to fulfil the objectives of the Directive. *He becomes manufacturer* -in the full sense-. **The system is therefore an apparatus** in the sense of the EMC Directive and **must comply with all its provisions.**

The EC declaration of conformity, as well as the instructions for use must refer to the system as a whole. It must be clear (e.g. by enclosing a list of all parts) which is/are the combination(s) that form(s) the system placed on the market for distribution and/or use. The system manufacturer assumes responsibility for compliance with the Directive, in particular with the protection requirements in all expected electromagnetic environments, and must therefore in accordance with chapter 10, provide clear instructions for assembly/installation/operation/maintenance in the instruction for use. One CE marking is sufficient, affixed just once on the main part of the system, if all parts are supplied as one unit.
Those parts of the system which are themselves compliant apparatus may, of course, be distributed and/or used outside the system.

If the electromagnetic environment in which the system is used is different from that intended by the system manufacturer, the system may be subject to EMC problems. The user, the assembler or the installer must therefore overcome these specific unforeseen EMC problems, (e.g. by following the procedure within 6.4.3) or else, purchase another system suitable for that environment.

6.4.3. System or apparatus with various configurations

Most often systems or apparatus are offered in different configurations, to perform different tasks. These configurations are variants of a complete or complex configuration. The system manufacturer (assembler or integrator) can follow the approach below, suggested as a way to simplify his tasks while fully complying with the EMC Directive:

The responsible person should attempt to define, from an EMC perspective, the configuration most likely to cause the maximum disturbance, or to be the most susceptible to possible disturbances. This configuration, often called the "worst case" should be defined, so that the other possible configurations are included in it in EMC terms.. Such a configuration is then brought into full compliance with the Directive, in accordance with article 10. The manufacturer then declares conformity and affixes the CE marking.

Once the *worst case configuration defined above* is in conformity, the manufacturer (assembler or integrator) can place on the market any of the possible variants or configurations *without further verification*, since they are included in it in EMC terms. They have better electromagnetic performance, i.e. they do not introduce new electromagnetic disturbances not covered in the worst case configurations(s) or do deteriorate the immunity compared with the (fully EMC compliant) worst case configuration(s). He then draws up and signs the EC declaration of conformity and affixes the CE marking to each variant.

The responsible person might want to add some new components that were not included in the original (fully EMC compliant) worst case configuration(s), from the EMC perspective, that were fully EMC verified. He may use either electromagnetically "relevant" or electromagnetically "irrelevant" components:

In the context of various configurations the following definitions apply:

An *electromagnetically relevant component* is defined as one that, due to its electromagnetic characteristics, is liable to cause or have its performance degraded by, electromagnetic disturbances such that it influences the EMC characteristics or the intended operation of typical assemblies into which it may be incorporated.

A *electromagnetically irrelevant component* is then defined as one that, due to its electromagnetic characteristics, neither is liable to cause nor have its performance degraded by, electromagnetic disturbances such that it will not influence the EMC characteristics or the intended operation of typical assemblies into which it may be incorporated.

It must be noted, that some passive-EM equipment may not be electromagnetically irrelevant in particular applications. Therefore, the classification of components as electromagnetically relevant or irrelevant is strictly related to the application and may change from application to application. (Some examples include: inductors, motors, cables). The effect of this phenomenon must be taken into account by the assembler of the system or apparatus.

If the manufacturer (assembler or integrator) later wants to add some new *electromagnetically irrelevant- components* to his configuration(s), that were not included in the original EMC "worst case" that was fully EMC compliant, he *is not requested to carry out further* verifications from the EMC point of view. He then signs the EC declaration of conformity and affixes the CE marking to the configuration(s).

However, if the manufacturer (assembler or integrator) later wants to add some new *electromagnetically relevant components* to his configuration(s), that were not included in the worst case(s) that were fully EMC compliant, he must then ensure that the new EMC *worst case configuration(s)* are in full compliance with the Directive.

6.5   Application of the Directive to installations

6.5.1   A common understanding of "Installations"

In normal usage the word "installation" is sometimes used to refer to an optional combination of several apparatus, to perform a specific task where the end-user is the person who decides which apparatus are used to construct this so-called "installation" and where the apparatus were not intended to be placed on the market as a single functional unit. **Such installations must be considered like those combinations described in chapter 6.4.1 which are commonly referred to as "systems" and treated as such.** They are not treated further in this chapter.

A good example of such an "installation" is a HI-FI installation composed of an amplifier, tuner, CD player and cassette deck, each of them separately CE marked and separately placed on the market.

6.5.2   Fixed "Installation" within the EMC Directive

6.5.2.1 General

Under Articles 1.1 and 2.1 the Directive applies to installations containing electrical and /or electronic components.

The application of the EMC Directive to installations is, from the experience obtained over the last four years, a very controversial issue. It is therefore important to present in this chapter an analysis on the applicability of the Directive, based on its spirit:

"Fixed Installation", in the broadest sense, is defined as *"a combination of several equipment, sys-*

*tems, finished products and/or components (hereinafter called "parts") assembled and/or erected by an assembler/installer **at a given place** to operate together in an expected environment to perform a specific task, but not intended to be placed on the market as a single functional or commercial unit".*

The Directive does not distinguish between different kinds of installations, but in order to avoid unnecessary burdens for manufacturers of parts and assemblers/installers, it is convenient to investigate which provisions of the Directive can be declared non-applicable without compromising the objectives of the Directive.

6.5.2.2   Application of the EMC Directive to fixed installations

In the installations defined in chapter 6.5.2.1, parts not intended to be placed on the market as a single commercial or functional unit may well be used, and it makes no difference whether they were placed on the market by the same or different manufacturers because none of these manufacturers will know the final electromagnetic effect the combination of parts in the installation will have; they can only assume responsibility for each individual part when placing it on the market.

EMC problems in apparatus when used in installations are solved on a case by case basis, by co-operation between manufacturers of parts incorporated into the installation, the user and on some occasions, an installation contracting company. The combined expertise of these parties results in the correct operation of the total installation, and also enables its integration into a network.

The installation must comply with the essential requirements of the Directive as defined in Article 4.

The person(s) responsible for the design, engineering, and construction (assembly and erection) becomes the "manufacturer" in the sense of the Directive, and assumes responsibility for the installation's compliance with all applicable provisions of the Directive, when taken into service. The EMC assembly instructions given by the manufacturer(s) of parts, and the whole method of installation has to be in accordance with good engineering practice within the context of installations, as well as installation rules (national, regional or local) that will ensure the compliance of the whole installation with the essential requirements of the EMC Directive. Such rules must, of course, be fully compatible with the Union Treaty and, in particular, cannot influence the design and manufacture of apparatus that are already in conformity with the EMC Directive.

Such an installation cannot "enjoy" free (physical) movement within the EEA market, and in respect of the EMC Directive there is no need for CE marking or an EC declaration of conformity or to involve a competent body. The manufacturer of the installation must provide clear instructions for operation and maintenance in the instructions for use in accordance with chapter 10.

6.5.3   Application of the Directive to movable installations.

Installations which are intended to be moved to and operated in a range of locations (e.g. the outside broadcast vehicle of a TV or radio station) may experience or cause changes in the electromagnetic environment. Such a movable installation has free (physical) movement within the EEA market, or within the EEA territory. Therefore such movable installations have to comply with the Directive like a system as described in chapter 6.4.

If such installations are, however, intended to substitute for, or to extend a fixed installation (e. g. for electricity generation or transmission in the high-voltage network) they have to be treated in the same way as a fixed installation in chapter 6.5.1.2. The temporary connections to the networks of such installations must be carefully planned, and installed by experts.

2.2.2   Definitions of systems and installations

From the guidelines, and also from the UK regulations implementing the Directive, it is clear that the Directive applies in different ways to systems and to installations, and the two have very specific differences in definition.

2.2.2.1   System

A system is defined (guidelines 6.4.2, simplified) as

> a combination of equipment and/or components put together by the system manufacturer and intended to be placed on the market for distribution as a single functional unit.

Key parts of the definition are that a system can "enjoy free physical movement" as a "single functional unit", and is "put together by one person (the system manufacturer)" who is responsible for EMC compliance of the system as a whole. A system fitting into this definition must comply both with the essential requirements of the Directive (article 4, and see para 2.1.1.1) *and* with the conformity assessment procedure (article 10), that is, the manufacturer must make a declaration of conformity using one of the routes detailed in section 2.3, and apply the CE Mark.

The guidelines make a number of further distinctions as to how such systems might be put together, and some examples are offered below:

- *systems made up only of compliant apparatus (guidelines 6.4.2.1)*: e.g. a computer system including peripherals and interconnecting cables, supplied as a package
- *systems made up of a combination of compliant and non-compliant parts (guidelines 6.4.2.2)*: e.g. a complex laboratory instrument such as an electron microscope or research equipment, comprising some bought-in parts (such as a PC, already compliant for a given environment) and some parts made in-house
- *systems with multiple variant configurations (guidelines 6.4.3)*: e.g. a bespoke power system using the same fundamental approach for each application, but with different combinations of power and control modules; the manufacturer certifies based on the worst case configuration
- *a mobile installation (guidelines 6.5.3)*: it has free physical movement, and can experience or cause changes in the electromagnetic environment, and is treated for compliance purposes as a system: e.g. an outside broadcast vehicle or a mobile emergency control centre.

2.2.2.2   Installation

A fixed installation is defined (guidelines 6.5.2.1, simplified) as

> a combination of several equipment, systems, products and/or components assembled at a given place to operate together in an expected environment, but not intended to be placed on the market as a single functional or commercial unit.

Here the key issue is that the installation is put together "at a given place" and cannot therefore "enjoy free physical movement". In contrast to the system, it is not supplied as a single functional or commercial unit, but is assembled on-site.

> EMC problems in apparatus when used in installations are solved on a case by case basis, by co-operation between manufacturers of parts incorporated into the installation, the user and on some occasions, an installation contracting company. The combined expertise of these parties

results in the correct operation of the total installation, and also enables its integration into a network.

An installation meeting this definition must still comply with the essential requirements of the Directive (article 4, and see para 2.1.1.1), but does *not* need to be concerned with the conformity assessment procedure, i.e. the declaration of conformity and the CE Mark. Consequently, the involvement of a competent body with fixed installations is not mandatory (but see section 2.3.2).

Examples of fixed installations might be a building management system or an office local area network for a specific building, a telephone exchange, a manufacturing plant, or a power station or substation. It may be argued that in some of these examples, the installation contractor does supply the complete installation "as a single functional or commercial unit"; but the defining factor seems to be that whatever its provenance, the installation is assembled on site from a collection of compliant apparatus and systems, rather than being manufactured elsewhere and shipped as a whole.

To avoid confusion between the expectations of the user of an installation and the hopes of system builders and installers, it is recommended that EMC requirements are made part of the contract between the parties. Although a customer cannot make the EMC Directive applicable to suppliers who are not covered by its scope, he can decide whether he wants the supplier to do all the EMC work and provide him with a guaranteed reliable (i.e. adequately immune) and interference-free "product", or he can decide that he will risk the delays, unreliability, and lost production which are implied by "solving EMC problems on a case by case basis". In the former case the supplier will need to do more work, and charge more, whereas in the latter case the original cost will be minimized but the lifetime cost may well be higher (refer to Appendix C for some case studies).

Since this area still suffers from ambiguity and confusion (1997 EMC Guidelines notwithstanding), and since the financial well-being of suppliers and users of systems and installations can be at stake when EMC is not addressed adequately, agreement on who does what, and who is responsible for what, as regards all aspects of EMC, is strongly recommended during the tendering and placing of contracts stages early in any installation project.

2.2.3   The "responsible person"

Despite the exclusion of fixed installations from the CE Marking requirements, they are still subject to the essential requirements of the EMC Directive – that is, they must be "good neighbours" to their adjacent environment, and must be able to withstand incoming interference to an acceptable degree for their normal functioning. Some entity has to take responsibility for achieving and ensuring this, and under the terms of the Directive this is the installation's "manufacturer". The guidelines state that this is "the person(s) responsible for the design, engineering, and construction (assembly and erection)" of the installation. If there is a prime contractor for the installation, this company's project manager would have to ensure that all EMC aspects are accounted for, usually by including suitable requirements in suppliers' contracts, but also by ensuring that correct installation methods are followed. Chapter 3 has more to say on this.

Where the prime contractor in the above definition is not clearly identifiable, some explicit contractual specification identifying who has EMC compliance responsibility will be desirable. As the guidelines say, EMC problems in installations are addressed through the combined expertise of the various parties on a case by case basis.

2.2.4   The "instructions for use"

Development of the EMC Directive has placed ever greater emphasis on the need for a supplier of a product to provide suitable instructions. As part of the definition of the essential requirements, clause 5(6) of the UK regulations [124] states that

> The information required to enable use in accordance with the intended purpose of the relevant apparatus must be contained in the manufacturer's instructions accompanying the apparatus

which requirement does in fact appear in Annex III of the Directive itself. The content of the instructions is not specified, but could reasonably be expected to include details of the intended conditions of use, general installation and assembly details, and any warnings on limitations of use. The purpose of these instructions is to ensure that no EMC problem is encountered when the product is used.

From the point of view of the system builder, the important issue is that the instructions can be (and are) carried out by the personnel concerned in assembling and operating the system, and do not contain any mutually exclusive or unreasonable provisions. The difficulty at the system design stage is that any EMC-specific instructions may not be provided until the product is supplied, so that without good liaison with a competent supplier it may not be possible to anticipate problems in fulfilling them. A further difficulty is that some suppliers of products which are deemed to be "components" regard themselves as outside the scope of the legislation and therefore not bound to supply EMC-related instructions at all. It may also be suspected, uncharitably, that they simply don't have the expertise to do this anyway. In these cases, sustained commercial pressure from their customers the system builders may be the only long-term solution.

2.2.4.1   Intended conditions

The most significant aspect here is likely to be the intended operating environment. The generic EMC standards differentiate between the industrial environment on the one hand, and the residential, commercial and light industrial environment on the other. Certain product standards offer other environmental definitions, and some standards allow significant relaxation in EMC performance if limits are placed on the operating environment, or if the product is only to be supplied to technically proficient customers. Particular examples of the latter are EN 61800-3 and EN 50091-2 for variable speed drives and uninterruptible power systems respectively. It is necessary to ensure that a product purchased for a system or installation will in fact be used in the environment for which it was intended.

2.2.4.2   Installation and assembly

The EMC of most equipment used within systems and installations will be affected to some extent by the way it is put together. The compatibility of electrical and electronic products will hinge on several installation practices:

- cable and connector type, length, layout and configuration;
- for screened cables, the quality of the screen itself and of its termination at either end, as well as what it is connected to;
- earth connections of enclosures;
- choice of enclosure type and design;
- proximity of equipment to each other, and to sources or victims of interference.

Any of these aspects may be significant for a particular apparatus; if it is, then the instructions for use should state this and should give directions for achieving compatibility. Typical instructions might be minimum segregation distances between different cable classes, instructions to connect cable screens in a particular way, or requirements to use a particular type of cable for a particular signal. Of especial note might be restrictions on cable length; some standards, including the generics, allow a relaxation in testing of ports to which only short cables will be connected. But this implies that the instructions must make clear that such a restriction is enforced.

If a particular product claims conformity to the EMC Directive's requirements, or to an EMC standard, with no especial instructions to the user or installer, it is reasonable to assume that such conformity is maintained however the product is installed or used. This of course is the preferred goal of any equipment supplier and, for that matter, of any system builder. But the need to achieve an economic design may dictate the acceptance of some control of build and installation methods.

2.2.4.3   Limitations on use

Some types of equipment will rely on operational limitations rather than, or as well as installation methods. Some examples could be:

- the need to control the proximity of portable radio transmitters; a particular case of this is the already-established practice of forbidding mobile phones in some parts of hospitals

- observation of static precautions because particular parts of the apparatus are especially sensitive to ESD

- possible degradations in performance to be expected in the presence of certain levels of interference (as allowed by the generic standards, provided the user is informed)

- the need to ensure an extended separation distance away from any radio receivers and other sensitive apparatus that could be affected by emissions from the equipment.

2.3   Routes to compliance

The EMC Directive allows three possible routes to compliance, one of which is only applicable to radio communication transmitters (Article 10.5) and requires the intervention of a Notified Body (NB). Since this is a rather specialized route it is not discussed further here. If a radio transmitter is used as part of a system it should be procured fully compliant with a copy of its Type Approval Certificate and can be treated as any other compliant item, with no need for NB involvement with the finished system or installation.

The other two routes are of direct interest. These are self declaration to harmonized standards, and use of the Technical Construction File. They are alternatives; it is not necessary to do both, although it is possible to take elements of each.

2.3.1   Self certification to standards

The Directive allows for standards which have been harmonized by publication in the Official Journal of the EC [123] to be used to give a "presumption of conformity" to the essential requirements. This means that if a manufacturer can find a suitable standard or set of standards that can be applied to the product, and are relevant for its intended environment, then he is able to apply these and having done so is free to make a declaration, apply the CE Mark and market the product. No third party involvement is needed.

The UK's EMC Regulations use the term "Self-Certification" but some EU Member States are uncomfortable with this, preferring the term "Self-Declaration" instead. It has been agreed at the EC that according to common parlance in the UK, the term "self-certification" has the same meaning as "self-declaration".

Since all EMC standards that have been listed in the OJEC are standards which prescribe test requirements and place limits on the outcome of the tests, it is normal for manufacturers to decide to do such testing. But it is not inevitable. Especially for small manufacturers of low volume products, the costs of the testing are such a burden as often to make it uneconomic to bring a product to market. Thus there has been intense pressure to find a formula which would enable the sensible implementation of the Directive without driving small companies out of business, but also without penalising those companies that wish to follow the requirements to the letter. This formula is enshrined in a document from the UK DTI entitled "Minimising the cost of meeting the EMC Directive" [126] and it hinges on an interpretation of the word "apply". The essence of this interpretation is that a company may apply a standard without testing if the company believes that, *were the product to be tested, it would pass*. Of course there are several caveats to this approach, not the least being that for many electronic products, actually making such a judgement is notoriously difficult. The final arbiter is naturally the test itself; but this approach puts the responsibility of taking the risk (and therefore saving the cost) of not testing squarely on the manufacturer.

From the point of view of the customer of such a manufacturer, there is clearly a risk involved in taking a declaration of conformity produced under such an approach at face value. This question is explored again under 2.3.3. Since under the EMC Directive a manufacturer assumes the legal responsibility for his whole product, including all the parts he purchased to build it, if he was relying on the proper compliance of a particular apparatus in order to make his own declaration of EMC conformity, it is advisable for him to have an idea at least of what tests were done and how. Existing case law in the UK suggests that actual evidence for the EMC conformity of the constituent parts of a system may be required, rather than merely their manufacturers' claim of conformity, to be able to mount a defence of due diligence in the face of a complaint of non-compliance of the system.

Merely passing an EMC test with an example of a product does not prove that the next one of that type to be manufactured is equally as compliant, and neither does the possession of an ISO 9000 approval certificate. A successful EMC test on a serially manufactured product only has any relevance for other units when the manufacturer has a quality assurance method in place that actually ensures continuing compliance. Such a QA method will involve EMC procedures for design, change control and concessions, modifications and variants, manufacture, documentation, as well as sample-based EMC testing. Chapter 3 goes into this issue in more depth.

2.3.1.1   Standards for products

For the manufacturer of individual products the standards route is fairly straightforward. There are many standards already harmonized for the EMC of products, and their number is increasing steadily. Even if there is not a standard that is directly applicable to a given product, the generic standards [69], [70] are available and can be applied in default to any product that will be used in a wide range of locations.

The main problem facing product manufacturers is actually in keeping up with the rate of introduction of new standards and the rate of changes to old ones. It is inevitable that as understanding of EMC technology and issues grows, so old standards (and especially the test methods that they specify) will become unsatisfactory and new ones or revisions will be needed. The widespread application of the old IEC 801 series of immunity test methods, for example, showed up many faults and inconsistencies in the originally specified methods and as a result new methods have been introduced. Similar changes are happening with the RF and harmonics emissions standards. Once a new version is published in the OJEC, it supersedes all earlier versions; but there is a transitional period during which the "presumption of conformity"[†] applies for either new or old versions. The Commission now publishes details of the transitional arrangements for each standard in the OJEC when it is listed.

The manufacturer's declaration of conformity must list the standard(s) that have been applied. The declaration does not have to be made available to customers, but it would be very unusual for a manufacturer to refuse to show it to a customer who specifically requests it; in many cases the declaration appears as part of the product documentation, although this is not mandatory. Even so, it is not always possible to tell from the declaration the full details of what is being claimed. For instance, a statement of compliance to EN 55 022 is not sufficient unless "Class A" or "Class B" is also stated. Compliance with the generic immunity standards involves the possibility of degradation of performance, and this degradation does not have to be stated on the declaration, nor even in the user instructions if it is what the user may reasonably expect. EMC-specific installation methods or operational restrictions do not have to appear on the declaration of conformity either.

2.3.1.2   Standards for systems and installations

In contrast with the situation for individual products, there are at the time of writing no harmonized standards which fully cover the EMC of systems and installations. Thus the system supplier who has to meet the EMC Directive's CE Marking requirement – that is, one who is not supplying an "excluded installation" (see 2.2.2.2) – has only two options: bend an existing harmonized product standard to fit the system, or use the Technical Construction File.

If the system is limited in size it is usually possible to apply a product standard to it as if it were a single item. For instance, both EN 55 011 [66] (for industrial, scientific and medical equipment) and EN 55 022 [68] (for information technology equipment) require a conducted emissions test on the mains port and a radiated emissions test at a fixed distance from the boundary of the product. The immunity tests may be a little harder to apply, especially for RF immunity, but are not impossible. The limiting factor on size is usually the capability of the chosen test house. This is closely linked to the cost of the test.

If the system is too large to test reasonably and realistically, or its ports and boundaries are widely distributed or ill-defined, it becomes impossible to apply the tests of a product standard and the Technical Construction File is the only option.

2.3.2   Using the Technical Construction File

The EMC Directive allows this option in cases where the manufacturer cannot, or is unwilling to, apply harmonized standards in full. The most likely reason for this is that

---

† The application of all applicable harmonized standards is said to give an apparatus *presumption of conformity* with the Directive's essential requirements.

harmonized standards are not available (although a suitable draft or non-harmonized version might be). This could be because of the nature of the product – as discussed above, it might be a system for which applying the usual test methods is impossible; or it might be because the product occupies an application niche for which no product standards have yet been developed but for which the generic standards are inappropriate, such as equipment for use down mines.

Another reason may be that the manufacturer wishes to certify a range of products against tests done on a worst case representative of the range. In fact, he can do this using the self-certification route if he is applying harmonized standards as far as the tests are concerned; as discussed above, it is not necessary to apply tests to every product variant that will be certified. But many manufacturers may be unwilling to take the responsibility of deciding which product to choose as the worst case representative, and will turn to external bodies for help and validation. Even so, it is not mandatory to use a TCF for this purpose; an EMC consultant's report in conjunction with the appropriate test results may be viewed as covering the risk of non-compliance in acceptable fashion.

2.3.2.1   The rôle of the competent body

A TCF, then, is used at the manufacturer's discretion when harmonized standards are not wholly applied, and in this case it is necessary to have independent overview by a "competent body". The competent body is a specific organization appointed by the authorities to provide this service for the EMC Directive. In the UK there are around thirty, and in the whole of Europe there are about one hundred. The competent body reviews the TCF documentation produced by the manufacturer, and then issues a certificate which must be included within the TCF. Only when this has been received can the manufacturer legally make the declaration of conformity, affix the CE Mark and market the product.

Although the purpose of the TCF is for use when standards are inappropriate or unavailable, the competent body is likely to be guided by the content of standards when assessing the TCF, and will be prepared to modify their strictures only when a reasonable technical rationale is presented. So for instance, a lack of emissions testing or relaxation in emissions limits is likely to be accepted only if it can be shown that the equipment will be used only in locations or circumstances where there will be no radio receivers to be affected. A TCF is certainly not a means of avoiding testing, and except in rare circumstances is likely to be more costly than self certification were that to be available, since the competent body has to assure itself that any tests that are carried out are performed in the correct fashion.

The advantage of the TCF approach is that the competent body accepts part of the liability for proper compliance with the essential requirements, and therefore the manufacturer's risk is lessened; it is much less likely that the enforcement authorities will challenge a competent body's opinion than they will a manufacturer's own self-declaration. Once the TCF has been approved, the manufacturer's responsibility is limited to ensuring that the equipment that is marketed is the same in all respects as that which is described in the TCF documentation.

2.3.2.2   TCF contents

The principal purpose of the Technical Construction File is to support the manufacturer's declaration of conformity to the essential requirements. As such its availability to enforcement officers is mandatory under the terms of the EMC Directive.

The content of any one TCF may vary from any other because no two types of product are exactly the same and their EMC characteristics may differ. The exact content and structure will be agreed between the manufacturer and the competent body assessing the file.

However, a general format for the document's contents can be offered and this has been agreed with the EMC Test Laboratories Association (and is intended to become a pan-European model through the Association of Competent Bodies) and is published by the DTI [125]. Table 2.1 gives a summary of the requirements. Even if the TCF route is not being followed, this structure is a useful rubric for an in-house document supporting self declaration.

Administrative details

According to the UK regulations, the TCF shall:

- be in material form
- be in English (if assessed by a UK competent body)
- give the name and address of the "responsible person" and if necessary of the manufacturer
- allow an enforcement body to identify the apparatus, and ascertain whether it conforms to the essential requirements
- state standards applied
- describe the suitable EM environment.

From the point of view of the enforcement officer, "ascertaining whether the apparatus conforms with the essential requirements" should simply be a matter of confirming that the product is as described in the TCF – for instance if it is a variant, confirming that the variant is included in the scope of the TCF – and that the marketing, user instructions and installation practice conforms with any conditions laid down by the Competent Body.

Identification of apparatus

The TCF can refer to a single item or to a range of variants which are technically similar. It must be possible to relate the documentation to the hardware that it describes, especially if a CB has based its assessment on inspection or testing of specific hardware; some assurance is needed that this hardware is representative of that which will be installed. Some reference to the build state of the apparatus in the document is essential.

Technical description

The purpose of the technical description is to make clear what aspects of the apparatus are relevant for EMC (and by implication, what aspects are not). Although some manufacturers interpret this requirement as meaning that the complete design of the equipment must be described, in practice it is usually possible and preferable to abstract only that information that is important for EMC. This is especially desirable if the apparatus is complex.

Rationale and design aspects

This section is the real meat of the TCF, as it describes *how* the apparatus meets the essential requirements.

A wide range of techniques can be included here, from use of specific components and design features through to purely procedural approaches. Examples of the latter might be an instruction to use the equipment only in a shielded room (if this is reasonable), or an instruction to observe separation distances from specific other disturbing or sensitive apparatus. Test apparatus, for instance, is allowed a dispensation in the UK regulations, but since there are no applicable standards, a TCF is necessary to use this dispensation.

Where compliance is based on use and proper installation of already compliant parts (for instance in the case of system integrators, see 2.3.3), it would be reasonable to expect to see a description of how the procurement and build procedures adequately control the EMC compliance of the parts used. However, a competent body is not expected to vet a manufacturer's quality system.

Test results

In most cases a TCF will include some testing. If standards were to be applied in full, a TCF would not be necessary, and therefore it is usual that some tests applied under a TCF will be non-standard. (Though it has been common to use a TCF to declare compliance against a standard that has not yet been harmonized, and therefore cannot be used for self-certification; if the TCF is subsequently reviewed, the standard might have been published in the intervening period.) They should be fully described within the TCF, along with some explanation of why they were carried out, and how they help to demonstrate conformance with the essential requirements.

The competent body certificate or report

It will be usual for the main body of the TCF to be written by the manufacturer in consultation with his chosen CB. The CB's role is to assert that the evidence contained within the TCF is consistent with conformity. Therefore it is quite possible that the CB's "deliverable" need be little more than a certificate stating that it agrees with the approach taken. Legally, a certificate has the same status as a report.

Alternatively, the CB report may include specific comments, including limitations and conditions which must be complied with if the certificate is to apply. These could include a limitation on the EM environment for which the equipment is deemed suitable, or that the equipment is only to be installed on a specific site, or limitations on the operational procedures to be used with the equipment.

2.3.3   Does "CE + CE = CE?"

The question of how to treat systems that are assembled from already-compliant apparatus has been controversial ever since the Directive was implemented. From the administrative point of view, it would be ideal if the CE Mark could be trusted to mean that any item bearing it would be compliant in any system into which it was assembled, and therefore that the system as a whole would be compliant, with no further action needed. Then, all that a system builder would need to do would be to ensure that they sourced only CE-marked parts, and they could declare compliance of their product without having to go through any more expensive testing or assessment. This approach has been encapsulated in the neat but fundamentally erroneous equation "CE + CE = CE".

Unfortunately for neat and tidy bureaucrats, life isn't like that. The CE Mark, in fact, says nothing of worth about the item at all. There are a number of reasons why the simplistic approach doesn't work:

Table 2.1 TCF contents

Identification of apparatus	• brand name • model number, or description of scope of variants • manufacturer/agent's address • functional description • photographs • operating environment • physical location, if an installation
Technical description	• functional block diagram • relevant technical drawings • interfaces with other products and devices • if relevant, description of variants and how they are related
EMC assessment	• operational and installation aspects that affect EMC
Rationale and design aspects	• explanation of how design and/or testing shows compliance with the essential requirements • design features specifically addressing EMC, and relevant component specifications (e.g., filters, shielded enclosures, cables) • procedures used to control variants • theoretical modelling • operational and installation practices
Test results	• list of EMC tests performed • test reports and description of methods • rationale for adequacy of tests • test results/reports on sub-assemblies, and how they relate to the whole equipment
Competent body certificate or report	• refer to the build state of the apparatus assessed • comment on technical rationale • states how design information was assessed • comment on procedures used to control variants • comment on relevant environmental, installation and maintenance factors • contain analysis of the tests performed • may include conditions

- the CE Mark, as said earlier in section 2.1.3, is merely a passport. It is not *intended* to make a technical statement about a product. It merely says that, if any of a number of directives apply, the manufacturer *declares* that the product complies with their essential requirements. It is entirely possible that a product may be CE Marked in respect of one Directive (say, the Low Voltage or Machinery Safety Directives) and not another (say, the EMC Directive) if that is how the supplier interprets the exposure of the product;

- because in the vast majority of cases the approach to CE Marking is self-certification, there is no guarantee backed up by a third party that the product does actually comply with anything, even if it carries the CE Mark. Anybody who relies on the CE Mark alone is placing themselves at the mercy of the supplier's interpretation of the requirements, his quality system and, in the last resort, of his integrity;

- the manufacturer is able to put a number of limitations on the conformity of his product. These can include a limitation on the environment in which it can legitimately be used, a set of instructions which must be followed if the product is to remain compliant, or an allowable degradation in performance when the product is subject to interference. Any or all of these may conflict with what the system builder requires of the product, but the CE Mark of itself says nothing about them;

- analogous to the confusion and misunderstanding surrounding the CE Marking of systems and installations, there is similar confusion regarding the exposure to the EMC Directive of components. There are several areas where a manufacturer could quite legitimately claim that he is making products which, as components, are outside the scope of the Directive and thus do not need CE Marking. Therefore the system builder is left unable to source CE Marked product anyway;

- in many cases, the integration of individual parts into a product radically changes the EMC characteristics of the part. Even if it was compliant when tested on its own or in a "representative" system, when used in a different way in a different system its own compliance may well not be guaranteed.

For all of these reasons, relying as many system builders have been doing on procuring only CE-Marked parts is a recipe for disaster. At least two companies in the UK have already suffered such a disaster, when they were prosecuted in 1997 for supplying non-compliant personal computers which had in fact been built out of CE-marked subassemblies. In mid-1999 the same fate befell two more PC companies. Due to the general confusion surrounding the CE Mark and because of ill-informed customer pressure to purchase only CE-Marked products, many component suppliers are looking to find ways to apply the CE Mark when it is unnecessary and may even be misleading – a far cry from the earlier days when the majority were looking to find ways to *avoid* CE Marking! In the UK regulations, applying a CE Mark to non-relevant apparatus is an offence, but this is not so in other countries. Such a practice further undermines any credibility the CE Mark may have been intended to have.

2.3.3.1 *The correct approach*

However, it is entirely reasonable for system builders to want to shift at least some of the burden of compliance onto their suppliers. What, rather than a blind insistence on purchasing CE Marked parts, should they be doing?

The necessary actions break down into:

- decide what are the EMC requirements that are imposed as a result of the operation of the system; these will be a maximum level of emissions and a minimum level of immunity dictated by the operating environment, along with a defined performance in the presence of interference

- check that each part that is built into the system is either irrelevant to these requirements or, if it is not, that it meets an equivalent or better specification; this will normally mean that it is compliant with a specific set of EMC standards, but it is necessary to be familiar with both the standards the system must meet and those that each part is said to meet before you can be sure that a match has been achieved

- check that any conditions that are imposed by the supplier in relation to meeting the claimed standards are acceptable in the context of your system

- finally, if there is any question as to whether your method of integrating the part into your system will affect its compliance status, assess whether or not you are going to have to include some further EMC testing in the programme, either on the system as a whole or on one or more of its sub-assemblies.

As can be seen, these actions require a fair amount of engineering knowledge and cannot usually be delegated to a purchasing manager with the instruction to "buy only compliant parts"! The management implications of this are spelt out more fully in Chapter 3.

2.3.4   The "Procedural Approach" to compliance

System builders producing small volume or one-off custom systems are faced with the problem of self-certifying or having a TCF approved for each new system, and where the costs of these systems are a few thousand pounds the financial burden of EMC compliance can seem unreasonable. There is a way out of this, the "procedural approach". This is similar to the approach described above to cover a range of products by testing one or two "EMC worst cases". Many system builders create completely different systems every time they take on a contract, but from an EMC point of view these systems often exhibit strong commonality.

For example, a control and instrumentation system builder may one day construct a system for a chemical process, and the next day construct a system for a car assembly line. These are quite different as far as functionality goes, but they are made of similar "parts" (measurement of temperature, flow, position, velocity, weight, etc., power supplies, motor drives, motors and pumps, computers and programmable logic controllers) and they are installed in a similar fashion (control cabinets with cables to the transducers mounted on the machine or process plant).

A "procedural approach" would only allow the use of parts which were known to have adequate EMC performance for the environment they were to be used in, ensure that the parts manufacturers installation instructions and limitations to use were followed, and would then ensure that EMC best-practices (described in some detail in this book) were always followed in design, assembly, installation, commissioning, servicing, and modification.

Some EMC testing (preferably on the "worst-case" examples) resulting from this procedure, on a reasonably regular basis, will still probably be required, but it avoids the need to test each system for conformity. This testing could be carried out by the system builder in his own factory, on the customer's site, or in a test laboratory.

It is possible to declare EMC conformity for a huge range of possible future systems, even though it is not known what they might be required to do or where they will be used, on the basis of a correctly constructed "procedural approach".

Although this approach can obviously save significant amounts of money and time, it must be recognized that it does involve other costs. A certain amount of EMC expertise is required to set it up and maintain it. Smaller companies may need to rely on the expertise of EMC consultants, whereas medium sized or large companies may find it more cost-effective to employ their own EMC experts to manage such a system. Where the systems concerned cannot be self-certified to harmonized EMC standards (for reasons described earlier), such a "procedural approach" must be made into a TCF and assessed by a competent body.

2.4   Enforcement and the future of the EMC Directive

2.4.1   The UK situation

Enforcement in the UK is carried out by Trading Standards Officers. TSOs have the sanction of court proceedings and withdrawal of the product from the market available to them, but have emphasized in their public statements that they view these as a last resort, and prefer to work confidentially with manufacturers to "encourage" them to achieve compliance. This is achieved by checking for the presence of CE Marking and associated documentation, and following this up with factory inspections and if necessary advice to modify an offending product within a fixed period. In most cases the mere threat of having to withdraw a product from the market is sufficient to make manufacturers bring their errant products rapidly into compliance, and several examples of this (sometimes occasioned by a competitor's complaint to their local Trading Standards Officer) are known. Thus, without several high-profile court cases, it is difficult to judge the extent to which the risk of non-compliance is affecting manufacturers' strategies.

In fact, the experience of many system manufacturers is that the principal pressure for compliance comes from their customers (as discussed earlier in this chapter). The chain of supply in systems projects often involves several levels of contractors; even if the final system does not require CE Marking, the prime contractor has some responsibility for meeting the essential requirements, and will normally place contractual requirements on its suppliers to ensure this.

2.4.2   Other countries

Enforcement of the EMC Directive in other countries within the EU is somewhat uneven. A 1997 review [14] identified enforcement activity in Sweden, Germany and Italy, and an article by a technical lawyer [27] has provided further information, summarized as follows.

Austria

A single national body is charged with enforcement of the legislation. The maximum fine is 35,000 schillings (approximately £1,920 or US$3,072).

Belgium

A single national body is charged with enforcement of the legislation. The primary means of enforcement is by prohibiting the non-compliant equipment from being sold. If the product is subsequently sold, a term of imprisonment of up to five years could be imposed.

Denmark

Local enforcement authorities are responsible for enforcement of the legislation. It is expected that the enforcement will be passive but products are being actively checked to determine whether they have the CE Marking on them. A fine can be imposed upon manufacturers for selling non-compliant equipment.

Finland

A single national body is charged with enforcement of the legislation. The maximum penalty is six months in jail or a fine of several thousands of markkas (approximately £3,000 or US$4,800).

Germany

The German version of the EMC regulations allows the enforcement authorities (the Federal Office of Post and Telecommunications, BAPT) to impose fines for infringing the requirements without taking companies to court. The German system has established 54 regional offices linked to a central database for "market surveillance". This processes around 1500 technical documents a month. In 1996, over 100 cases were investigated, resulting in fourteen companies receiving administrative fines, five others receiving formal warnings with fines, and eight more receiving formal warnings.

Iceland

The exact position in Iceland is unknown but the enforcement will be by a single national authority. It is also understood that there will be some active surveillance of the market.

Ireland

The Irish legislation is closely modelled on the English legislation. However, a single national authority will be responsible for enforcement. The maximum penalty for non-compliance is 1,500 punts (approximately £1,530 or US$2,448).

Italy

Apparatus that does not comply is liable to seizure and confiscation. The maximum penalty for sale of non-compliant equipment is 90,000,000 lire (about £35,000). A lower fine of 30,000,000 lire applies where the offence relates to the failure of a manufacturer or importer to affix the CE marking to relevant apparatus. A person who sells apparatus that does not carry CE marking when it should risks a maximum fine of 18,000,000 lire.

The Italian government has agreed with several EMC test houses to undertake testing for enforcement, without charge.

Luxembourg

A single national authority (the regular police!) is responsible for enforcing the legislation. There are no specific penalties laid down in the legislation for failure to comply.

The Netherlands

A single national body is responsible for enforcing the legislation. The penalties are variable and may be fixed individually by the court on a case by case basis.

Norway

A single national body is responsible for enforcement of the legislation. There is some active enforcement of the legislation through random sampling. In appropriate circumstances, the authorities can order the supplier to undertake a product recall of non-compliant products. This compulsory product recall is the main means of enforcement.

Portugal

A single national body is responsible for enforcement of the legislation. The maximum penalty for enforcement is 3,000,000 escudos where a company breaches the legislation (approximately £11,550 or US$18,480). The maximum penalty where an individual breaches the legislation is only 500,000 escudos (approximately £1,920 or US$3,072).

Sweden

Sweden was the first country to remove products from the market for EMC Directive non-compliance. Three electric motor power drive converters, from large multi-national organizations, were tested, found non-compliant, and the Commission was notified. Sweden's policy has been to focus on products which are known to have high emissions; other products such as electric tools and fluorescent lamps have also been under investigation.

The main penalty provided by Swedish legislation is withdrawal of product from the market. In appropriate circumstances, the Swedish authorities can order the recall of all previously supplied non-compliant products.

2.4.3   The SLIM initiative

In 1996 the European Commission launched the SLIM initiative (Simpler Legislation for the Internal Market) with the purpose of reducing the legislative burden for European industry. The EMC Directive was reviewed under SLIM in the summer of 1998 and the review committee (just ten members, drawn from several European countries) presented their report to the Commission in the autumn. The recommendations covered the following ground:

- basic principles of the Directive;
- installations and large machines;
- conformity assessment procedures;
- standards;
- EMC requirements in other Directives;
- impact of the Commission guidelines.

In fact, many of the recommendations are non-controversial. The immunity requirements should be more fully and explicitly addressed in the Directive, but should not extend to functional safety. A strategic review panel for EMC standards is advised, looking at the relevance and applicability of all existing EMC standards, and at the necessity of a new mandate to CENELEC and ETSI with the aim of producing fewer and more useable standards. No change to the conformity assessment procedures is recommended. The Directive should be reviewed in the light of the guidelines, to incorporate specific definitions, the EMC analysis process, the procedure for application to installations, apparatus and systems with various configurations, and any other areas that could be usefully transferred.

The most interesting recommendation, and that which is most relevant to the subject of this book, is that "in the absence of complaints, installations and large machines should not be subject to assessment tests". For clarification, the definitions of installations large and small, large machines and networks should be added to the Directive. It is made clear that fixed installations should not need CE Marking, nor a declaration of conformity nor the involvement of a Competent Body. The Directive should allow installations to be constituted either by CE marked apparatus, or by apparatus with CE Marking and parts without, or by no CE marked parts at all. The Directive should be amended to say that compliance of a fixed installation with the essential requirements is ensured by following the EMC assembly instructions given by the manufacturer of the constituent parts and using good engineering practice in the method of installation.

None of this seems to take us any further forward than the existing interpretations that attach to the Commission's 1997 guidelines. And the likely timescale for implementation of the SLIM recommendations, some of which necessitate modifications to the wording of the EMC Directive, stretches into several years. For these reasons, the general reaction of industry at the end of 1998 has been to pay little attention to the outcome of SLIM for EMC, and to carry on in much the same way as before.

Chapter 3

Management of systems EMC

In the Castle the telephone works beautifully of course, I've been told it's going there all the time, that naturally speeds up the work a great deal. We can hear this continual telephoning in our telephones down here as a humming and singing, you must have heard it too. Now this humming and singing transmitted by our telephones is the only real and reliable thing you'll hear, everything else is deceptive. There's no fixed connexion with the Castle, no central exchange which transmits our calls further. When anybody calls up the Castle from here the instruments in all the subordinate departments ring, or rather they would all ring if practically all the departments – I know it for a certainty – didn't leave their receivers off. Now and then, however, a fatigued official may feel the need of a little distraction, especially in the evenings and at night and may hang the receiver on. Then we get an answer, but an answer of course that's merely a practical joke. And that's very understandable too. For who would take the responsibility of interrupting, in the middle of the night, the extremely important work up there that goes on furiously the whole time, with a message about his own little private troubles?

– Franz Kafka, The Castle

EMC does not happen by itself, whether we are looking at a product, a system or an installation. As with any other aspect of the process, it must be managed. This chapter looks at the tools that project managers need to ensure that EMC is properly accounted for throughout the project.

The level of detail and activity to which EMC control is taken will depend on the degree to which EMC affects the outcome of the project. A space system, for instance, will have many more critical aspects than a building management system and will justify more effort in the planning. This chapter therefore presents the control activity in its fully-developed form; some areas can be reduced or omitted as circumstances dictate.

3.1   The EMC control plan

A schematic form of the structure of a system development plan with respect to EMC is shown in Table 3.1. The control plan has to fit into this development structure; its initial release creates a rubric for the EMC aspects of the programme including basic design guidelines, while subsequent updates incorporate refinements as the programme progresses.

3.1.1   The EMC control board

It is essential to have a management structure defined for the EMC activities. This is achieved by establishing an EMC control board or committee, whose purpose is the effective execution, on schedule, of the EMC control plan. Typically, it will be chaired by the prime contractor's project manager, or a deputy with specific EMC expertise. A large part of the control board's activity will be concerned with sourcing equipment to

Table 3.1 EMC deliverables in the system development programme

Programme milestones	EMC deliverables
Project inception: establish EMC control board	EMC performance specifications: 　System interface requirements 　Known operational environment(s)
Requirements definition review	Define system level EMC requirements: 　Define module level emissions and immunity requirements and procurement specification (choose standards to apply) 　Define power system compatibility requirements
Preliminary design review	Complete EMI environmental assessment for radiated and conducted requirements Define power distribution and earthing/isolation architecture
Submit first EMC control plan	Assign cable design engineering function and responsibility for module and system interconnections: define wiring layout and shielding considerations Preliminary test for power line conducted emissions to define extra filtering needs Identify specific EMC threats and victims
Critical design review	Review engineering model test data Verify module-level and equipment-level requirements Define and resolve outstanding EMC issues
EMC compliance test	System tests (if necessary) for system-level EMC assurance Pre-shipment or in situ conformance tests

the correct specifications, so the board must have procurement authority. Given that normal practice in system development is to proceed via a series of review stages, progress to the next stage being allowed only on successful review of the previous one, the control board must also have authority in the review process. Finally, it must have the ability and the resources to define and test the various module- or equipment-level requirements that will ensure total system-level compatibility.

3.1.2   Identifying EMC issues

A checklist for the EMC issues to be addressed is a good starting point for any control plan. Various sources of such checklists exist, for instance [73], [116], [6]. Our own checklist for managers is given in Appendix A. The following bullet-point list is derived from prEN 50 174-2 [73], regarding IT installation planning within buildings:

- nature of the building
- nature of the power distribution system
- catalogue of disturbing sources
- customer requirements for security of performance
- structure of existing and/or future earthing network
- layout and earthing of cable conduits/ducts/trays/raised floors.

3.1.2.1   The environment

A crucial initial stage in the EMC control of any systems project is to assess the

Table 3.2 The content of the control plan

	Content
EMC programme management	Procurement responsibilities, reporting structure, control of design changes Planning and scheduling the control programme, including resources, co-ordination and review procedures
System level performance and design	Definition of the intended environment Definition of critical circuits and modules System and sub-system design issues: Controlled earthing scheme, structural bonding, wiring layout, shielding and termination practices, corrosion control etc.
Sub-system performance and verification	Allocate EMC performance at the equipment level, via specific standards or tailored requirements Integrate the test results from equipment level EMC tests, to analyse and determine the effects of acceptable degradations
EMC analysis	Prediction of intra-system compatibility based on known sub-system characteristics, together with solutions for predicted or actual interference problems
System level verification	System-level test plan, including rationale for selection of critical sub-systems for safety margin demonstration, performance criteria and operating configuration for in situ compliance tests

electromagnetic environment that will be enjoyed (or suffered!) by the system. This will in turn define the degree of rigour that will be needed by the system's components in dealing with various phenomena, in the levels of stimulus or limits of emissions that are applied, and in the amount of degradation of functional performance allowed for each function for each disturbing phenomenon. The various product and generic standards that are available to some extent present such a picture of the environment, but only in very broad terms. For a closer match for any given project, a more specific review of the intended environment is needed.

For most commercial and industrial systems projects, this review can use as its basis the international standard IEC 61000-2-5, "Classification of electromagnetic environments" [87]. The advice given in this standard is expanded in Table 4.1/2, presented in detail in Chapter 4. Other application environments – for instance military, space, automotive, railway – emphasize different EM phenomena and most have explicit documents detailing these. Using Table 4.1/2 as a guide, the environment for a given installation can be tabulated, and this will provide a clearer picture of the overall requirements that will be needed for individual apparatus and for the installation practices that are used. In some cases, the requirements may be relaxed over those inherent in the generic or product standards, in others they will be more onerous.

3.1.2.2   System interface requirements

The system components interface both with each other and with the greater environment and must exhibit electromagnetic compatibility in each case. Since interference is propagated through the interfaces (including the enclosure as an interface, in the same manner as cable connections), identifying these and listing the environmental phenomena that may apply to each is a necessary step towards specifying the compatibility levels and tests that will be necessary.

3.1.2.3 Module level emissions and immunity

Once the interface requirements have been defined based on the environment, then the compatibility requirements for each module can be listed, and hence the test compliance specification for each module. These can be separated into emissions control and immunity control. Bear in mind that RF emissions control is intended to protect victim radio receivers, while RF immunity is intended to guard against aggressive radio transmitters: there is a large difference in amplitude (microvolts per metre versus volts per metre) between these. In many cases the requirements are covered by compliance with the generic or product standards, but in some cases there will be extra or more onerous needs. This would be the case if for instance portable radio transmitters and/or RF production equipment (such as dielectric welders) are known to be used in close proximity to the apparatus, a situation which is specifically excluded from consideration in most harmonized immunity standards.

3.1.2.4 Power system compatibility

The mains power supply is a fundamental interface to many if not all functional modules in a system, and compatibility with disturbances that exist on it – and control of disturbance generation – is necessary for all such modules. IEC 61000-2-5 lists a number of interference phenomena that are associated with propagation via the power supply. Many of these are low frequency phenomena, some of which (such as frequency and voltage variations within operational limits) may well be regarded as functional rather than EMC specifications. The power supply connection also acts as a conduit for high frequency disturbances which are much harder to predict and deal with other than by rigorous filtering at the module supply interface.

The basic characteristics of disturbances on the supply may be similar throughout the system. But if the supply is broken into a number of different segments within the system, and if each segment has some degree of filtering or suppression isolating it from others, then the supply interface EMC requirements may well be different for different clusters of components.

3.1.2.5 Earthing, bonding and isolation

An important tool in EMC control for any system is a controlled equipotential bonding scheme. In general, using any part of the mechanical structure as a *functional* electrical current path is to be avoided. All return paths should be carried in their relevant cable harnesses. Where functional connection via the metalwork is necessary for weight or topological reasons, the current path through the structure must be established and controlled, and the impedances across significant paths (and across the whole of the applicable frequency spectrum) have to be analysed. Whilst this is possible for some closed systems such as aircraft and spacecraft, it normally isn't for distributed systems in buildings and industrial plant.

Use of the structure for interference (as opposed to functional) earthing is often necessary, and this brings in considerations of electrical bonding and corrosion control. Bonding is needed for the management of interference current paths and to minimize voltage differentials across the structure both for interference and safety purposes.

3.1.2.6 Cabling issues

System components are interconnected at their interfaces, and these interconnections are typically made by electrical cables. The effect of the cable engineering on EMC issues has historically often been neglected, but it should form an important part of the overall EMC control plan. Good cabling design can enhance the barrier between the system and external interference, while poor design can worsen it. All aspects of cable design are relevant: routing, segregation, allocation of signal classes, choice of signals in each bundle, choice of conduit, not to mention the selection and termination of the cable itself. The decision as to whether to use shielded or unshielded cable should be made at this stage, once the characteristics of each interface have been detailed as already discussed.

Other considerations may dictate a prior choice of cabling: for instance, structured cabling in a building services system may already be in place, so that the choice of cable layout, segregation and shielding may not be available. In this case, the control plan is driven backwards, towards specification of the interfaces based on the known limitations of the cabling, rather than vice versa.

3.1.3   Identifying and sourcing critical parts

Systems are composed of interconnected modules. There are two aspects to the criticality of a module in this context: functional criticality, and EMC criticality. That is, how important is a module to the operation of the system, and how important is it to the EMC of the system. To take a couple of near trivial examples: the emergency stop switch is absolutely critical to both the safety and the functioning of a process control system, but because it is entirely electromechanical it has no EMC importance whatever. Vice versa, we can visualize an electronically controlled or fluorescent lighting unit which has no effect on the daytime running of the system and perhaps only marginal effect at night, yet which creates unacceptable interference perhaps both inter- and intra-system when it is switched on.

Table 3.3 Functional criticality

Functional category	Description
Safety critical	EMI problems could result in loss of life, property or the system
Mission critical	EMI problems could result in injury, damage to the system, loss of or delay to operation, or performance degradation which unacceptably reduces operational effectiveness
Non critical	EMI problems could result only in annoyance, nuisance or minor discomfort, or acceptable and temporary loss of performance

Table 3.4 EMC criticality

EMC category	Description
EMC benign	Will not cause or suffer from interference; has no effect on the EMC performance of the system (e.g. passive switches, indicators)
EMC relevant	Will not of itself cause or suffer from interference, but may have an effect on the EMC of related items; e.g. cables, connectors and enclosures
EMC critical	Active electronic equipment, other interference sources such as motors and lighting equipment

Categories of functional criticality can be divided up as shown in Table 3.3. These will determine the weight that is given to the electromagnetic immunity of each item (see also 3.4.1.4). EMC criticality (Table 3.4) is a classification of whether and to what extent EMC issues need to be considered for this item. Partly, it is a measure of the "electromagnetically benign" nature of the part (see section 2.1.1.4). But there are some items which though they are not "electromagnetically active" still need consideration from the EMC viewpoint, since they affect the compatibility aspects of other equipment. Cables are a good example of this.

The interplay between functional criticality and the EMC phenomena which may degrade various functions is discussed in more detail in Appendix B.

3.1.3.1   Control of procurement

The non-benign system components which are bought in must be subjected to some overview at the purchasing stage. Section 3.4 discusses this in greater detail.

3.1.3.2   Change control system

It is in the nature of purchasing that parts become obsolete or unobtainable and have to be replaced by other parts. To control the impact this has on EMC, there should be a system which protects the compliance status of bought-in parts. This can typically be a review of the aspects described in section 3.4 of this chapter (declaration of conformity, assembly instructions etc.) which is triggered whenever a change in specification occurs. Obviously this is not necessary for EMC-benign parts, and therefore the parts list for each system must identify which parts are relevant or critical, so that the change control system is only invoked when necessary. This should be no more than is already done within a company's manufacturing system for other purposes: in such cases the extension to EMC aspects is only administrative.

3.1.4   Control of assembly and installation

Maintaining EMC requires observance of a number of specific assembly and installation practices. These may include:

- implementation of protection zones in the installation building
- bonding of the structural components of racks, cabinets and ducting
- layout and connection of apparatus, filters and surge protectors within cabinets
- layout, segregation and routing of cables within the system
- proper termination of screened cables.

The EMC control plan should show how this is to be achieved. It would be usual to invoke in-house company practices, provided these were adequate, though in many system companies that have had no previous EMC experience these will have to be imported or built from scratch. Even if there is an existing body of in-house practice, this will not always be enough to ensure a fully compliant result, and may in some cases work against it. Two further possibilities exist:

- the installation instructions for particular apparatus mandate different, extra and on occasion mutually exclusive requirements with respect to the company practices;
- testing of a complete or partial system reveals an area in which performance is non-compliant.

In either case, the control plan must provide for a flexible and informed response to such situations, and a means of ensuring that changes to the build state which could affect EMC aspects are reviewed.

3.2   The EMC test plan

Testing a system or installation in situ is described in more detail in Chapter 10. The project schedule will require an idea of the tests that are needed, what they will be applied to and how long they will take. This is the purpose of the test plan.

An EMC test plan is a vital part of the specification of a new system. It can act as the schedule for in-house testing, or can be used as a contractual document in dealing with an external test house. Although it should be prepared as soon as the project gets underway, it will generally need revision as the system itself develops and especially in the light of actual test experience on different parts of the system.

The content of the EMC test plan will be determined by the specification of the electromagnetic environment to be withstood and the functional degradation permitted for each immunity test phenomenon, and the maximum levels of emissions not to cause interference problems with other apparatus. It will also be determined by the requirements of the applicable harmonized standard(s) if these are being used, or by the competent body if the TCF route is being followed. Typically it will need to cover the subject areas shown in Figure 3.1.

Content of the test plan

- Description of system under test
- Statement of test objectives
- The tests to be performed and their schedule
- Functional performance criteria for immunity
- Criteria for determining monitoring/injection points
- Description of ancillary equipment, simulators & software
- Details of the test set-up
- Evaluation of test results

Figure 3.1 The content of the test plan

3.2.1   Defining the configuration to be tested

The most important first stage in the test plan is to define what the system (or collection of modules) is that will be tested. Obviously this should bear some resemblance to that which is to be certified; but it is not always possible or reasonable to test the exact system that will enter service. Very often it is required to test a configuration that will be used to represent the EMC performance of several other options. In this case, a study of the justification for the representativeness of that configuration will be needed.

"Configuration" in this context refers both to the make-up of individual modules in the system, and to their layout and interconnections. Equally, it applies to their mode(s) of operation, which in turn will lead on to the performance criteria for acceptable operation to be applied in the immunity tests. All of these aspects should be fixed in the test plan.

3.2.2   Defining the tests to be done

Frequency ranges and test equipment will be determined by the standard(s) in use, although there may be a desire to extend the chosen standard's coverage. Most standards have specific requirements for test equipment, for example CISPR 16-1 on instrumentation. An external test house will be able to determine the instrumentation that they will need to use to cover the required tests; otherwise, that is the manufacturer's responsibility.

The number of ports to be tested, and the number of different functional modes of operation to be tested, directly influence the test time. The applicable standard(s) may define which ports should be tested. It may be possible to test just one representative port and claim that it covers all others of the same type.

The point at which the test is made can be critical, especially for testing cable emissions or immunity in large systems, and in the application of electrostatic discharge. Some pre-testing is usually needed to find the most appropriate point, and the results of this should be noted.

Both the manufacturer and any subsequent assessment authority should know why it was decided to apply tests to particular points on the EUT. These may be specified in the chosen standard(s), such as the mains lead for conducted emissions. But for example the choice of ESD application points should be supported by an assessment of likely use of the equipment and/or some preliminary testing to determine weak points. A decision not to test emissions or immunity of certain connected signal or I/O leads may rest on an agreed restriction of the allowable cable length that may be connected to the ports in question. The use of a voltage probe rather than a coupling network for supply line emissions measurement may be due to insufficient current rating of the available coupling network.

3.2.3   Testing system modules

Part of the test plan is likely to include tests on individual components of the system; this might be covered by the procurement specification on each module, but in some cases the system builder will find himself responsible for such tests.

Some military/aerospace tests require that each module is tested in conjunction with the other items to which it will be connected in the final installation. Besides not affecting the outcome of the test, these items should also offer the appropriate RF terminating impedance to the connected cables. Thus the description of the ancillary equipment set-up must be detailed enough to allow the RF aspects to be reproduced.

The basic description of the EUT must specify the model number and which (if any) variants are to be tested under this generic model type. If it can only be tested as part of the whole system, e.g. it is a plug-in module or a computer peripheral, then the components of the system with which it will be tested must also be specified. The test results must not be compromised by a failure on the part of other system components.

If the EUT can form part of a system or installation which may contain many other different components, a representative system configuration must be defined for test purposes. The criteria on which the choice of configuration is based, i.e. how to decide what is "representative", must be made clear.

3.2.4   Testing the whole system

Because of interactions between modules, and because of layout issues, testing of individual modules on their own rarely gives adequate information about the EMC of the whole system. A full-system test is usually necessary for at least some phenomena, especially the RF-related ones. An important aspect in such tests is the detail of the physical layout of what is to be tested.

Layout is defined in general terms in the various standards, but the instructions given in these will have to be interpreted to apply to the particular EUT in question. Critical points are distances, orientation and proximity to other objects, especially the ground plane. The final test report should include photographs which record the set-up, as well as sketch drawings showing relevant distances.

Cable layout and routing has a critical effect at high frequencies and must be closely defined. A cable which is run close to the ground plane and in the opposite orientation to the measuring antenna, will radiate far less than one which is suspended in free space and aligned with the antenna. Types of connector and cable to each relevant module should be specified, if they would otherwise go by default, as the termination affects the coupling of interference currents between the module and the cable.

The system may benefit from special software to fully exercise its operating modes; if it is not stand-alone it will need some ancillary support equipment. Both of these should be described and calibrated or declared fit for purpose. If the support equipment is not to be subject to the tests it can be interfaced via filtering or by separation distance which will reduce fortuitous emissions and isolate it from disturbances applied to the system. This filtering or separation arrangement will need to be specified.

If there are several different operating modes then it may be possible to identify a worst case mode which includes the majority of operating scenarios and emission/susceptibility profiles. This will probably need some exploratory testing. Choice of mode has a direct influence on the testing time. The rate at which a disturbance is applied or an emission measurement is made also depends on the cycle time of the specified operating mode. If the system only emits, or is susceptible, during a particular part of its operation then this must be synchronized with the test cycle (Figure 3.2); or the operation cycle can be "patched" to run continuously.

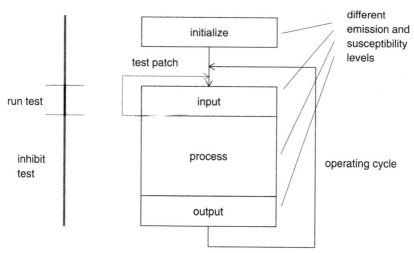

Figure 3.2 Synchronizing the test with the operating cycle

The order in which tests are applied and the sequence of operating modes should be specified; the results of one test may unintentionally (or intentionally) set up conditions for the next, or it may be possible to damage the EUT by overtesting, especially with high energy surges – these should be left till last as a precaution.

3.3   Documentation

Companies who CE Mark their products via the self-certification route for both the EMC Directive (EMCD) and safety Directives such as the Low Voltage Directive (LVD) and Machinery Safety Directive (MSD) are declaring that each item meets the essential protection requirements of each Directive. To be able to do this, there should be a documented rationale which supports it. Although the LVD and MSD explicitly require that a Technical File can be produced to an enforcement officer (the EMCD does not), the principal reason for generating CE Marking documentation is internal: it provides a means by which the responsible signatory can confirm that their signature on each Declaration of Conformity will be valid.

For this reason, companies develop CE Marking files for each product, which bring together in one place all information needed to support the declaration. Although not legally required, it would be typical and most cost effective for these to cover all applicable Directives. The contents of such a file for EMC can be loosely based on the recommendations for an EMC Technical Construction File, adapted as necessary.

3.3.1   Documentation for in-house use

The general practice in firms which follow quality procedures within the ISO 9000 structure is to maintain design and production documentation that describes and controls each product. The CE Marking document should simply be an addition to this practice, and will typically refer to much of the material produced anyway as a matter of course.

3.3.2   Documentation for a TCF

Section 2.3.2 and Table 2.1 describe the layout and general content of the Technical Construction File. The detail will be determined by the requirements of the Competent Body. To re-iterate – since there is still much confusion over this – *legally*, the Technical File for MSD and LVD purposes is not the same type of document as the EMC Technical Construction File. *Practically*, similar types of information (relating to safety on the one hand, EMC on the other) would appear in each.

3.4   Purchasing

Many final apparatus contain complex electrical and/or electronic items which have been purchased from other suppliers, for example:

- Finished products may contain bought-in sub-assemblies such as computer boards, or complete units such as power supplies, PLCs, computers, motor drives, panel meters, instrumentation and control modules, etc. (some of which may be finished products in their own right).
- Finished systems and installations are usually constructed from bought-in finished products, and systems, such as computers, telecommunications gear, instrumentation and control equipment, machinery, etc.

Compliance of the final apparatus depends upon the EMC performance of the bought-in items. Liability for non-compliance cannot easily be passed on to the supplier of a non-compliant item. Even where this may be possible, contingent losses such as product recall costs or harm to brand-image may well prove impossible to recover from suppliers or their insurers.

Where a final apparatus is found to be non-compliant by reason of the non-compliance of an incorporated item, enforcement agencies are likely to take action against both the final manufacturer and the supplier of the item.

The correct way to ensure that incorporated items do not compromise the compliance of the final apparatus is not to rely on CE marking, but instead to ensure that their EMC and safety engineering performance is adequate. The recommendations below are straightforward for engineers to adopt, being similar to the process they go through to ensure that functional performance is adequate. These recommendations make it quite easy to achieve due diligence for the final apparatus whilst also minimising development and manufacturing costs and timescales and reducing business risk.

Even so, some companies may not feel that they have the resources to follow these recommendations. As long as the sum total of the hazards and risks of their products to users are low in the global sense (i.e. only a few people exposed to low levels of hazards) they may be able to demonstrate due diligence without going through all of them, but in such cases it is recommended that their local enforcement agency should be asked to endorse their approach.

3.4.1   Determining the EMC specifications for an incorporated item

3.4.1.1   Electromagnetic stresses on the final apparatus

To begin with it is necessary to decide which EMC and safety standards the final apparatus needs to comply with, taking into account its physical and electromagnetic environments.

This may not be as simple as choosing harmonized standards from a list, because harmonized EMC standards may not adequately cover the actual electromagnetic environment, and harmonized safety standards may not adequately cover all the essential health and safety requirements for the physical environment. Other standards may have to be employed, and/or unique specifications written, to ensure that the final apparatus meets the essential requirements of all the relevant Directives.

If, say, an operator is expected to use a walkie-talkie radio handset whilst controlling a machine, the generic industrial EMC immunity standard (EN 50082-2) will not be tough enough to cover the level of exposure to VHF or UHF fields expected for the control panel. Neither of the early generic immunity standards (EN 50082-1:1992 and EN 50082-2:1995) include tests for the AC supply surges caused by distant lightning, or by the small dips and dropouts normally experienced on AC supplies, so additional EMC standards to cover these environmental conditions may need to be applied.

3.4.1.2   Foreseeable extremes and misuse

For the EMC of non-critical functions it is enough to consider the normal operating environment of the apparatus. But for the EMC of all critical functions it is necessary to consider all reasonably foreseeable situations, even if they have a low probability. This includes the probability that an operator or visitor will use a mobile radio device (e.g. cellphone or walkie-talkie) in areas where their use is banned.

Where electromagnetic interference can cause a safety hazard or increase risk, such possibilities are covered by safety Directives and *not* by the EMC Directive, so they should figure in all risk analyses under these Directives. An example here is the possibility of interference with a PLC controlling an industrial robot, causing it to "go wild" and operate outside of its programmed range. It is known that some robot manufacturers do not consider this safety risk when creating the technical documentation required by the MSD, despite only guarding for the robot's programmed range, and despite deaths known to have occurred in Japan due to this very problem.

3.4.1.3   Physical and electromagnetic stresses on an incorporated item

Having determined the physical and electromagnetic stresses for the final apparatus, the specifications for the items to be purchased may be derived.

Sometimes incorporated items are protected from the external environments to some degree (e.g. a shielded metal enclosure can protect against physical impacts and reduce field strengths), but sometimes they are exposed to higher stresses (e.g. an item mounted near to a variable-speed motor drive may suffer locally intense electromagnetic fields).

The resulting engineering specification for the purchased item will ideally be a list of harmonized EMC standards, but may have to include modifications to them, (e.g. field strength increased to 30V/m in the VHF band to cope with 4 watt VHF walkie-talkies no closer than 0.5 metres). Other standards may also need to be added, for instance surge testing to EN 61000-4-5 and/or EN 61000-4-12 at a defined level, and/or relevant IEC or ISO safety standards that are not harmonized.

IEC 61000-2-5 [87], and Table 4.1 and 4.2 of Chapter 4 in this book which is developed from it, will be found to be very useful by people who are not EMC experts, in helping to assess electromagnetic environments. EMC site surveys are recommended where safety-critical issues are concerned, although these may fail to detect electromagnetic events with a low probability of occurrence, or changes to the environment after the survey is carried out, so some estimation is generally required for some phenomena.

3.4.1.4   Functional performance requirements

To complete the engineering specification for the EMC of a purchased item, the functions that the item performs (or that depend upon its correct operation) are analysed for their criticality (see section 3.1.3 and Appendix B).

Safety functions are allowed no significant degradation of performance over the whole range of electromagnetic stresses, including those caused by misuse, overload, failure of another item, and low-probability environmental extremes.

Where the degradation of a function may cause significant financial loss (such as loss of production), or embarrassment to a project (such as a satellite launch being delayed) it may be decided to treat it as if it were safety-critical. Less critical functions may be allowed temporary degradations of performance during transient stresses. Monitoring, reporting, and alarm functions often fall into this category, as long as they automatically recover after the event.

The use of a product is important when deciding criticality of functions. Some DC power supplies actually switch their output off whilst they are experiencing transient overvoltages, whereas others will ride through such transients without significant deviation of their output voltage. Both of them may legally claim that they meet the relevant generic EMC immunity standard, since these allow any amount of temporary degradation of performance during transient tests – and do not mandate that this degradation is specified by the manufacturer to the user, as long as it is what he may "reasonably expect".

Where the power supply is feeding lamp and indicator circuits it may be acceptable (although annoying) for it to respond incorrectly during a transient, as long as it recovers afterwards. But where the power supply is feeding a circuit involved in critical functions (e.g. a PLC, relays, contactors, pneumatic solenoids, etc., controlling machine operations) it is obviously important to choose a power supply which rides-through the transient, especially as some premises have been logged as experiencing several hundred transients on their mains supplies every day.

3.4.1.5   Emissions may be too high

Harmonized emissions standards allow emissions to occur, and these may be too high in situations where sensitive apparatus is nearby. This is especially important in some scientific and medical situations, usually where sensitive measurements are involved. How many machinery manufacturers who, when asked to install a waste crushing and packaging machine to a hospital, would automatically ask what there was on the other side of the wall that their contactors and motor drives might interfere with (such as a life support ward using sensitive physiological monitoring apparatus)?

Although continuous RF emissions from 150kHz to 1GHz are covered by harmonized emissions standards, emissions outside this range are not, yet may need to be controlled to prevent interference problems which would upset the user (or third parties) and prevent compliance with the protection requirements of the EMC Directive. Transient emissions are another example of a phenomenon which is generally not much controlled by harmonized standards, yet is known to cause problems in real life.

3.4.1.6   Emissions can add up

The total of the electromagnetic emissions from a number of incorporated items will exceed their individual emissions. In some cases this will result in a busier emitted *spectrum* without any increase in emitted *levels*, but in other cases the emissions from the various units will be so close together in the spectrum that they will measure as higher emissions levels.

Increases in emitted levels are most likely to occur when a number of similar items are incorporated into the final apparatus. For identical items whose internal electronic operations are not synchronized together (such as motor drives) ten of them may be crudely assumed to increase the total emissions by 10dB (square root of ten). When items employing digital processing or switch-mode power convertors have their respective internal electronic operations synchronized by a "master clock", ten of them may be crudely expected to give an emissions level increased by 20dB.

Where a number of items are incorporated in one enclosure with a single power lead, the emissions standard that applies to the final product may be exactly the same as the emissions standards that apply to the incorporated items – one of the reasons why the CE + CE approach to EMC compliance (section 2.3.3) is unreliable. These crude rules of thumb help determine the emissions specifications for an individual incorporated unit to be calculated on the basis of the number of those identical units which are to be fitted in a single enclosure.

3.4.1.7   Specifying the item

Once all the above considerations are complete, it is possible to write a complete engineering specification for the EMC performance of an incorporated item. This should include all the electromagnetic stresses it is to withstand, the amount of functional performance degradation allowed during the application of those stresses, and the amount of electromagnetic emissions it must not exceed.

In many cases this specification will be able merely to list harmonized EMC standards to describe the stresses and electromagnetic emissions. As long as critical functions are not involved the specifications for functional performance degradation may not be onerous for the item supplier.

The specification should be sent to the favoured item suppliers for their replies, pointing out that actual *independent evidence* of conformity with the specification will be preferred. Sales people will readily supply an EU Declaration of Conformity, but this is not *evidence*, so some education of suppliers' sales people is to be expected. (How suppliers' evidence may be judged is discussed in section 3.4.2.)

3.4.1.8   Negotiating with suppliers

Suppliers may not be able to meet the specification, or may not be able to provide all the evidence that is required. Negotiations may ensue, leading to the acceptance of a reduced specification or reduced amount of evidence. It may also prove possible to alter the design of the apparatus to accommodate the specifications of standard items.

All engineering is compromise, and the great advantage of following these recommendations is that the designer of the final apparatus will be working with *known* compromises rather than invisible or unexpected ones. Murphy's Law (to which all other physical laws are subservient) guarantees that an unknown engineering compromise will cause the worst possible problems at the worst possible moment, so these recommendations may be thought of as an anti-Murphy defence.

It is almost always commercially best to use items with adequate EMC and safety performance, rather than to purchase items which are (or may be) inadequate and deal with the resulting issues. Material costs may increase, but since it costs less to deal with problems at earlier stages of integration the final apparatus should benefit from least overall cost, and improved financial margins.

A final requirement is to make sure that the agreed safety and EMC specifications (and the agreed requirements for evidence that they have been achieved, as discussed below) are written into the purchasing contract accepted by the supplier of each item.

Suppliers who follow the prevailing culture of high specifications, low cost, and CE marking, without being able to provide acceptable evidence of actual performance, know that in the final analysis the law is "buyer beware". So it will be appreciated that following these recommendations tends to limit the number of suppliers to those who have shown that they can actually satisfy their customers' real engineering needs.

3.4.2   Checking suppliers' evidence of EMC performance

The real EMC performance of an item is unknown until evidence of engineering performance and quality control has been seen, and checked to be satisfactory (ideally by comparison with its purchasing specification determined as described above).

Not many suppliers yet provide as a matter of course the functional performance specifications achieved by their items during EMC immunity tests, so it may be necessary to pursue this vital data. Items for which the necessary evidence is not available (for whatever reason) should not be purchased, unless it is intended to put the final product through further EMC and safety compliance tests, and unless contingency costs and timescales have been allowed for remedial work and re-testing.

If potential suppliers claim design confidentiality issues as a reason for not providing evidence, insist on a trusted third party report which confirms that the item meets all the engineering specifications for EMC without revealing any (supposed) secrets. Such reports need not be expensive or difficult for a supplier to obtain, *if* he actually has the evidence he claims.

3.4.2.1   Checking Declarations of Conformity

A suppliers' Declaration of Conformity (D of C) cannot in general be considered to be actual evidence, although it may be possible for small companies making low volume apparatus with no safety implications to rely on them (check with the local EMC enforcement officers).

Even so, Declarations of Conformity are useful as a guide to the intended use of the item and the competence of its supplier. Things to look for in a D of C include whether it lists the EMC standards required by the engineering specification for the item. It may prove difficult to judge whether items are suitable if they list different standards. Some standards, such as EN 61800-3 for the EMC of motor drives and EN 61131-2 for the EMC of PLCs, cannot be applied to the final apparatus and so may be of little help.

It is also worth checking whether the D of C actually covers the item concerned (and not something else), and is clearly signed and dated by the supplier's Technical Director or equivalent. Dates which are only a few days old, for items which have been on the market for many months, must be suspect.

Also check for any inappropriate or unreasonable warnings, limitations to use, or attempts at disclaimers, such as "Do not use this product if it causes interference" or "May stop working when interfered with", neither of which are unknown. Products not intended for safety-critical application (such as ordinary PLCs) should make this clear, and this may appear on their D of C, but do not rely on it being so obvious. Legally, manufacturers are only required to state such restrictions in their "instructions for the user". Figure 3.3 shows an example of the sort of Declaration of Conformity which is commonplace and quite legitimate.

3.4.2.2   Problems to watch for concerning standards

It is impossible to discuss the full range of EMC standards here. There is often a lot of confusion over the generic EMC standards – with suppliers choosing those that make it easier for their CE marking, rather than providing the engineering performance that their customers actually need.

Remember that it is the function and user environment of the final apparatus that governs which standards apply to it, rather than the technology it incorporates. This can lead to a number of problems with the standards applied to incorporated items, some of which are described below. For instance, a washing machine which incorporates a microprocessor has to use EN 55014 (the EMC emissions standard for household appliances), and cannot use EN 55022 (the EMC emissions standard for information technology).

Note that functional safety may require tests to additional EMC phenomena to be considered, and/or alterations to the range or levels of the test standards that have been applied for EMC Directive compliance, as discussed elsewhere in this book.

Declaration of Conformity

Manufacturer/importer name

Address

certifies that the following apparatus conforms with the protection requirements of Council
Directive 89/336/EEC on the approximation of the laws of the Member States relating to
electromagnetic compatibility:

Name of product

(description of variants)

EMC standards applied:

> **EN 50081 part 1 : 1992 (Electromagnetic compatibility –
> generic emission standard – part 1: residential, commercial
> and light industry)**
> and
> **EN 50082 part 1 : 1992 (Electromagnetic compatibility –
> generic immunity standard – part 1: residential,
> commercial and light industry)**

Signed: (signatory)

Date of Issue: XX/XX/XX

Figure 3.3 Example declaration of conformity

3.4.2.3   Problems to watch for concerning the generic EMC standards

There are two sets of two generic EMC standards [69], [70], covering emissions and
immunity for two different environment classifications, making four generic EMC
standards in all:

- EN 50081-1: the tightest emissions standard, for residential, commercial
  and light industrial environments. This is equivalent to EN 55022 Class B,
  VDE 0891 Class B, CISPR 22 Class B, and broadly similar to EN 55014-1,
  EN 55011 Class B, and FCC Part 15 Class B.

- EN 50081-2: a more relaxed emissions standard for (heavy) industrial
  environments. This is broadly similar to EN 55011 Group 1 Class A, and
  EN 55022 Class A.

- EN 50082-1: immunity for residential, commercial, and light industrial. The
  use of issue 2: 1997 is preferred over the original 1992 issue, especially as
  it completely supersedes the 1992 issue on 1st July 2001.

- EN 50082-2: the toughest immunity standard, for (heavy) industry
  environments.

The best items for general uncontrolled use, or where the user's environment may not be very well defined, are those that meet the toughest standards for emissions and immunity: EN 50081-1 and EN 50082-2. The best items will also meet IEC 61000-4-5 for surges at level 2 (light industrial) or 3 (heavy industrial), and the mains dips and dropout tests of IEC 61000-4-11. Although surge, dip, and dropout tests are not included in the early generics we know that they do occur in real life. Standardising on such items makes the selection of items and their use in custom engineering projects much easier.

Items declared using EN 50081-2 are sometimes sold for incorporation in apparatus intended to be used in light industrial and commercial environments – but their emissions are too high for these environments and their use would necessitate additional EMC work and probably EMC testing of the final apparatus, for due diligence to be achieved. Similarly, items declared using EN 50082-1 are often sold for incorporation in (heavy) industrial environments, where their immunity will be too low without additional EMC work and probably some testing of the final apparatus, for due diligence.

Despite the practice being impossible to justify, some items are declared using EN 50081-2 and EN 50082-1, the easiest of all the four generics – but this means they are too noisy for residential, commercial, and light industrial environments and not immune enough for heavy industrial environments, so they cannot be used anywhere without significant additional EMC work, plus (probably) some testing of the final apparatus, for due diligence.

3.4.2.4   The implications of EN 55022 Class A

Items which may be classed as information technology or telecommunications equipment (ITE), e.g. computers, modems, printers, VDUs, keyboards, etc., are allowed to apply the Class A EMC emissions limits in the product-specific EMC standard EN 55022 for use in the commercial and light industrial environments – though not for domestic use.

But almost all other EMC emissions standards require tighter limits for commercial and light industrial environments (usually equivalent to EN 55022 Class B). So when an item which meets EN 55022 Class A is incorporated into final apparatus that is not allowed to declare compliance using EN 55022 because it is not ITE, such items can cause excessive emissions and lead to non-compliance with the relevant EMC emissions standard.

The reverse situation is also true, as described above: immunity is inadequate for use in the heavy industrial environment. This is a common problem when integrating computers and peripheral devices into industrial control systems or other more stringent applications.

3.4.2.5   The implications of EN 55011

Items declared using EN 55011 are known as "ISM" (industrial, scientific and medical) equipment: which means they may generate and/or use electromagnetic energy to achieve their main function. Examples include dielectric heaters such as wood gluers, plastic sealers and welders; induction heaters; RF stabilized electric welders; spark erosion machines; magnetic stirrers; and diathermy equipment (both medical and cosmetic).

EN 55011 allows *very high, and even unlimited levels of EMC emissions* at specified frequencies (the so-called "free radiation frequencies", listed in Table 1 of the standard – the most important are 13.56MHz, 27.12MHz, 40.68MHz, 2.45GHz,

5.8GHz and 24.125GHz), which could cause considerable immunity problems for other equipment, and even health hazards for their operators. As the standard says, "special measures to achieve compatibility may be necessary where other equipment satisfying immunity requirements is placed close to ISM equipment".

When incorporated in final apparatus that cannot utilize EN 55011, ISM items can cause excessive emissions which lead to non-compliance, and may require significant additional EMC work and probably some testing of the final apparatus for due diligence to be achieved. (But note that use of EN 55011 does not *necessarily* mean high emissions – other apparatus which does not generate high levels at the free radiation frequencies may also declare compliance against this standard.)

3.4.2.6   Checking assembly and installation instructions

For an item to actually achieve the EMC performance that its test and other evidence implies, it is necessary to assemble or install it fully in accordance with its supplier's detailed instructions. This is very important for EMC, which can easily be compromised, for example by the use of the wrong type of cable, or the incorrect use of a "pigtail" on the screen of a cable. It is also very important for safety, which can easily be compromised by inadequate mounting or ventilation.

Suppliers who cannot (or do not) provide detailed assembly and installation instructions relevant to safety and EMC should be treated with suspicion: it may be possible to assume that an item which is devoid of such instructions does not need any special installation precautions, but it would be wise to seek specific assurances. If the supplier cannot provide them, their products are best avoided.

A big problem for many one-off engineering projects is that assembly and installation staff do not usually follow suppliers' detailed instructions, preferring to use what they have considered to be "best practices" (often unchanged for the last twenty years). Many modern EMC installation best-practices run counter to typical electrical practices, as can be seen by a review of IEC 1000-5-2.

Suppliers' instructions should be checked for inappropriate or vague limitations or instructions, e.g.:

- "Do not use this product if it causes interference."
- "If interference occurs, fit filter and/or fit product in shielded box."
- "This product may require manual reset after transient interference."

Assembly and installation instructions should also be checked to see if they specify expensive or exotic cables or connectors, additional filters or shielding, or unusual environmental conditions. These can significantly affect the overall project cost and timescales, a good reason for carefully reading an item's assembly and installation manuals *before* making the decision to purchase it, rather than after as is more common.

3.4.2.7   Checking test results and certificates

With a little experience, suppliers' test reports can be more revealing than they may have realized. Full test results from an accredited test laboratory make the most convincing evidence. "Accredited" means that their measurement integrity, understanding of the standard, quality systems, and independence have all been checked and approved by a government-appointed accreditation body, giving a useful degree of confidence in their testing and results.

EMC tests are notoriously inaccurate, with even world-class laboratories experiencing differences in measurement on the same test item of 6dB (that is, +100%

or –50%). A similar issue concerns the interpretation of safety standards, with one laboratory safety manager usually managing to achieve several points of difference from another on any harmonized safety standard. Accreditation helps to reduce differences in interpretation, though it does not necessarily mean a test is more "accurate". When we say "accuracy", we are really referring to a known and declared value of measurement uncertainty. Accredited labs (in the UK, at least) are required to know their uncertainty (sic), but do not have to give it in their test reports except in certain circumstances – although many do.

Test laboratories can only be accredited for specified test standards – so although we tend to say "Accredited Laboratory" what we really mean is "Laboratory which is accredited for the test standards concerned". So don't take for granted the logo of the accrediting body on the test laboratory letterhead: check whether the laboratory actually is accredited for all the tests covered by its report. An accredited lab's report should identify explicitly any test results which it reports that are not covered by its scope of accreditation.

Full test results should include: the exact identification of the model (and version) tested; detailed sketches or photographs of the test set-ups; lists of the test equipment used and their calibration dates; whether the item passed or failed the test; and the signature or identification of the test engineer. EMC reports should include emissions graphs and result tables showing the margin under the limit lines, and the functional performance criteria for the immunity tests.

3.4.2.8   Checking test set-ups

Proper EMC test reports will include sketches or even photographs of the test set-ups employed, and descriptions of how the tests were conducted. These should be checked for the following:

- Do they agree with the supplier's detailed EMC installation instructions? Watch out especially for the use in the tests of special types of cables or connectors, or ferrite clamps or additional earth bonds, which are not mentioned in the installation instructions.

- Do the test set-ups relate to how you intend to use the item? Check especially for a lack of some of the external cables (cables usually create the biggest EMC problems, so leaving them off usually gives better EMC results).

- Were the emissions consciously maximized, and immunity consciously minimized, by the test procedures, methods, and set-ups?

In all EMC test reports, make sure that there are no comments along the lines of "the product was modified and passed the test .....". It is not uncommon for the supplier's engineers to apply remedial measures to products during testing, which an aware test engineer will fully record in the test report. What can then happen is that these remedial measures get "forgotten" when the items are manufactured.

3.4.2.9   Checking EMC Technical Construction Files (TCFs)

Where a product has been declared compliant with EMC Directive by using a TCF rather than harmonized standards, it will be valuable to read this – or at least, the Competent Body's report – along the above lines. It is not uncommon to find a number of comments in TCFs where the assessor has decided not to declare the product non-compliant, but nevertheless has serious concerns. Such warnings are often along the lines of: "The supplier should make clear to the customer certain specific installation requirements and limitations to use...."

3.4.2.10 Suppliers' quality control

The fact that a supplier has had an example of an item tested for EMC performance using acceptable standards, and it has passed, of itself proves nothing about the EMC or safety performance of any of the other units of the same type and/or model. For that, a quality control system is necessary.

Even where a supplier has a BS EN ISO 9000 quality system in place, by itself this is no guarantee that the standard items supplied to the manufacturer of the final apparatus have any EMC or safety performance. All it means is that the company is audited against its quality manual, so it is important to discover what their quality manual says about maintaining the specified EMC performance in production. To control the EMC performance of standard products a supplier must have controls over design changes and production concessions, unit build standard, repairs, refurbishment, and upgrades, at least as far as all EMC issues are concerned.

Even with all these controls in place a number of elements are still uncontrolled – especially the performance of the components that they buy in – and this makes it necessary for suppliers to have a sample-based testing policy for EMC. The better the suppliers' controls over its design, purchasing, production, and repair, the lower need be its rate of sample testing.

Companies with a "supplier approval" procedure will find it quite easy to add the necessary additional requirements to ensure that the EMC evidence provided by the supplier stands some chance of being representative of the items actually purchased.

3.5   Maintenance, upgrades and enhancements

A system is rarely fixed for life: technology "push" and customer "pull" will create opportunities and desires for changes, and the natural degradation of physical structures with time will have a further impact on EMC.

3.5.1   Maintenance

The EMC Directive says very little about ongoing maintenance. The supplier's responsibility is to ensure that he supplies compliant apparatus, and the CE Marking procedures are all geared to the act of "placing on the market". In the UK Regulations the user has a responsibility to "take into service" compliant apparatus, but there is very little understanding of the legal position of a product or system that was compliant when it was first used but ceases to be so due to degradation over time.

Such legal niceties raise no dilemma for safety- or mission-critical systems as might be found in the aerospace, military, transport or medical sectors. If a system relies on EMC to be fit for purpose, then it is reasonable to expect it to remain compatible for the duration of its operational life. Thus for these systems much more than for those which are merely covered by the EMC Directive, through-life EMC related maintenance is an essential part of the control plan. Examples of aspects which should be addressed are shown in Table 3.5, along with ways in which they might be controlled.

Note that some failure mechanisms (such as open circuit filter capacitors or surge protection devices) are impossible to detect by functional checks, and some way of testing the system's immunity to "synthetic" interference is needed. Variations on the theme of the IEC 61000-4-4 electrical fast transient burst test can be very useful here (see section 10.2.3 for a more detailed description).

Table 3.5 Items to check in periodic maintenance

Fault	Test
Screened enclosures	
Corrosion between mating surfaces	High current bond resistance check
Damaged conductive gaskets	Visual
Loss of torque in fasteners	Periodic re-tightening of fasteners
Cables	
Failure of screen-to-backshell joint	High frequency screening effectiveness test
Earth faults due to trapped or punctured insulation	Earth continuity and isolation checks (can be designed into system)
Filters	
Development of high impedance to earth	Bond resistance check
Open-circuit failure of filter capacitors	High frequency filtering effectiveness test

3.5.2   The relevance of upgrades or enhancements

Systems already in the field are often subjected to mid-life upgrades, or changes to the installed configuration. These can take place for a number of reasons:

- to extend the system's existing capability: for instance, adding more line cards to a telephone exchange
- to include new functions: for instance, monitoring a new process variable in a production plant
- to improve the performance or correct deficiencies in existing functions: for instance, a software bug fix
- to implement a repair: for instance, swapping a faulty board.

With the exception of a repair by replacement of an identical component (and not all apparently similar components are identical in EMC terms), each of these may have EMC implications. As far as the application of the EMC Directive is concerned, the UK regulations include reconditioning and "modification which substantially alters the EMC characteristics" in their definition of "manufacture", but exclude repairs. A slightly different approach is proposed in Chapter 7 of the EC Guidelines, which introduces the concept of "as-new" apparatus; this is apparatus already taken into service "which is subject to an industrial operation that implies a substantial modification in order to obtain identical (or similar) performance as ... new apparatus placed on the market at the same time". The general principle is that the EMC Directive is re-applied only if the apparatus is to be considered "as-new" and if it is going to be placed on the market again.

Since in the context of installations the re-configured system or installation will not be "placed on the market" again, the conformity assessment procedure for CE Marking will not apply. What is still applicable is the system builder's original responsibility to ensure that the system or installation meets the Directive's essential requirements. To do this for any but trivial modifications will need some level of EMC analysis. It is the responsibility of the person making the changes to do this, who could be either the system builder or the user. In the worst case this could require re-testing of the reconfigured system, but possibly less onerous alternatives could include documentary proof of the compliance status of newly-introduced items, or their benign nature, or a technical analysis which shows that the emissions and/or susceptibilities which are created are negligible.

The same analysis process should be done for upgrades to systems where EMC is an essential part of fitness for purpose, as in safety-critical systems. The difficulty in many cases lies in deciding how relevant a modification is to EMC, and therefore what level of analysis should be undertaken. This difficulty is reduced if the original project documentation included a clear description of the EMC-criticality of all parts of the system, so that changes to any part can be judged against this earlier knowledge.

The trigger for the upgrade analysis process should be built into the change control procedures for the whole system, as described in section 3.1.3.1 and 3.1.4.

3.6   Training

It is sometimes assumed that technical personnel know all that is needed to know about the company's products without any formal induction or guidance; they learn on the job and if they make mistakes, well, as long as they are corrected no harm is done.

This attitude seems to prevail with respect to the straightforward functions, operation and assembly of the products. It is even more prevalent when it comes to EMC, despite the fact that many if not most of a company's technical staff will have had little or no background or experience in EMC problems and how to deal with them.

On the shop floor and at installation traditional assembly methods must adapt to take account of good EMC practice. This may well require a change in thinking for some staff, particularly those whose attitude is "we've been doing it this way for years, we know best", or "can't see this making any difference". The rest of this book discusses such developments in practice.

3.6.1   Training and awareness check list

The most cost effective way of bringing this knowledge in-house is by providing suitable training courses for critical personnel. The following is a suggested guide to structure a company-wide EMC awareness and training programme. It takes typical modules from course providers and applies them as necessary for each job function.

3.6.1.1   Project managers

Modules:
- Introduction to EMC
- Application of the EMC Directive
- EMC control planning
- Overview of design and installation factors affecting EMC

3.6.1.2   System designers, installation and commissioning engineers

Modules:
- Introduction to EMC
- Interference fundamentals and coupling mechanisms

- Earthing and bonding
- Cable layout and termination
- Shielding techniques
- Segregation of equipment
- Filtering and surge protection

3.6.1.3 Procurement and quality control

Modules:

- Introduction to EMC
- Overview of installation factors affecting EMC
- Use of standards for specifying EMC of parts
- Assessing suppliers' evidence of compliance
- EMC change control

3.6.1.4 Installation and assembly technicians, maintenance and field service engineers

Modules:

- Introduction to EMC
- Overview of installation factors affecting EMC
- In-house EMC-specific installation practices
- Controlling the EMC effect of modifications

3.6.1.5 Sales and marketing management and personnel

Modules:

- Introduction to EMC
- Application of the EMC Directive
- The marketing benefits of following a correct approach to EMC
- Assessing the customer's electromagnetic environment and the permissible functional degradation allowed during disturbance events
- Marketing and selling products only into the environments, and for the intended use, where they will be EMC compliant

3.6.1.6 General and financial management

Modules:

- Introduction to EMC
- Application of the EMC Directive
- The financial benefits of following a correct approach to EMC

3.6.1.7 Documentation and manual writers

Modules:

- Introduction to EMC
- Application of the EMC Directive
- Creating technical documentation that shows due diligence in EMC compliance
- Informing the user of limitations to use and installation methods

Chapter 4

Interference sources, victims and coupling

The general interference phenomena were catalogued in Chapter 1, section 1.1.1. This chapter will look in more detail at the type of source that will produce these phenomena, and the apparatus that may be affected by them, and then goes on to discuss the modes of coupling. The subject of mains harmonics is covered in a separate section. As an endpiece, the interrelationship with safety issues is also covered.

In the context of EMC legislation, it is usual to refer to a piece of equipment in relation to its environment. The environment contains all other sources of disturbances, and also contains all other apparatus which might be affected by the source equipment's own disturbances. Control may therefore be exercised either on the equipment's ability to create disturbances, or on its immunity from external disturbances, or both; the coupling paths are defined as the totality of the equipment's interaction with its environment.

4.1   Phenomena in the electromagnetic environment

The source document for environmental electromagnetic phenomena is IEC 61000-2-5 [87]. The tables on the following pages are based on that standard. They are classified into continuous phenomena and transient phenomena. Several types or higher levels of transient phenomena may occur only occasionally, and these have been separately discussed in the second part of Table 4.2.

A great deal of research has been conducted into the prevalence of these phenomena, which has been distilled into IEC 61000-2-5. Some of the more useful source papers can be found in [16], [29], [30], [140]. Most of the basic immunity test standards in the IEC 61000-4-X series include a description of the sources of the phenomenon they are dealing with, and IEC 61000-4-1 is itself an overview of the immunity test methods.

Not all of these phenomena are covered by the product or generic standards which are applied under the EMC Directive: in fact, it's probably fair to say that the majority are not (see Table 1.1 on page 4). This can be a blessing, if you are a product manufacturer faced with days of EMC compliance testing, or it can be problematic, if you are a user who needs to have confidence in the performance of an item of equipment when faced with a particular electromagnetic threat. In the latter circumstance it is usually necessary to discuss the detailed performance with the equipment supplier. The following tables, and the standards they refer to, may offer a helpful framework for such discussions.

Table 4.1 Continuous phenomena

The electromagnetic phenomena, their principal sources or causes, and some examples and comments	Basic standards allowing assessment of the environment	Basic test methods for emissions	Apparatus particularly susceptible to the phenomena	Basic test methods for immunity, and compatibility levels
AC or DC supply voltage variations (slow variations) Most supply variations do not exceed 10%, although in some parts of some countries (even in the EU) the official figures for supply tolerance should not be accepted without question. In certain industries, excessive variations may be caused by very heavy loads, such as arc furnaces, welding, electrolysis and electroplating, and other continuous loads with varying current demands.	IEC 1000-2-5		All normal electronic equipment considered susceptible at > 10%.	IEC 61000-4-14 IEC 61000-4-28, and -29 1: 3% Vnom 2: 10% Vnom 3: 15% Vnom
AC supply phase unbalance Caused by asymmetrical loads, or large single-phase loads.	IEC 1000-2-5		Three-phase equipment which relies upon phase balance (e.g. AC motors, transformers). Neutral cables may overheat. Contact breakers may trip or fuses blow unexpectedly.	IEC 61000-4-14, and -27 1: 2% Fnom 2: 3% Fnom
DC supply ripple Caused by AC rectification, or battery charging. A consideration for equipment that operates directly from rectified AC or batteries that are charged during its operation.	IEC 1000-2-5		All normal electronic equipment considered susceptible at > 10%.	IEC 61000-4-17

Table 4.1 Continuous phenomena (continued)

Phenomenon	Refs	Test method	Effects	Levels
Harmonics and interharmonics of the AC power supply (waveform distortion) Non-linear loads: - static frequency and cyclo-converters, induction motors, welding equipment - AC-DC power converters (e.g. adjustable speed drives or multiple single-phase electronic power supplies) – transformers driven into saturation • Strongly-disturbed networks (e.g. in steelworks) can exceed 10% THD. • Supply network resonances can create very high levels at certain frequencies. • Supplies in developing countries (e.g. China) can have very heavy levels of harmonic distortion indeed.	IEC 1000-2-4 IEC 1000-2-5	IEC 61000-3-2 (≤16A/φ, LV) IEC 61000-3-4 (>16A/φ, LV) IEC 61000-3-6 (MV or HV) IEC 61000-3-9 (interharmonics)	Power converters and other electronics, which use zero crossing, peak, or slew rate of supply waveform AC/DC power supplies can output lower than expected voltages due to supply waveform distortion Power factor correction capacitors, distribution transformers, cables, AC motors, and switchgear can overheat Excessive acoustic noise and vibration in electromechanical equipment	IEC 61000-4-7 and -13 1: 4% THD 2: 8% THD 3: 10% THD
AC or DC magnetic fields Medium and high-voltage supply distribution, heavy power use. Principal sources are power conductors and magnetic components Audio-frequency magnetic fields also exist near audio power amplifiers and induction loop systems All cables carrying currents, whether analogue or digital, leak magnetic fields to some extent, and this may be important for very sensitive circuits	IEC 1000-2-7 NRPB-R265 [136] IEC 1000-2-5	No basic IEC or EN test method. Search coil method: Annex A of EN 55103-1	CRT-based displays and computer monitors start to suffer visible image degradation at >1A/m but can be shielded to achieve 20A/m or more Microphones, hearing aids with inductive loop pickup, loudspeakers, Hall effect and other magnetic transducers can produce erroneous outputs	IEC 61000-4-8 1: 3A/m 2: 10A/m 3: 30A/m 4: 100A/m
AC or DC electric fields Medium and high-voltage supply distribution, heavy power use. Principal sources are power conductors	NRPB-R265 [136] IEC 1000-2-5		Unshielded sensitive or high-impedance analogue circuits or transducers	1: 0.1kV/m 2: 1kV/m 3: 10kV/m 4: 20kV/m

Table 4.1 Continuous phenomena (continued)

Phenomenon	Basic standards (IEC)	Emission standards	Victims	Test standards / levels
Signalling voltages on the AC power supply Ripple control (100Hz to 3kHz) and power-line carrier systems (3 to 95kHz) used by electric utilities. Signalling in end-user premises (95 to 148.5kHz) • Supply network resonances can create very high levels at certain frequencies.	IEC 1000-2-1 IEC 1000-2-2 IEC 1000-2-12 IEC 1000-2-5	IEC 61000-3-8 (outside Europe) EN 50065 (in Europe)	Power converters and other electronics, which uses the zero crossing, peak, slew rate, or other characteristics of the supply waveform.	IEC 61000-4-13 (to 2.4kHz only) 1: 5% Vrms 2: 9% Vrms (0.1–3kHz only)
Conducted interference DC to 150kHz in all conductors (voltages and currents) Industrial electronics (power semiconductor devices such as rectifiers, thyristors, IGBTs, FETS, etc.), leakage currents of RF filters and other earth currents, VLF and ELF radio transmitters. This phenomenon is most likely to be observed in installations using large amounts of power. • 50V differences in earth potentials are allowed by UK electrical safety regulations and occur in some premises. • Practical experiences in Sweden show that the following levels of 50Hz conducted interference may be expected: 1V in protected environments, 250V in unprotected installations, 500V in outdoor installations associated with HV switchgear.	IEC 1000-2-1 IEC 1000-2-2 IEC 1000-2-4 IEC 1000-2-5 IEC 1000-2-6 IEC 61000-2-12		Long wave and medium wave radio receivers. Analogue telephone systems. Sensitive instrumentation (e.g. temperature, flow, weight), audio, and video.	IEC 61000-4-16 1: 1Vrms 2: 3V 3: 10V 4: 20V X: special (case by case)
Conducted interference above 150kHz in all conductors (voltages and currents) Most importantly from the RF fields generated by fixed and mobile radio transmitters and some ISM equipment (especially Group 2 of EN 55011). Also coupled into conductors from synchronous (clocked) digital circuits and semiconductor power converters. May apply less, or only in certain frequency bands, or not at all, to equipment and all cables used in screened rooms (depends on the screening performance of the room).	IEC 1000-2-3 IEC 1000-2-5	EN 55011, 55013, 55014, 55015 or 55022	Radio receivers. Digital control and signal processing. Sensitive analogue instrumentation (e.g. temperature, flow, weight), audio, and video. Analogue telephones.	IEC 61000-4-6 1: 1V (7mA) 2: 3V (21mA) 3: 10V (70mA) 4: 30V (210mA)

Table 4.1 Continuous phenomena (continued)

Phenomena	Basic test methods for emissions		Apparatus particularly susceptible	Basic test methods, and compatibility levels
Radiated interference above 150kHz Most importantly from fixed and mobile radio and TV transmitters, and some ISM equipment (especially equipment covered by Group 2 of EN 55011). Also from synchronous (clocked) digital circuits and semiconductor power converters such as PSUs and AC motor drive inverters. May apply less, or only in certain frequency bands, or not at all, to equipment and all cables used in screened rooms (depends on the screening performance of the room).	NRPB-R265 IEC 1000-2-5	EN 55011, 55013, 55014, 55015 or 55022	As above. High levels of radiated interference can cause sparks and ignite flammable materials and atmospheres.	IEC 61000-4-3 1: 1V/m 2: 3V/m 3: 10V/m 4: 30V/m
Radiated interference above 150kHz from high-voltage power lines Due to: - corona discharge in the air at the surfaces of conductors and fittings; - discharges and sparking at highly stressed areas of insulators; - sparking at loose or imperfect contacts.	BS 5049-1 CISPR 18-1	BS 5049-2, -3 CISPR 18-2, -3	Radio receiver.s	

Table 4.2 Transient phenomena

The electromagnetic phenomena, their principal sources or causes, and some examples and comments	Basic test methods for emissions	Apparatus particularly susceptible	Basic test methods, and compatibility levels
Voltage fluctuations, dips and short interruptions on AC and DC power supplies Load switching and fault clearance in LV power supply networks, transient current changes in connected equipment	IEC 61000-3-3 (≤16A/φ from LV supplies) IEC 61000-3-5 (>16A/φ from LV supplies) IEC 61000-3-7 (supplied by MV or HV) IEC 61000-3-11 (conditional connection to public LV supply, <75A/φ)	All digital systems can fail if their supply voltage briefly drops below minimum. Specific devices and circuit techniques are available for automatic recovery but are not universally used Analogue signal processing can also fail, but will generally recover when the supply quality is back to normal	IEC 61000-4-11 (AC) and -29 (DC) Dips can vary from 30 to 100% of Vnom

Table 4.2 Transient phenomena (continued)

Conducted and radiated fast transient bursts Arcing during initial opening of switch contacts feeding an AC or DC load, worst with inductive loads Applies to conductors connected to these loads, conductors connected to AC or DC supplies, and (to a lesser extent) conductors which may be in proximity to the cables mentioned above	EN 55014 Discontinuous disturbance (Household and similar apparatus only)	All digital systems can fail when affected by fast transient bursts. Specific devices and circuit techniques are available for automatic recovery but are not universally used Analogue signal processing can also suffer errors, but will generally recover after the burst	IEC 61000-4-4 1: 500V 2: 1kV 3: 2kV 4: 4kV

Voltage surges on AC and DC power supplies and all long cables (including telecomms)
 	
Load changes in LV power supply networks, especially reactive loads such as power factor correction capacitors and resonating circuits associated with switching devices (e.g. thyristors)
 	
The ring wave phenomenon described by IEC 61000-4-12 is mainly applicable to equipment connected to AC supplies in certain countries (such as the USA)
		Semiconductors in off-line electronic circuits (e.g. switch-mode power converters) and all semiconductors connected to long cables, are the most prone to suffering actual damage from differential (line-to-line) surges
		
		All electronics can suffer actual damage from CM surges (line-to-ground) if they have inadequate creepage, clearance, or insulation resistance, at any point where the surge voltages exist
			IEC 61000-4-5 (unidirectional)
			and IEC 61000-4-12 (ring wave)
			1: 0.5kV CM
			   0.25kV DM
			2: 1kV CM
			   0.5kV DM
			3: 2kV CM
			   1kV DM

Conducted damped oscillatory wave
 	
Power switching with re-striking of the arc (duration measured in seconds)
		As above, with greater possibilities of actual damage due to the longer duration and hence greater energy of the surge	IEC 61000-4-12 (damped oscillatory wave)
			1: 0.5kV CM
			   0.25kV DM
			2: 1kV CM
			   0.5kV DM

Table 4.2 Transient phenomena (continued)

Electrostatic discharge (personnel discharge, machine discharge, or furniture discharge) Tribocharging of personnel, workpieces (including some liquids, dusts, and vapours), and unearthed metalwork Direct and indirect discharges can occur. Equipment used in controlled-ESD environments (e.g. semiconductor assembly areas) may be exempted from some or all ESD requirements (case-by-case basis)		All digital systems can fail when affected by electrostatic discharge events. Specific devices and circuit techniques are available for automatic recovery but are not universally used Analogue signal processing can also suffer errors, but will generally recover If the discharge current couples directly into conductors or devices in connectors, cables, keyboards, etc. it can destroy internal circuitry	IEC 61000-4-2 Basic test methods only exist for personnel discharge 1: 1kV 2: 4kV 3: 8kV 4: 16kV
Low probability of occurrence, and unusual environments			
AC power supply frequency variation Major faults in supply networks		Real-time clocks operating from the supply frequency Processes in which the rates of production are related to supply frequency, e.g. an induction motor driven machine may get unacceptably out of step with a DC motor or stage timed from a more stable source	IEC 61000-4-14 and -28 1: 2% of nominal frequency 2: 3% of nominal frequency
Short duration AC or DC voltages (in all long signal, control, telecommunication, and data cables) Associated with faults in areas of heavy power use (especially earth faults)	Meshed earth systems can reduce this exposure (see IEC 61000-5-2:1998).	All semiconductors connected to long cables are most prone to suffering actual damage from these surges if they have inadequate creepage, clearance, or insulation resistance, at any point where the surge voltages exist These are generally common-mode (CM) voltages at the frequency of the power supply, and can equal the full supply voltage for the time taken for fault clearance (e.g. by fuses) (especially where the earth is not equipotential – typical of older installations)	

Table 4.2 Transient phenomena (continued)

Phenomenon / source	Note	Victim	Standard / test levels
Voltage surges on AC and DC power supplies and all long cables (including telecomms) Lightning, and field collapse in loads with large stored energies (e.g. large motors, superconducting magnets)	Note that exposed sites can suffer from surges of up to 10kV several times a year	Semiconductors in off-line electronic circuits (e.g. switch-mode power converters) and all semiconductors connected to long cables, are the most prone to suffering actual damage from differential (line-to-line) surges All electronics can suffer actual damage from CM surges (line-to-ground) if they have inadequate creepage, clearance, or insulation resistance, at any point where the surge voltages exist	IEC 61000-4-5 (unidirectional) and IEC 61000-4-12 (ring wave) 2: 1kV CM, 0.5kV DM 3: 2kV CM, 1kV DM 4: 4kV CM, 2kV DM
Conducted voltage surges due to fuse operation Fuse operation		As above	IEC 61000-4-5 1: 0.5 times V_{peak} 2: 1 times V_{peak} 3: 2 times V_{peak}
Conducted damped oscillatory surges on power lines and all other cables Switching of isolators in HV/MV open-air stations, particularly the switching of bus-bars		As above	IEC 61000-4-12 1: 0.5kV CM, 0.25kV DM 2: 1kV CM, 0.5kV DM 3: 2kV CM, 1kV DM (2.5kV CM for substation equipment) 4: 4kV CM, 2kV DM
Short duration (pulsed) magnetic fields Lightning and fault currents in earth conductors, supply networks, traction systems Mainly applies to equipment used in electrical plants and switchyards		CRT-type VDUs may suffer momentary image movement Hall effect and other magnetic transducers may suffer temporary output errors	IEC 61000-4-9

Table 4.2 Transient phenomena (continued)

Phenomenon		Effect	Standard / Levels
Radiated (damped oscillatory) magnetic fields MV and HV switching by isolators Mainly applies to equipment used in high-voltage substations and switchyards		CRT-type VDUs may suffer momentary image movement Hall effect and other magnetic transducers may suffer temporary output errors	IEC 61000-4-10
Radiated pulsed fields near gas-insulated substations HV/MV disconnect switching in gas-insulated substations (rise time around 10ns) • A 25mm gap in an SF6 switch was stressed to break-down at 80kV and gave the following maximum fields: At 2 metres distance: 340 V/m/ns and 608 A/m/ns; At 10 metres distance on the other side of a plaster-board wall: 11 V/m/ns and 29 A/m/ns	The duration of the pulsed fields is generally such that a rate of change figure of 10V/m/ns translates into a field strength well in excess of 10V/m	Likely to have a more catastrophic effect on digital circuits and software than on analogue circuits Sensitive circuits (whether analogue or digital) could suffer actual damage from these pulsed fields Equipment intended to be exposed to these pulsed fields will generally need to be designed specially	1: 100 V/m/ns 2: 300 V/m/ns 3: 1000 V/m/ns 4: 3000 V/m/ns 5: 10,000 V/m/ns
Radiated short duration (pulsed) fields HV/MV disconnect switching in open-air substations (rise times around 100ns), and due to lightning ground strikes (rise times between 100 and 500ns)		As above	1: 30 V/m/ns 2: 100 V/m/ns 3: 300 V/m/ns 4: 1000 V/m/ns 5: 3,000 V/m/ns
Radiated short duration (pulsed) fields under overhead lines Where the lines carry pulse currents (due to HV/MV disconnect switching in substations or lightning), (rise times around 1μs)		As above	1: 3 V/m/ns 2: 10 V/m/ns 3: 30 V/m/ns 4: 100 V/m/ns 5: 300 V/m/ns
Direct lightning strike Exposed equipment not fully protected by LPS		Catastrophic effects on all electronic components, and many electrical and structural elements	1% of strikes exceed 200kA, see BS 6651 and IEC 61312-1

4.1.1   Examples of radiated field threats

4.1.1.1   Examples of LF magnetic fields:

- 100A/m has been seen 10m from steel rolling mill DC drive cables (±8kA), and > 1000A/m at ≤ 1m.

- A 1.1kHz 800kW steel billet induction heater has been seen to emit 100A/m at 1m.

- A 230kHz 400kW steel tube welder with a coil diameter of 50mm can create 40A/m at 0.25m from its coil.

- A 50Hz 6MW copper billet heater generated 430A/m at a distance of 1m.

- A 700Vdc 60kA electrolytic process can create 15kA/m at operator position.

- At ground level under an overhead 400kV line: 32A/m.

- At ground level above an underground 400kV line: 160A/m.

- Commercial premises under-floor heating can create 160A/m at floor level, 16A/m at 1m height above floor.

- 8A/m has been seen at floor level in a multi-storey office, above a sub-floor carrying cables from distribution transformer to switch-room, and up to 2A/m at desk height.

- A 1kW water pump has been seen to emit 800A/m at 10mm distance, and 3A/m at 400mm, whereas an 18kW motor emitted 6A/m at 200mm.

- A TIG welder has been seen to emit 800A/m at the surface of the welding cable and surface of its power supply, and ≤ 160A/m at the operator's position.

4.1.1.2   Examples of LF electric fields:

- A 490kHz 8kW steel tube heater with a coil diameter of 60mm has been seen to emit 100V/m at 0.3m from its coil.

- A 20kHz 1.5kW induction cooker hob generated 28V/m at 250mm.

- 1kV/m = outdoors under 30kV lines, or indoors under 765kV lines.

- 10kV/m = outdoors under 400kV lines.

- 20kV/m = outdoors under 765kV lines.

4.1.1.3   Examples of RF fields:

- MHz-operation dielectric heaters of 3 to 15kW have been known to create 300V/m at the operator's position (this is not regarded as safe!).

- Hand-held walkie-talkies and cellphones can generate 30 V/m field strengths at distances of 400mm and 250mm, respectively (greater fields at smaller distances).

- A 1200kW medium-wave broadcasting station generated 32V/m at 0.5km.

- A wire-type spark erosion machine generated the equivalent of 0.02V/m field at 1m.

4.1.2   Continuous radiated threats from radio transmitters

The distances given below assume free-space radiation and a fall off with distance proportional to 1/r. Local field strengths can be doubled by reflections and resonances in nearby metal structures. In practice, near-field effects and scattering or absorption by objects along the path mean that actual fall-off with distance is proportional to $1/r^n$, where n varies between 1.3 for open country and 2.8 for heavily built-up urban areas (EN 55011 suggests an average value of $n = 2.2$). Table 4.3 is therefore conservative.

General background EM radiation at a representative site has been found [15] to be between 0.1 and 1V/m, dominated by long, medium and short wave signals.

Table 4.3 Distances versus radiated field strengths

Total emitted RF power, and type of radio transmitter	Proximity for 1V/m	Proximity for 3 V/m	Proximity for 10 V/m	Proximity for 30 V/m
0.8W typical (2W maximum) hand-held GSM cellphone, and 1W leakage from domestic microwave oven	5 (7.8)m	1.6 (2.5)m	0.5 (0.8)m	0.16 (0.25)m
4W private mobile radio (hand-held) (e.g. typical VHF or UHF walkie-talkies, TETRA mobiles)	11m	3.6m	1.1m	0.36m
10W emergency services walkie-talkies, and CB radio	16m	5.0m	1.6m	0.5m
20W car mobile cellphone, also aircraft, helicopter, and marine VHF radio-communications	25m	8m	2.5m	0.8m
100W land mobile (taxis, emergency services, amateur); paging, cellphone, private mobile radio base stations	54m	18m	5.4m	1.8m
1.0kW DME on aircraft and at airfields; 1.5kW land mobile transmitters (e.g. some CBs)	210m	70m	21m	7m
25kW pulsed marine radars (both fixed and ship-borne)	850m	290m	89m	29m
100kW long wave, medium wave, and FM radio broadcast (Droitwich is 400kW)	1.7km	580m	170m	58m
300kW VLF/ELF communications, navigation aids	3km	1km	300m	100m
5MW UHF TV broadcast transmitters	12km	4km	1.2km	400m
100MW pulsed – ship harbour radars	55km	18km	5.5km	1.8km
1GW pulsed – air traffic control and weather radars	170km	60m	17km	6km
10GW pulsed – some military radars	550km	180km	55km	18km

4.1.2.1   Attenuation of field strength by buildings

The attenuation of a double-brick wall in the UK may be assumed to be one-third (10dB) on average, but can be zero at some (unpredictable) frequencies. The attenuation of a typical steel-framed building can be much better than this except in the region 50 to 200MHz, depending on position within the building. [51] gives attenuation values for seven different types of building common to the US.

4.1.2.2   A note on radars

The peak value of emitted power depends on the type of radar and its pulse characteristics. Radar fields are line-of-sight, and the very high powers of ground-based radars are considerably attenuated by geographical features such as hills or the curvature of the earth. Fixed radars are normally aligned so as not to include people or buildings in their main beam.

4.2 Coupling

Source, coupling path and victim

Electromagnetic compatibility invariably has two complementary aspects. Any situation of incompatibility must have a source of interference emissions and a victim which is susceptible to this interference. If either of these is absent, there is no EMC problem. There is also a third factor: there must be a coupling path between source and victim, which can be through a direct connection, through proximity, or through radiated energy. The same equipment may be a source in one situation and a victim in another.

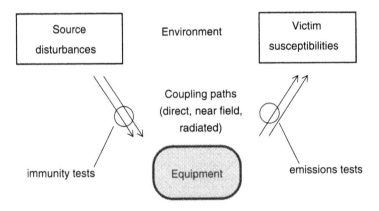

Figure 4.1 The three parts of an EMC situation

The various possibilities for the nature of the coupling path are explored below.

4.2.1 Direct coupling

4.2.1.1 *Coupling via power or signal lines*

The easiest method of coupling to visualize is when there is a direct connection between the source and the victim. The most typical example of this would be via the power supply (Figure 4.2(a)). Disturbances present at the supply port of the interference source are fed onto the power supply conductors, along which they propagate, and from which they enter the power supply port of the victim. The power supply network itself is usually regarded as a passive player in this scheme. The disturbances can be propagated along it either differentially, that is between its conductors, or in common mode, that is along all its conductors with respect to a remote earth reference (see 4.2.4). In most coupling situations, both modes occur, though one is usually dominant.

The total coupling path can be modelled as a noise source with its own impedance Z_S feeding into the characteristic impedance Z_0 of the power network treated as a transmission line, which then feeds the load impedance Z_L of the victim (Figure 4.2(b)). Clearly the level of disturbance which ends up at the victim will be a function of these impedances, which are complex and dependent on frequency. For the mains supply network in particular, Z_0 is heavily affected by loads which are attached at other points in the network; between 2Ω and 2000Ω depending on frequency and time of day can be observed, although the characteristic impedance of the copper wiring itself is

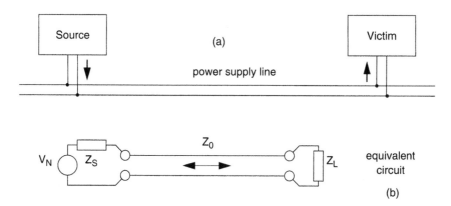

Figure 4.2 Coupling via the power line

reasonably stable and predictable. The impedance of 50Ω/50µH between each phase and earth, which has been developed by CISPR to cover the frequency range 9kHz to 30MHz for conducted emissions tests, is an average value which was determined by measurements across a wide range of installations in different countries. (Note that this is quite different from the standardized low frequency source impedance between phases of $0.24 + j0.15 \, \Omega$ which is used for flicker tests.)

Whether or not the mains network should be treated as a transmission line depends on the length of wiring between source and victim, and the frequency of interest. For longer distances than a quarter wavelength, transmission line parameters are essential; shorter than this, the impedances can be represented by a lumped approximation. At or above 30MHz, this implies that distances longer than 2.5m should be treated as transmission lines. At, say, 1MHz, the appropriate distance is 75m.

The amount of disturbance power that is fed into the network from the source, apart from being determined by the available voltage, is also affected by the impedance match between the source and the line. Maximum power transfer occurs when the source presents a conjugate match to the line; in the context of the CISPR impedances quoted above, that is when the source impedance is 50Ω plus whatever capacitance is necessary to resonate with 50µH at a given frequency. The same will be true of the victim. Of course, in real life interfering sources and victims do not present such a perfect match, and so maximum disturbance coupling rarely occurs. The purpose of interference suppression filters at the mains ports of equipment, and for that matter anywhere along the power network, is to deliberately increase the mismatch and therefore reduce the power transfer even further, usually by several orders of magnitude. The degree of mismatch is affected by the impedances presented either side of the filter, which is why real life filter attenuation performance rarely resembles that which is presented in manufacturers' data. This subject is covered further in section 8.1.3.

Coupling via the power network is not the only example of direct coupling, though it is the most significant because there is usually no specific relationship between the source and its victim, so that the systems designer has no way of predicting the exact interaction between them. Signal and control lines between different equipment can also carry disturbances, and the same coupling analysis can apply to them.

4.2.1.2   Common impedance coupling

A second form of direct coupling occurs between two nominally separate modules which share a common impedance. The most frequent example of this is a common earth or power supply connection. Figure 4.3 shows two modules each of which is earthed to a single busbar, which is then taken to the system earth reference by a further wire. Module 1 may have internal noise V_N which is coupled to its earth wire, and referenced externally to the system earth usually by stray capacitance C_S or by connections to other modules. The current produced by this noise voltage flows not only in Module 1's earth wire, but also in the common earth wire that connects the busbar to the system reference. Since this wire has a finite impedance (see section 5.2.1 on page 112) a further noise voltage V_{int} is developed across its length which then appears between the busbar and the system reference, and hence is coupled to the earth connection of Module 2.

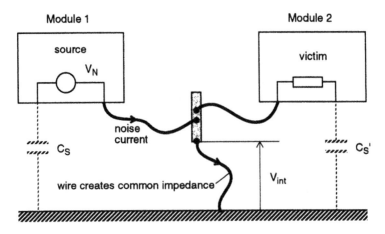

Figure 4.3 Common impedance coupling

This mechanism occurs in all such interconnection cases, and of course very often the actual interconnections are more complex and may involve several modules, which are interconnected to each other via other cables as well as the earth. If problems do not arise despite this, it is because the noise voltages generated are small, the modules are well-designed with respect to EMC, and because the earth conductor impedances although non-zero are negligible at the operating frequency of the equipment.

Conductor impedances are not negligible across the whole spectrum of frequencies of interest to EMC, though. Not only that, but the collection of impedances which appear in any given disturbance current path are usually made up of both inductance (due to the earth wires) and capacitance (due to stray coupling between the module and its surrounding structure). Such a mesh of interconnected L and C results in a number of resonant peaks in the total circuit impedance. If any of these peaks coincides with a particularly significant disturbance or susceptibility frequency, the effect of the coupling path is magnified.

To minimize the effects of common impedance coupling:

- the coupled network must be carefully designed so that high interference currents do not flow in impedances which are connected to circuits which may be sensitive to the noise, and

- the impedance of unavoidable common connections should be kept as low as possible over as wide a frequency range as possible; this is one reason for the use of short, wide earth bonding straps rather than wires.

4.2.2   Near field (inductive and capacitive) coupling

Modules do not need to be directly connected for interference coupling to take place. Whenever current flows in a conductor, a magnetic field exists around it; whenever a voltage appears between two conductors, an electric field is present. Each of these field phenomena is capable of inducing an interfering signal in a second circuit that is coupled to the field. The principles involved are deliberately used in transformers and capacitors; dealing with a near-field EMC coupling problem is, in principle, nothing more than controlling the unwanted transformers and capacitors in the structure. These are sometimes called "strays" or "parasitics" and become very significant at higher frequencies.

4.2.2.1   Magnetic or inductive coupling

The voltage induced in a conductor by a current flowing in another conductor in close proximity (Figure 4.4) is

$$V_N = -M \cdot di/dt \qquad\qquad (4.1)$$

where M is the mutual inductance of the coupled conductors. This voltage appears in series with the desired signals in the victim circuit, and is unaffected by the circuit impedances. The mutual inductance is determined by the separation of the conductors, the length along which they are in proximity, their geometry, and the presence of any magnetic screening around one or other conductor. The most usual situation in which magnetic coupling is significant is when several circuits are run together in a single cable loom. In this situation the circuits are in close proximity for the length of the loom and their mutual inductance is high. This is one reason for recommending segregation and separation of different cable classes (see section 7.4.2). The quantitative effects of separation distance are shown in Figure 4.6.

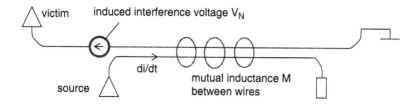

Figure 4.4 Magnetic field coupling

However if a return circuit is run adjacent to the signal (or power) conductor, the magnetic field due to the signal current is nearly cancelled by the magnetic field due to the equal and opposite return current. The greater the distance from the conductor pair, the more completely is the field cancelled. This is why signal and return, or power and return, pairs[†] should always be run together in any cable loom. The mutual inductance between cable pairs is inversely proportional to the log of the square of their separation

---

† Or triples for 3-phase delta supplies, or quads for 3-phase star supplies.

distance, whereas that between two individual wires carrying different circuits is inversely proportional to the log of the separation distance directly.

A further situation in which magnetic coupling is significant is when a magnetic component is involved, such as a transformer or motor winding. The magnetic field around the core of such a component can be much higher than around a conductor carrying an equivalent current, but it falls off much more rapidly, inversely proportional to the cube of distance. Problems are usually only significant if a cable carrying a sensitive circuit runs directly past such a component, or if a magnetically sensitive device such as a sensor is placed right next to it. A small amount of separation will show a very large improvement in this situation – for instance, a change in separation distance from 2cm to 20cm would show a 60dB ($1000 \times$) reduction in coupling.

Magnetic screening is difficult to achieve: most conducting materials such as copper and aluminium are by themselves largely transparent to the magnetic field and do not offer significant attenuation at low frequencies (see the discussion on skin depth in section 6.2.1.2 on page 139). A magnetically permeable material such as steel or mu-metal is more effective but requires bulk for effective absorption and is difficult to handle. The only realistic solution to many magnetic screening problems is to surround the offending or victim conductor with a conductive tube along its length (Figure 4.5), within which an equal but opposite current is allowed to flow[†], which has the effect of negating the magnetic coupling with fields outside the tube. If the tube is perfectly coaxial with the inner conductor and the currents are exactly opposite, the magnetic coupling is nil. This is one of the main principles by which coaxially screened RF cables work, but it is only imperfectly effective for other structural geometries – the twisted pair being the next closest approximation.

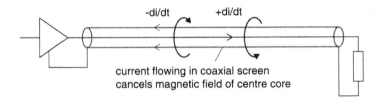

Figure 4.5 Magnetic field attenuation

4.2.2.2 Electric or capacitive coupling

The complement to magnetic coupling involves capacitance between structures and is mediated by the electric field between them. When a voltage difference appears between two conductors an electric field is created; this field induces a voltage on the victim conductor of

$$V_{in} = C_c \cdot Z_{in} \cdot dV_s/dt \qquad (4.2)$$

where V_{in} is the voltage induced on the victim circuit of impedance Z_{in}, by an interfering voltage V_s (of negligible source impedance) coupled through a mutual capacitance C_c. Notice that (a) in contrast to the magnetic field coupling, the severity of electric coupling depends on the victim's load impedance, so that high impedance circuits are much more susceptible; and that (b) the source and victim circuits need to

---

† This is the principal reason behind the advice to bond cable screens at each end, explored more thoroughly in section 7.2.4.1.

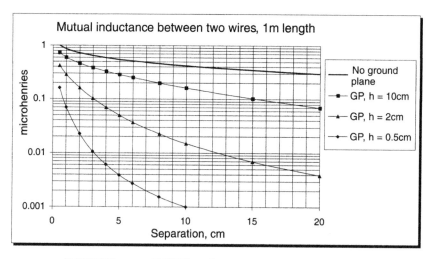

Two wires, no ground plane, $D/l \ll 1$

$$M = 0.002 \cdot l \cdot (ln(\frac{2l}{D}) + \frac{D}{l} - 1)$$

Return current through ground plane

$$M = 0.001 \cdot l \cdot ln(1 + (\frac{2 \cdot h}{D})^2)$$

M in microhenries

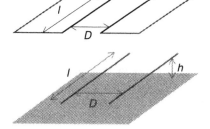

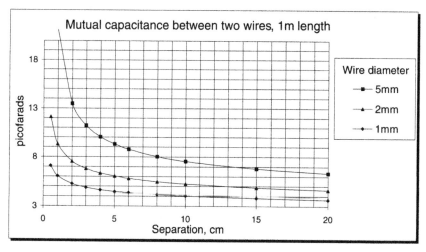

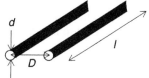

$$C = \frac{0.0885 \cdot l \cdot \pi}{acosh\frac{D}{d}}$$

C in picofarads, $\varepsilon_r = 1$

(All dimensions in cm)

Figure 4.6 Spacing between conductors

be referenced together. This is easily seen in Figure 4.7 when the circuits share a common 0V or earth reference, but E-field coupling occurs even between circuits that are totally isolated. In this case, there are two coupling capacitances involved, one between the source and victim nodes, and one between the reference points for each circuit.

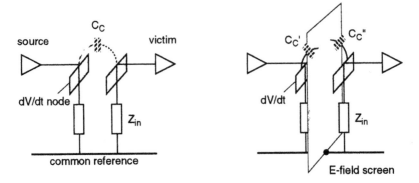

Figure 4.7 Electric field coupling

The mutual capacitance appears between two voltage-bearing circuit nodes, through which the current is irrelevant, rather than between two current-carrying conductors. Of course, these may be one and the same structure, such as a pair of wires, but the distinction between current and voltage is important. The capacitance is affected by separation distance, geometry (in particular the area of overlap of the two nodes), the nature of the dielectric between the nodes, and the presence of any electric field screening between them. Since area is significant, capacitive coupling tends to be greater between large objects than between small ones, but this is often mitigated because larger structures tend not to carry high levels of dV/dt. Where they do, such as with switchmode power converter heatsinks, capacitive coupling is a serious threat.

Screening against electric fields is very much simpler than against magnetic fields. Any conductive material will act as a barrier to the electric field; the lower the resistance of the barrier the better, but even a material of a few ohms per square (such as nickel paint) will severely attenuate the field. A partial screen can be effective, although some of the field leaks around the edges of the screen. The electric field is easily distorted even by dielectric materials, which makes for difficulties in measurement when this is needed, since any measuring probe will affect the field it is supposed to measure. The important aspect of E-field screening is that the screen itself must not carry any interfering voltages, or these will capacitively couple (C_C" in Figure 4.7) to the circuit that is being protected. Thus it must be connected to a point of low interference potential; this is naturally assumed to be "earth", but the difficulty with practical implementation is that many "earths" actually carry significant interference voltages, or there may be voltage differentials between nominally identical earth points. In detailed system screening design, the point of connection of the screen must be chosen carefully. A conducting structure that is unconnected to any potential is of no use as an E-field screen, though it may have other purposes.

4.2.3   Radiated coupling

So far we have considered magnetic and electric field coupling in isolation. At DC and low frequencies this is quite acceptable – it is known as the "quasi-static approximation". But any varying electric field between conductors requires a current to cause the voltage to change, and any changing current will develop a voltage differential. Thus AC fields are inherently composed of both electric and magnetic components, and as the frequency rises so it becomes more difficult and less meaningful to treat them separately.

At a sufficient distance from the structure which is carrying the radiating currents and voltages, the magnetic and electric components resolve themselves into a propagating electromagnetic wave. The two component vectors are at right angles to each other and to the direction of propagation, and lie in a plane surface which can be visualized as expanding outwards from the radiator in all directions. In free space, at any point on this plane the ratio of electric to magnetic components is constant and equal to $120 \cdot \pi$ or 377Ω, which is known as the *impedance of free space*. The amplitudes of the two components will vary in different directions away from the source as a result of the geometry and phases of the various radiating elements.

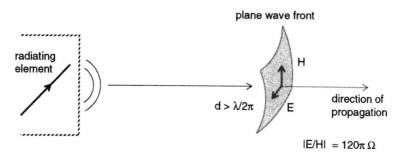

Figure 4.8 The plane wave

The "sufficient distance" beyond which the plane wave region of constant impedance begins can be derived from Maxwell's field equations and is given by $\lambda/2\pi$, or roughly one sixth of a wavelength. Examples of transition distance would be 1.6m at 30MHz, 48m at 1MHz, 16cm at 300MHz and 48km at 1kHz. Beyond this distance in the *far field*, coupling is radiative, that is the propagating electromagnetic wave induces voltages and currents in the victim structure as if it were acting as an antenna. The interference potential of the wave can be expressed equivalently as a power density (watts per square metre or milliwatts per square centimetre), or as an electric field strength (volts per metre), or as a magnetic field strength (amps per metre). Conventionally, in the frequency range of interest for EMC the electric field strength is quoted.

The radiated coupling both from a source and to a victim then hinges on the effectiveness of either as an antenna. The structures of non-radio electrical or electronic products are rarely designed with this purpose in mind, and they are inefficient converters of radiated energy at most frequencies, which is fortunate; though it is also possible that, on occasion, a particular arrangement of elements can result in a high antenna efficiency at a particular frequency. Efficient antennas have arrays of elements which are intentionally laid out so that the currents and voltages are in desirable phase relationships at the resonant frequency of the structure, and therefore give maximum transfer of energy at that frequency.

Conversely, good EMC design to minimize radiative coupling consists of deliberately arranging the mechanical layout to stop resonances, damping those which are unavoidable and ensuring that the induced currents and voltages do not couple well with the internal circuit operation.

Stopping resonances is difficult, especially given that most enclosures are rectangular in shape, so that a resonance will exist when each major dimension is a multiple of a half or quarter wavelength. In the range from 30 to 300MHz, any system enclosures of typical size will exhibit many such structural resonances – see Table 4.4 later. Damping them is more successful, and normally occurs anyway with the introduction of parasitic smaller structures (such as PC boards or internal cables) into the main structure; such damping can be enhanced by the deliberate introduction of RF absorptive material such as ferrite sleeves on relevant cables. The main weapon in good EMC design consists of separating the unavoidable interference currents that exist in the structure from the operational currents and voltages that exist in the circuit. This is the purpose of screening, segregation, earth layout, and filtering.

4.2.4   The modes of coupling

One of the most important aspects in EMC is to understand the distinction between the possible modes of coupling. The basis for this distinction is the idea that two separate circuit paths can coexist in the same set of conductors. One of these is the circuit that was intended by the designer – signal and return, or power and return, along which the desired signal currents flow, "differentially", that is in opposition to each other.

The other is the parasitic circuit that is formed between this desired circuit and the structure within which it is located. This is called the "common-mode" circuit, because the currents in the conductors are all flowing in the same direction. Figure 4.9 illustrates these different modes for a generalized apparatus with a mains power supply and a signal line. The arrows in the figure imply emissions; their direction could quite easily be reversed to imply susceptibility.

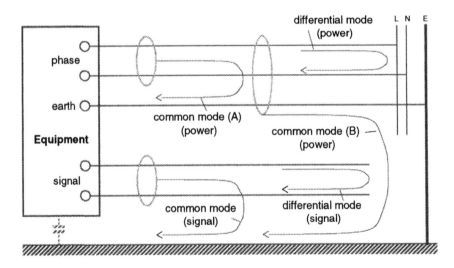

Figure 4.9 Differential and common mode concepts

4.2.4.1   Conducted differential mode

Conducted coupling in differential mode on the power supply is probably the simplest to visualize. Interference appears between the phases (L and N, DC + and –, or P1/P2/P3) of the power supply and is carried into or out of the equipment by the phase conductors only. Placing a filter in line with these conductors is the conventional way to attenuate this noise. Typical sources of emissions are switch-mode power supply or switching converter currents, and of incoming interference, fault- or lightning-induced surges are the most common. Conducted emissions tests on the mains port will measure half the differential component on each phase (see section 10.1.2 for more detail).

A similar mechanism operates on the signal lines, although here the threat and method of dealing with it is different: since in general signal connections are point-to-point, they do not have the opportunity to pollute a wide area as is the case with power connections. Differential currents in signal lines are principally concerned with the signals themselves. If these are DC or low-frequency (e.g. for sensor or audio connections) then their interfering capability is low; filtering is straightforward and is aimed principally at improving immunity to induced signals. If the signals are wideband (data or video) then there will be both interference capability and susceptibility issues, and filtering is more difficult. Either screened cables or special protection measures at the interface are needed. However, the main problem with such signal lines is radiated rather than conducted coupling (see 4.2.4.3).

4.2.4.2   Conducted common mode

The power supply phase conductors, and the signal line conductors, also carry conducted interference in common mode. In this case the interference does not appear *between* the conductors. It appears on each conductor with reference to a third point, and the interference currents flow in a loop which includes this third point. In the case of the power supply, there are two possibilities for the third point: it can be the safety earth wire (common mode (A) in Figure 4.9) or it can be the external structure (common mode (B) in Figure 4.9). Although the safety earth is usually connected to the external structure at some point, there are differences in these two modes. The most notable difference is that in the first, the currents remain within the power cord – flowing down the phase conductors and returning via the safety earth, whereas in the second, all conductors *including the safety earth* carry common-mode current which returns via a separate path. Clearly for Safety Class II apparatus (without a safety earth wire) only the second mode applies.

Sources of common mode emissions are much harder to visualize and predict, and also harder to control. They are usually associated with internal high frequency functions within the equipment (such as microprocessor clocks, and also switch-mode supply oscillation) which are not intentionally or directly coupled to the power port but nevertheless appear there through stray coupling. Similarly, incoming common mode interference (typically fast transient burst noise, and radio frequency signals) is coupled into the internal circuits via such parasitic paths. Simple filtering between the phase conductors has no impact on this mode of interference. Mode (A) coupling can be filtered by a common mode choke and parallel capacitors between phase and earth (which worsen the leakage current pollution of the earth network), as is usually found in most bought-in mains filter units. Conducted emissions tests measure the L-E and N-E voltages separately; if there is no differential component then common mode (A) signals are indicated directly. Mode (B) coupling cannot be dealt with by capacitive filtering since there is nowhere to connect parallel capacitors to; common mode chokes in all lines including the safety earth can help, but otherwise the only successful solutions involve structural remedies to the equipment itself.

Common mode noise in the signal lines is equally significant. The comments above regarding filtering are equally applicable, with the added problem for wide-band signals that capacitive filters will affect the wanted signal as much as the interference. As well as stray coupling within the equipment, another source of emissions occurs for wide-band signals: through leakage between the cable and the environment, some signal current "escapes" from the differential circuit and returns through the structure, thus resulting in a common mode component of the signal. The degree to which this occurs is predictable from the cable parameters, and cables for wide-band applications may quote their performance in this respect as "longitudinal conversion loss", discussed further in section 7.1.2.1.

Conducted common mode interference is more of a problem in general than differential mode, because its coupling paths include physical structures which are normally not designed for the purpose. Consequently,

- its effect is diffficult to predict and control;
- it can change with time because of uncontrolled structural changes;
- it can pollute a variety of unrelated equipment;
- the currents can flow within a large and uncontrolled loop, increasing their potential for radiated coupling.

4.2.4.3   Radiated differential mode

The same conceptual circuits of Figure 4.9 can be used to visualize radiated coupling. Because the efficiency of a coupling structure peaks when it is close to a quarter- or half-wavelength in dimension, at the lower frequencies the major mechanism for radiated coupling as discussed in 4.2.3 is via cables, which are generally longer than other elements of the system. Table 4.4 relates the half-wavelength dimension to frequency and to typical sizes of structures. From this it is clear that equipment enclosures are likely to dominate the radiative mechanism above 100–200MHz; long cables will dominate it below 30MHz. If the coupling is between two structures in close proximity, stray capacitive and inductive (near-field) effects dominate and wavelength plays a less important part, and the coupling can be expected to occur over a wide range of frequencies.

Table 4.4 Frequency versus half-wavelength

Frequency	0.5·λ	Typical structures
10MHz	15m	Long cables
50MHz	3m	Medium cables, large cabinets
100MHz	1.5m	Short cables, medium cabinets
300MHz	50cm	Medium cases, 19" rack enclosures, internal wiring
600MHz	25cm	Small cases, PCBs

In differential mode in cables, the return current path is known and (unless the circuit has been poorly designed) it will be in close proximity to its send path. This means that the magnetic fields due to the currents from each conductor will tend to cancel each other, and the electric fields from the voltages on the conductors will tend to be concentrated between them. Radiated coupling is therefore minimized when the conductor pairs are kept as close together as possible, as in a cable loom, and best exemplified by a twisted pair. Nevertheless, at high frequencies the small area that remains between the conductors is still capable of radiated coupling and this puts a limit on the amplitude and frequency of wide-band signals that can be transported along cables of a given geometry.

Within equipment it is often harder to maintain the optimum geometry for all circuits; internal wiring runs and the constraints of PCB layout tend to compromise the rule of minimum enclosed loop area. Thus radiated differential coupling with the internal circuits takes on a greater importance, particularly for frequencies where the structures are resonant, typically above 100–200MHz. A screened enclosure creates an intentional barrier to this coupling mode.

4.2.4.4   Radiated common mode

The common mode current path, by its very nature, breaches the optimum layout requirement of close proximity of signal and return. Often there is no control over cable routing, and therefore no control over the enclosed area of the common mode loop, which can become very large. Only when a cable is run for its whole length against a structure that intentionally carries the common mode return current, such as a properly bonded cable conduit, is the coupling loop area controlled to a minimum. In other circumstances the cable acts as a reasonably efficient antenna, with maximum efficiency at its resonant frequency. Worse, it is entirely possible (and indeed normal) for common mode currents to flow in the screen of a screened cable. The design of the screen should ensure that internally-generated common mode currents do not transfer to the outside of the screen and that conversely, incoming interference currents that inevitably flow on the outside do not transfer to the internal circuits. The quality of the screen and its termination at either end are crucial factors in this respect.

Radiated common mode coupling is not limited to cables, although because of their length they assume a great importance. Any conducting structure will carry common mode currents and will act as a radiator or receptor. Current flows and voltage differentials occur on the outside of metal enclosures, and at the frequencies at which they are resonant, they are most efficient at radiating them. Since metal structures can't be avoided in most systems and installations, good EMC design requires care that they are not excited by interference currents generated by the system, and that the currents which are developed by external interference sources are not then transferred to the system's operational circuits.

This description of common mode coupling between structures and circuits, which occurs both within cables and within the equipment itself, should emphasize that *limiting transfer between modes* is an important goal of interference control. The *transfer impedance* of a given structure is a measure of this parameter and is discussed in more detail in the next few chapters, on earthing, enclosures and cabling.

4.2.5   Protection measures

In the context of electronic products, EMC protection can be applied at the circuit level, at the interfaces, and at the enclosure. The desired degree of protection can be obtained by trading off techniques at each of these levels against each other to achieve a cost-effective optimum.

The system designer does not have this luxury. System components are "black boxes" whose internal workings cannot be modified, and whose interface parameters are already fixed, which is why EMC aspects are so important in purchasing (section 3.4). The only variables the system designer has to work with – assuming that the black boxes themselves are already specified – are the interconnections between the component modules, the physical layout of the modules, and the possibilities of electromagnetic containment. All of these impact the coupling to, from and between modules, rather than the performance of the modules themselves.

Subsequent chapters in this book will deal with each of these areas in detail. Here we simply enumerate the measures which can be taken at the system level to minimize the effects of the coupling paths that have just been discussed. These can be conceptualized as:

- creating zones of different degrees of interference potential (partitioning the system);
- erecting barriers between the zones;
- applying protection at the electrical interfaces across the barriers;
- deliberately designing the earth reference network as a sub-system in its own right.

4.2.5.1   Zoning and barriers

The concept of zoning is very simple: the operating environment of a system or installation is partitioned into a number of different classes of zones, within which different levels of EMC protection apply. Two zones is the minimum for this approach; three is quite reasonable, but more than three would be unusual and only necessary for special cases. Figure 4.10 shows the concept graphically.

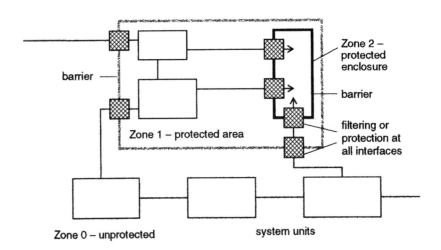

Figure 4.10 The concept of zoning

In this example, **Zone 0** is essentially the unprotected environment – the "outside world", so to speak. In an industrial control system this would be the shop floor, or in a building management system it would be the outside, or parts of the building without electrical services if this were a typical office or hotel, say. The significant aspect of this zone is that *no special EMC precautions* are taken within it. Equipment located in Zone 0 must comply with all the EMC requirements for emissions and immunity that are applicable for the general environment.

Zone 1 is a protected area within the general environent of the system. Within this area lower levels of immunity can be tolerated, as can higher emissions, since the effect of the zoning is to provide some level of isolation from the environment. Alternatively, a higher degree of reliability is offered to equipment with a standard level of compatibility. The nature of the protection can vary, but two typical approaches would be to create an area within which the earthing system is properly controlled, and/or to screen the walls, floor and ceiling of a single room. A Zone 1 area might be used for the control room of an industrial plant, or a safe room for the main servers of an office computing network.

Zone 2 is a smaller area of even greater protection. It could be located within Zone 1, as shown in Figure 4.10, or more rarely it could be an area within Zone 0 which has a greater degree of protection than Zone 1. The most usual approach to implementing a Zone 2 area is to provide a fully shielded enclosure such as a racking cabinet. Within this could be housed very noisy or very sensitive equipment such as motor drives or medical or scientific instrumentation – not both together, of course!

The zoning principle requires that the boundaries, or barriers, of each zone are defined, and that the protection offered by each zone is quantified. The screening effectiveness of an enclosure, for instance, could define the difference between Zone 1 and Zone 2. Given a known or assumed level of disturbance and susceptibility in the external environment (Zone 0), the corresponding levels within each zone can be determined by the attenuation across the zone boundary. This can then be used to define the requirements for the equipment that will be located in the zone.

It should be clear that whatever services, electrical or otherwise, cross the boundary must be treated to maintain the required attenuation. This is the function of interface protection (filtering and suppression), discussed next.

The various EMC phenomena may not all be attenuated equally at a zone boundary. For instance, a controlled earthing regime may not be as effective as a shielded enclosure for reducing field strengths, but can reduce the amplitude of transients and surges within the zone. It is entirely reasonable, and good practice, to match the zoning philosophy and techniques to the needs of the equipment within the zones and the environment in which they exist.

4.2.5.2   Interface protection

The attenuation across a zone boundary must be applied to the electrical services to the same degree as is offered by the physical arrangement, such as screening or earthing. Filtering and/or transient suppression across the appropriate frequency range for every unscreened cable that crosses the boundary is necessary. This is not difficult to do for power cables, but signal and control lines may pose a greater problem. For this reason it can often be helpful to partition the whole system so that the minimum number of difficult-to-treat cables cross the zone boundaries.

Screened cables do not need filtering, although surge protection may occasionally be required on internal conductors. However, it is essential to bond the screen to the earthing structure at the boundary (see section 7.2.4.2 on possible exceptions to this rule), whether this is a screening wall or an earthing bar. This ensures that the interference currents on the screen are returned to the earth rather than penetrating from one zone to another, which would compromise the attenuation otherwise offered at the zone boundary. This bonding requirement applies not only to electrical screens but to any conductive service that crosses the boundary, such as pipework, conduits or ducting.

4.2.5.3   Earthing

A well designed earth structure can contribute greatly to reduced interference coupling within a protected zone. This is because the earthing controls the common mode currents that were discussed in section 4.2.4. Interference currents that remain do not generate large potential differences across the zone, so that each item of equipment within the zone is subject to lower electromagnetic stress. Conversely, noise currents developed by equipment within the zone are returned directly to the source and do not propagate significantly beyond the zone structure.

In order to achieve this function the earth structure must be regarded as a component of the system in its own right, and designed as such. This is discussed in Chapter 5.

4.3   Mains harmonics

4.3.1   Their causes and problems

Non-linear loads generate harmonic currents which in turn create harmonic voltage distortion due to the finite (and complex) impedance of the mains supply network. These harmonic currents can overload neutral conductors and distribution transformers, cause nuisance tripping of protective devices, and create EMC problems. Harmonically distorted mains waveforms can cause a variety of reliability problems with electronic apparatus.

Examples of non-linear loads include:

- Transformers and DOL induction motors (usually fairly small contributors);
- Rectifier systems used in electrolytic processes;
- Fluorescent lamps;
- The DC power supplies of any electronics (e.g. adjustable-speed motor drives, information technology and telecommunication equipment, etc.) whether linear or switch-mode;
- Three-phase power convertors (six-pulse, twelve-pulse, etc.);
- The DC power supplies of microwave magnetrons and klystrons, and other radio-frequency generators;
- Arc welding, arc furnaces, electric smelting, etc.

Where the mains supply suffers from severe waveform distortion, even linear loads such as resistive heaters draw harmonic currents which then pollute their supply networks.

Of all the above it is the electronic DC power supplies that are causing the most concern these days, due to the ubiquity of electronic devices such as TV sets in domestic premises (soon to be increased by the use of variable speed drives for many appliance motors); computers and computer technology in commercial buildings; and adjustable-speed drives in industry. In most cases the building mains and earthing networks were installed when large harmonic currents were not a consideration.

Traditionally, harmonics was only a concern for larger systems and installations, particularly for power generation and distribution and heavy industry. But the modern proliferation of small electronic devices, each drawing perhaps only a few tens or hundred of watts of mains power, and usually single-phase (e.g. personal computers), has brought the problem of mains harmonics to almost every type of system and installation.

Triplens

Emissions of "triplen" harmonics (multiples of 3: 3, 6, 9, 12, etc.), or triplen harmonic distortion of the incoming supply, add constructively in Neutral conductors and are known to reach 1.7 times the phase current in some installations [26]. Single-phase non-linear apparatus has a much greater propensity to emit triplen harmonics than three-phase equipment, which is partly why the problem of triplen harmonics in neutral cables and transformers of systems and installations is a particularly modern one. Other non-triplen harmonics have phase rotations which are either faster or slower than the rate of the fundamental, and can even rotate backwards. These can cancel out to some degree leaving a complex ripple current (a detailed analysis is usually required to determine the degree of cancellation).

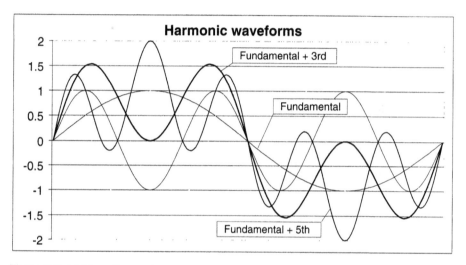

Figure 4.11 Addition of harmonics

4.3.1.1   *Transformer and Neutral conductor overload*

Harmonic currents in the Neutral conductors present serious reliability risks, and even safety risks, where Neutral conductors have not been suitably dimensioned. Many modern installations use Neutral conductors of the same cross-sectional area (csa) as their associated phase conductors, and some (usually older) buildings are known to use half-size or smaller Neutral conductors, with obvious problems. In some countries (e.g. South Africa) it is possible to find buildings wired with quarter or even eighth-sized Neutrals.

As well as creating problems with Neutral conductors, these excessive "zero-phase" currents cause excessive zero-phase flux in delta-wound transformers, leading to possible overheating (often only at particular locations within the transformer). Most distribution transformer manufacturers will provide formulae which allow correct dimensioning on the basis of known harmonic currents.

4.3.1.2   *Unexpected tripping*

Overcurrent protection devices (fuses, MCBs, etc.) can trip even though the phase current appears to be much less than their rating. The real issue here is that the protection device is working properly, but the electrician's AC ammeter will read low

(−30% is not unusual) unless it is a true RMS type. The danger is that the electrician will fit a higher-rated protection device to prevent "nuisance tripping" and the cables will then not be adequately protected, resulting in increased fire safety hazards.

4.3.1.3   Conductor skin effect

Harmonic currents cause significantly greater ohmic heating in conductors than do 50 or 60Hz currents. This is due to the effects of skin depth (section 6.2.1.2) at higher frequencies. For example the density of the current in the cross-section of a copper conductor at the 7th harmonic of 50Hz (or the 5th harmonic of 60Hz) is reduced to around 32% (one skin-depth) at a depth of around 5mm from the outer surface. Since higher frequency currents are forced to flow through less of the conductor's cross section, they experience greater resistance and hence produce greater heating effects. This is a greater problem for larger diameter cables. A similar problem of increased thermal losses also occurs in the windings of distribution transformers, and possibly also in their magnetic cores.

4.3.1.4   Effects on electronics

Electronic circuits that rely upon zero-crossing detection can mis-operate, due to harmonically distorted waveforms. Consequences range from increased acoustic noise and/or energy inefficiency, through worsened reliability, to "explosive dis-assembly".

Electronic power supplies may provide much lower unregulated voltages than expected, since they usually rely on charging to the peak voltage of the sinusoidal mains. Harmonic waveform distortion (mostly due to the use of such power supplies) results in "flat-topping" and the power supplies don't charge up to as high an unregulated voltage as would be expected from a voltmeter reading of the supply. Consequently the regulated outputs of these power supplies may fall below their minimum specified levels, with unpredictable results for the functionality of the equipment, when the RMS mains voltage is only slightly below nominal.

4.3.1.5   Motor failure

Direct-on-line motors operated on distorted waveforms will try to rotate at the speed of phase rotation of the various harmonics. They will try to run at 3 times the rate of the fundamental for the 3rd harmonic, 6 times for the 6th harmonic, etc. At the very least the harmonic distortion of the supply waveform will cause a decrease in efficiency and a rise in winding temperature. Higher levels of distortion can cause serious overheating of the motor with consequent shut-down (assuming full protection is in place), excessive acoustic noise and vibration, damage to bearings (which can wear-out at as little as 10% of their nominal life), and dis-assembly of enclosures and mountings due to high levels of vibration. At even higher levels of harmonic waveform distortion motors may even refuse to start, drawing such high levels of third harmonic currents (the slip speed for the 3rd harmonic is much greater than for the fundamental) that their protective devices open.

4.3.1.6   Voltage distortion due to resonances

System resonance effects at the harmonic frequencies can create areas of the power distribution network where the voltage is more heavily distorted than elsewhere, and/or has significant over- or under-voltage. Also, some areas of the network can suffer from much higher levels of current than elsewhere in the network, at a few harmonic frequencies. At 50 or 60Hz most systems do not suffer from resonances, but at the higher frequencies of the harmonics, with the levels of capacitance increasingly seen due to the use of EMI filters in modern apparatus, it is becoming increasingly possible for power networks to resonate at harmonic frequencies.

4.3.1.7   Power factor correction

Power factor correction capacitors can fail, sometimes explosively (either the large ones used by the electricity suppliers or in the distribution room, or the smaller ones fitted to electrical apparatus and fluorescent lamp ballasts). This is due to the much higher harmonic currents they experience because of their lower impedance at higher frequencies, leading to overheating. Sometimes harmonics can cause overvoltage breakdown.

4.3.1.8   Radiated interference

Harmonic currents in mains supply cables give rise to emissions of electric and magnetic fields from the supply cables themselves, and also from the earth network which suffers from increased levels of harmonic currents due to the effects of the distorted supply voltages on the earthed capacitors in RF filters. These emissions especially cause problems for audio induction loop systems, and can also cause problems for VDUs, audio systems in general, and other sensitive equipment.

4.3.1.9   Increased earth leakage

Harmonic distortion of mains voltage causes protective earth conductors to carry higher levels of harmonic leakage currents, due to the lower impedance of EMI filter capacitors at these higher frequencies. These currents can cause the earth leakage to exceed safety norms. They also create potential differences between different parts of the earth network and hence give rise to common mode LF conducted disturbances on signal and power lines.

4.3.2   Harmonic solutions

It is important to make sure that all electrical measuring instruments in use by electrical engineers measure true RMS with a bandwidth of at least 2kHz, otherwise inaccurate readings of voltage and current will result, with possible unreliability and even safety consequences. Test equipment is also available from a number of suppliers that will display harmonics as individual bar graphs or readings.

To deal with mains harmonics it is often necessary to involve the local electricity supplier, especially where the problem is due to waveform distortion on his supply. Waveform distortion in the UK should typically be held to < 4% THD (total harmonic distortion) according to the electricity supply industry's guidelines, but might in future be relaxed to 8% throughout Europe due to the increasing difficulty experienced by power generators and distributors in maintaining a pure sinewave. The problem is compounded because European countries have different supply topologies, some of which are more susceptible to harmonic pollution than others. In some countries waveform distortion can be much greater than 4% THD: in parts of mainland China it is reputed to be so bad that the mains waveform is a reasonable square wave.

4.3.2.1   Standards

Standards for harmonic emissions from apparatus include the older IEC 555-2 /EN 60555-2, which applies to domestic electrical equipment, and IEC/EN 61000-3-2 (which has superseded it), IEC/EN 61000-3-4, and IEC/EN 61000-3-6.

IEC/EN 61000-3-2 [88] becomes mandatory on 1st January 2001 for all apparatus offered for sale in the EU that operates from public low-voltage supplies (400/230 Vac) and draws under 16 Amps per phase. This standard is an attempt to control the ever-increasing levels of harmonic pollution on the public mains supplies.

IEC/EN 61000-3-4 is for all equipment running on public low-voltage supplies which draws up to 75 Amps per phase. This will probably never be a mandatory EMC standard, instead it provides a number of standard forms allowing the hopeful user of a new apparatus to negotiate with his power supplier on the basis of the harmonics emitted by the new apparatus. Only equipment which meets IEC/EN 61000-3-2 (regardless of its current consumption) is to be allowed unfettered connection to the public mains supplies without permission from the supplier of the electricity.

IEC/EN 61000-3-6 provides a set of application forms, similar to IEC/EN 61000-3-4, except that it applies to apparatus supplied from MV or HV.

4.3.2.2   Remedies

Remedial measures for systems and installations include:

- Oversizing HV, MV, or LV/LV transformers;
- Use star power distribution for different applications by means of different feeders or transformers;
- Use an adequate csa for the neutral conductor, possibly even larger than the phase conductor;
- Equalize the load sharing between phases (does nothing for harmonics but reduces the heating effects in the neutral);
- Use only apparatus that conforms to IEC/EN 61000-3-2 or the appropriate sections of IEC/EN 61000-3-4 or -6;
- Fit resonant filters "tuned" to the harmonic(s) which are causing the problems. These may be series or parallel resonant, depending upon the effect required;
- Fit active harmonic correction equipment [40]. These are now available from several manufacturers, in a variety of sizes, and are usually intended to provide local correction for a machine or process, or a floor of information technology equipment. Active harmonic correction equipment is connected in parallel with the mains supply and draws current when the voltage waveform is too high compared with a pure sine-wave, storing the energy in internal capacitors and releasing it to the supply when the voltage waveform is too low. Apart from inevitable inefficiencies (minimized by their use of switch-mode technology) these units do not consume power. They protect the "upstream" mains distribution network from the harmonic currents flowing in the "downstream" network, and may possibly increase the harmonic currents in their downstream network. They also help to preserve supply waveform quality, but may not be able to be used where the incoming supply is significantly distorted – check with the manufacturer.

4.4   EMC versus safety

A discussion of EMC protection would not be complete without mentioning the potential conflicts that may exist between EMC protection measures and safety protection measures. Safety of electrical equipment is not often compromised by EMC protection measures, but there are a few areas where this can happen if no thought is given to the system design. Particular issues are:

Effects of filters and suppressors:

- earth leakage currents
- residual voltage on capacitors
- fusing of filters and surge protection devices (SPDs)

Effects of screened enclosures:

- creepage and clearance distances between conducting parts
- over-temperature and ventilation.

4.4.1   Filter problems

4.4.1.1   Earth leakage

It is usual for mains filters to include capacitors between each phase and the safety earth. As discussed in section 8.2.2.2 on filtering, these capacitors suppress common mode noise emitted by apparatus and also reduce the amplitude of incoming transients. Since there is an impressed voltage between the live conductor and earth of 230V at 50Hz (in Europe), a current will flow through the live-connected capacitor, and under certain fault conditions through all capacitors, given by

$$I \quad = \quad 230 \times (2\pi \cdot 50 \times C) \text{ amps} \qquad\qquad (4.3)$$

This translates to $80 \times C$ mA if C is in microfarads and a 10% increase is allowed for the effect of voltage variation above nominal. For reasons of protection of personnel against electric shock under fault conditions, safety standards limit the current that is allowed to flow continuously in the earth conductor. Different standards have different maximum values, and the values depend on the way that the equipment is connected to the mains source. Table 4.5 shows these values for a few of the most important equipment safety standards. The commonly-used figure of 0.75mA for portable earthed equipment results in a maximum capacitor value of 4.7nF between each phase and earth, and many mains filters incorporate this value.

Table 4.5 Earth leakage current limits in some equipment safety standards

Standard	Class I, portable	Class I, stationary	Class II
EN 60335-1 household	0.75mA	3.5mA	0.25mA
EN 60950 ITE	0.75mA	3.5mA	0.25mA
	Sinusoidal	Non-sinusoidal	DC
EN 61010 meas't & control	0.5mA	0.7mA	2mA
	Type B	Type BF	Type CF
EN 60601-1 medical	0.5mA earth leakage		
Patient leakage	0.1mA	0.1mA	0.01mA

From the systems point of view, it will often be the case that several items of equipment will be paralleled onto one mains supply. There is the real possibility that, even though an individual apparatus remains within the leakage current limits, the combination will not. However, the limit for portables does not apply to permanently-wired apparatus, so when such a combination is permanently wired in place the problem may be avoided. For instance, from Table 4.5, five appliances which meet the portable limit of EN 60950 can be paralleled if they will be connected in such a way as to render them "stationary". Heavy industrial equipment which has its own power supply does not fall under the rules of these standards, which can be fortunate, since it often needs much higher values of common mode capacitance to meet emissions requirements. On the other hand, patient-connected medical apparatus is only allowed 100µA of earth leakage current. This normally precludes any common mode capacitance in the filter at all, since such a low level can easily be generated by stray capacitances and leakage paths within the apparatus itself. Leakage current limits for certain medical systems can turn out to be particularly onerous when it comes to meeting the associated EMC requirement.

When a system is being built from a number of items of equipment which are known to be individually compliant both for safety and for EMC, the leakage current question must be properly addressed in the design. Special installation requirements apply where the leakage current exceeds 3.5mA, according to the IEE Wiring Regulations [114] section 607 and also EN 60950 [75] section 5.2.5. This provides a number of options for connecting single items of equipment to a final circuit, depending on the leakage current magnitude:

- 3.5mA < I_{lkg} ≤ 10mA: permanent connection to installation, or connection by means of plug-socket combination to BS EN 60309-2 (BS 4343)
- I_{lkg} > 10mA: permanent connection to installation (preferred), or connection by means of plug-socket combination to BS EN 60309-2 (BS 4343) plus additional 4mm^2 protective conductor, or by multicore with two protective conductors totalling 10mm^2.

EN 60950 (safety of information technology equipment) requires a warning label for leakage current greater than 3.5mA, and under no circumstances is it acceptable for earth leakage to exceed 5% of the total input current per phase.

Options for a high-leakage final circuit

If the final circuit is expected to carry a leakage current greater than 10mA, the following options are available:

- high integrity protective single conductor not less than 10mm^2, or
- separate duplicate protective conductors of not less than 4mm^2, or
- duplicate protective conductors totalling not less than 10mm^2 (minimum 2.5mm^2), or
- earth monitored circuit, or
- circuit supplied by double wound transformer, or
- "modified" ring final circuit.

The modified ring circuit is the most widely used [26]. The ring circuit's protective conductor must have a minimum cross-sectional area of 1.5mm^2, connected in the form of a ring with each leg connected separately at the distribution board. The final circuit must have no spurs and only *single* socket-outlets may be used.

If the installation is designed to supply a large installed base of IT equipment or other items with significant leakage currents, both the requirements for final circuits and the integrity of the distribution circuit protective conductors must be considered. Furthermore, RCDs (residual current devices) protecting a circuit should have a rated residual current of at least four times the total earth leakage, to prevent nuisance tripping.

4.4.1.2   Residual voltage

The capacitors connected across the phases in mains filters can take quite high values, up to several microfarads being not uncommon in high power units. These capacitors must withstand and carry the phase voltage continuously. Apart from mandating a particularly reliable form of construction (which is normally specified by quoting a part to IEC 60364-14), if the supply is suddenly interrupted with the load disconnected, the instantaneous phase voltage remains on the capacitor. This voltage can be anywhere between zero and the peak value, depending on the supply phase at the instant of disconnection. With high values of capacitance, if such a voltage becomes accessible to personnel a severe shock hazard can exist.

The typical case in which this could happen is with pluggable apparatus which can be switched off on the load side of the filter; the hazardous voltage can remain between the live and neutral pins of the plug when it is removed from the supply socket. For permanently wired systems apparatus, the threat is really only applicable to service and maintenance personnel. Even so, it should still be addressed. The normal solution is to ensure that a bleed resistor is wired across the phases to discharge the capacitors within an acceptable time when the supply is disconnected.

4.4.1.3   Fusing

Filter components are not perfect and can fail just like other components in the mains supply circuit of equipment. Additionally, transient suppression devices (typically, in mains circuits, metal oxide varistors or MOVs) can be connected both across phases and from phases to earth, to meet the increasing requirement for surge immunity. If these are subject to a high level surge they may fail short circuit, putting a low impedance path across the supply.

To meet this threat it is vital to include fusing of the proper value on the supply side of any filter or surge suppressor (see section 9.3.6.7 on page 232). An un-fused unit will present a potentially serious fire hazard in the equipment. Any bought-in items that already comply with safety standards should include this protection (but check!), but the principle must also be applied to any other discrete filtering or suppression that is put in place to implement the zoning concept within the system

4.4.2   Screened enclosures

4.4.2.1   Creepage and clearance

The mechanical design of many electrical products relies on separation distances between live and accessible parts to maintain safety against shock. When such products are placed within screened enclosures, it is necessary to ensure that such separation distances are not compromised. A typical problem might occur when a product in a plastic case with normally sufficient separation between hazardous and safe circuit connections, is bolted to a metal frame inside a screened enclosure; the separation distances from each set of connections to the metal frame may not now be sufficient. A related issue occurs when a plastic case is conductively coated for screening purposes.

The original designer may well have relied on the internal case geometry being insulating to maintain safety clearances. Adding a coating can breach these clearances by providing a new conducting path. The dangers of coating also extend to the possibility of flakes of coating becoming detached due to environmental stresses such as vibration or temperature extremes, and lodging in areas where they may bridge critical insulation.

Screening is not the only situation in which clearances may be affected: a poorly designed or installed mains filter or surge suppressor can create similar issues.

4.4.2.2   Thermal effects

To be fully effective a screened enclosure must be totally sealed with respect to electromagnetic penetration. This means that ventilation openings are anathema to the EMC engineer, who will bend every effort towards minimising or preferably eliminating them. It is easy to carry this process too far, and design a well-screened enclosure which overheats due to internal dissipation and may create a fire hazard or cause insulation to degrade. Liaison is necessary between the EMC design aspects and the thermal design aspects of an enclosure. EMC-screened ventilation openings are available (such as honeycomb panels) which are highly effective, if somewhat more expensive to implement.

It is also advisable to take into consideration the thermal properties of high current filters; any series choke will have appreciable resistance which is in series with the supply and which will cause additional dissipation. Very often this is the limiting factor in filter design. Expect a filter choke to run at elevated temperature, and check that it does not create a hazard in the worst case ambient temperature. Bear in mind that chokes are normally rated for RMS current, and that many electronic devices (particularly switch-mode converters) draw current with a high crest factor, which increases the I^2R loss over that expected with sinusoidal currents.

Chapter 5

Earthing and bonding

Give me where to stand, and I will move the earth.

– Archimedes, 3rd c. B.C.

Confusion about the earth reference has bedevilled EMC ever since the subject came into being. It has especial relevance to systems, where the concepts of functional earth, safety earth, RF ground reference and the interconnection of metallic parts are all interrelated and have become the objects of entrenched and not necessarily correct practices and opinions over the years. A major part of the problem has been that different practices are appropriate for different purposes; even if the purpose is purely EMC, frequency range is an important parameter, and its relevance is only imperfectly understood. This chapter will discuss the principles behind the purpose of earthing and will then look at how these are best implemented.

5.1   The concept and practice of earth

It's OK, we're earthed!

5.1.1   The purposes of the earth connection

First of all, it's necessary to decide what we mean by the term "earth". Difficulties immediately arise through the alternative American usage of the word "ground" to mean essentially the same thing; but because the words are often used interchangeably and in different contexts, confusion easily results. Dr Jasper Goedbloed [4] has argued cogently that both terms should be swept aside in favour of the more accurate "reference", leaving the word "earth" purely for application to the safety function. In this book we shall attempt to stick with the word "earth", only because it is the more familiar, and to indicate either explicitly or by context what function is meant.

5.1.1.1   Safety earth

The purpose of the safety earth is to guarantee personnel safety under fault conditions. The IEE Wiring Regulations (BS 7671, [114]) define "earthing" as

> Connection of the exposed conductive parts of an installation to the main earthing terminal of that installation.

Earthing ensures the provision of a low impedance path in which current may flow under fault conditions. Exposed conductive parts are those conductive parts of equipment which may be touched and which may become live in the case of a fault. The earthing connection prevents such live parts from reaching a hazardous voltage. Equipotential bonding, in this context, is a means of electrical connection intended to maintain various exposed and extraneous conductive parts at substantially the same potential under both operational and fault conditions. The protective conductor (typically colour coded green-and-yellow) provides this means and also connects the conductive parts to the installation's main earthing terminal. The prospective touch voltage within the installation is then the product of the impedance of the protective conductor and the earth fault current.

Equipotential bonding therefore creates an equipotential zone within which exposed and extraneous conductive parts are maintained at "substantially" the same potential. It is worth noting that although the voltages within such a zone may be safe, they are not necessarily, and not even usually, zero. Continuous currents from various sources, including equipment earth leakage, are likely to be flowing, even in a "healthy" circuit. Allowable earth leakage levels from individual items of equipment are covered in section 4.4.1.1 on page 97. An "equipotential" zone may protect people but may not protect equipment or wiring. Moreover, within a large installation there may be more than one "equipotential" zone for safety purposes and the voltages existing between them may well be large and undefined; special precautions need to be taken for signal wiring (such as local area networks) that may cross such zones, to prevent them importing potentially hazardous voltages.

Protection against electric shock is typically provided by earthing in conjunction with automatic disconnection of the supply. For this purpose, it is vital to co-ordinate selection of the protective device (for instance, fuse, circuit breaker or RCD) with the installation's earth fault impedance, to ensure that disconnection occurs sufficiently rapidly to prevent the touch voltage from rising high enough to cause a shock. The sizing and hence resistance of the earth protective conductor will therefore be determined largely by the prospective fault current available from the rest of the system. In the case of protective multiple earthing (PME), the protective conductors may take substantial continuous currents, even when the supply is off, as a result of circulating currents in the bonding network, and therefore may need over-rating. Since the concern is currents at low frequencies, it is resistance rather than inductance which determines the conductor impedance; this is not the case for high-frequency earths, as we see later.

5.1.1.2   Functional earth

In order for an electrical circuit to interface correctly with other equipment, there must be a means both of relating voltages in one equipment to those in another, and of preventing adjacent but galvanically separate circuits from floating.

This is the purpose of the functional earth and it must be distinguished from the safety protective earth. Earth conductors which are used for functional purposes only need have no requirements for sizing according to safety but should be coloured cream (according to BS7671) to identify them. Even so, because of the threat of circulating currents and potential differences between earthing zones, there may be other practical constraints on the widespread use of functional earthing on large systems.

Signal circuits of equipment should normally be specified for a maximum common mode voltage, which will be the voltage that appears between different parts of a functionally-earthed system. If this voltage is likely to be exceeded, implementation of a Common Bonding Network (CBN, see 5.1.3.3) in the system is recommended. If this is impractical or inadequate, isolated circuit interfaces are the normal solution.

5.1.1.3 Lightning protection earth

In building installations, there is a further important safety-related earthing function, and this is to provide a return connection for currents induced by a lightning strike. In many respects this is the *only* correct use of the term "earth", since this function is normally provided by ensuring a low-impedance connection throughout the building fabric to the literal earth on which the building sits. Since lightning potentials are built up between the cloud structure of a thunderstorm in the atmosphere and the surface of the Earth, connection to earth is the correct way to complete the circuit in the shortest manner.

Several standards for lightning protection have been published by the various standards bodies (e.g., [111]) and the reader is referred to these for detailed advice (Table 5.1). Section 9.2 reviews the main principles.

Table 5.1 Standards for lightning protection

	Protection of structure	Protection of contents	Risk assessment
IEC	IEC 61024–1:1990 IEC 61024–1–1 guide A	IEC 61312–1:1995	IEC 61662:1995
CENELEC	ENV 61024–1:1995		
BSI	BS 6651:1992	BS 6651:1992 App C	BS 6651:1992

5.1.1.4 EMC earth

The EMC earth has the sole purpose of ensuring that interfering voltages are low enough compared to the desired signal that incorrect operation or excessive emission does not occur [38]. It has no explicit safety or operational function. Because of this, and because of the wide frequency range over which it must work, earthing for EMC usually takes advantage of distributed structural components that are part of the whole system – typically, chassis members, enclosure panels and so on. The value of an EMC earth is directly related to its physical geometry, as we shall see in 5.1.2 and 5.1.3. This means that design and implementation of an EMC earthing system is not restricted to the electrical engineering discipline alone – it must also involve constructional aspects, that is, the mechanical designers and installers.

5.1.2 Definitions of the EMC earth

For EMC purposes, we can distinguish three almost hierarchical definitions of the earth:

- An equipotential area or plane used as a system reference;
- A low impedance path for currents to return to their source;
- A low transfer impedance path to prevent common mode currents converting to differential mode.

5.1.2.1   The equipotential area

A highly desirable feature of an earth is that there should be no potential difference developed between two points that are geographically separate. Thus connections to each point can be regarded as connections to the same potential. In the context of EMC, no noise voltages exist between these points and therefore no interference is injected into a system that is "earthed" only to these two points.

Such a perfect equipotential structure is in fact impossible, but it can be approximated by a highly conducting plane. Section 5.2.1 shows that wires have inductive impedance which renders them anything but equipotential, whereas a plane does not exhibit inductance, only resistance. The backplate of a systems equipment cabinet is an example of such a plane, and the meshed earthing structure of a computer room is an approximation to it. (Compare this description with the comments in 5.1.3.4.)

5.1.2.2   The low impedance return path

A perfect equipotential structure is impossible as soon as currents are allowed to flow in it. Any practical structure has a non-zero impedance between any two points, and when a current passes through such an impedance a voltage must be developed across it. In practice, the EMC engineer is always trying to minimize the interference voltages developed in the earth, and therefore the prime requirement for a good earth design is that it should offer the minimum impedance to whatever interference currents are flowing [44].

When a circuit is designed to use the "earth" as its functional return path – by deliberately connecting the returns of the source and load to the earth structure rather than to each other – then the path between these points must have the minimum impedance. Figure 5.1(a) shows this situation, where I_{ret} flows between points A and B on the earth structure. Clearly, if the circuit is to be at all controlled, the current path between A and B must be as carefully designed as that between C and D which carries the signal current, otherwise an unknown potential difference between A and B is injected directly into the circuit. Usually, the earthing structure is outside the control of the circuit or equipment designer, if only because external interference currents are easily induced in it, and therefore designing the circuit to use the earth as a return path is strongly discouraged.

If the earth is under the designer's control, then the method of Figure 5.1(a) can be successful, provided that the earthing structure does offer a low impedance path to the return current. But the more usual technique of Figure 5.1(b) is to be preferred, where a separate conductor is provided for the return current, which can be paired with the signal conductor thereby minimizing the total circuit inductance. If the circuit must be "referenced" to earth this can be done at one point, thereby preventing signal currents polluting the earth and vice versa. (The question of which point to connect to earth may not be trivial.)

5.1.2.3   The low transfer impedance path

The earth structure in Figure 5.1(b) still has a function with respect to the circuit that it

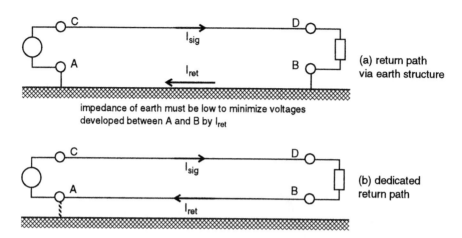

Figure 5.1 Earth as a return current path

is "host" to, and this is arguably its most important function. It should be so designed as to provide a low *transfer impedance* (Z_T) to the circuit [37]. Transfer impedance is defined as the voltage developed within the victim circuit divided by (interfering) currents flowing within the source. Low Z_T means that interference currents flowing within the earth – which are, for all practical purposes, unavoidable – transfer minimum interference voltages to the circuit. And vice versa, currents within the circuit do not create significant interference voltage differentials within the earth structure. Since the components of the earth are passive, the coupling is reciprocal, and a structure which has low transfer impedance in one direction will also be good in the other.

The essence of transfer impedance is *geometry*. It is mediated by near field (inductive and capacitive) coupling between the structure and the circuit and therefore whatever affects the coupling parameters, particularly separation, orientation and coupled area, will also affect the transfer impedance. Actually predicting the value of transfer impedance in a given situation is rarely feasible – the coaxial screened cable being the main exception to this.

It is though, possible to rank particular structures in order of reducing transfer impedance, as shown in Figure 5.2. A circuit on its own (a) cannot be described in terms of transfer impedance; there is no structure to transfer from, so all induced interference current flows within the circuit itself. The simplest structure to visualize is an earth wire in parallel with a circuit (b). This can be improved upon by mounting the circuit against a flat plate (c); the wider the plate, and the closer the proximity of the circuit to it, the lower is Z_T. This will be recognized as an alternative description of a printed circuit board on a metal chassis, but the same principle applies to a cable laid against a cabinet wall, for example. The transfer impedance is dominated by coupling from the edges of the plate; the further away these are from the circuit the better, but of course this is limited by available space.

Folding the edges of the plate up to surround the sides of the circuit (d) allows the edge effects to be reduced. This describes a conduit or a U-shaped chassis. Covering the conduit or chassis with a lid reduces or (if the lid makes continuous contact) eliminates the edge effects. The minimum transfer impedance is reached when the circuit is made coaxial with a cylindrical outer structure (e). Although this is largely impractical for equipment, it is straightforward for cables; the lowest transfer impedance for coaxial screened cable is achieved with a solid outer sheath, and braided cables compromise this to a certain degree, but the geometry of the coaxial cable is optimum.

All these drawings show the earthing structure as being continuous while it is surrounding or adjacent to the circuit. This is the crucial requirement for such a structure; if the earth currents are disrupted by discontinuities across the direction of flow – for instance, slots in a metal panel, or gaps in a cable conduit – the transfer impedance rises dramatically.

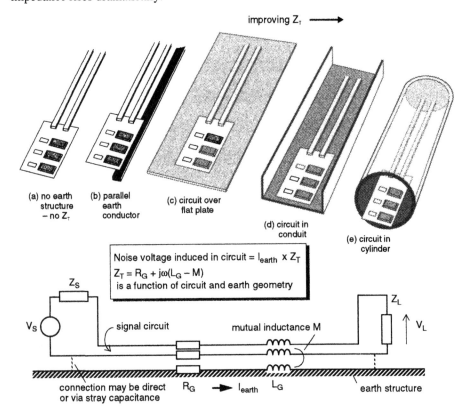

Figure 5.2 The transfer impedance of structures

5.1.3   Earthing techniques

As far as electrical signals and power (and their associated electromagnetic disturbances) generated and consumed within a building are concerned, the connection of the building to the local earth mass (earth, soil) is immaterial. But a coherent earth-bonded structure within a building is still required for safety, signal integrity, equipment reliability, and EMC within that building.

As the frequencies used by modern electrical/electronic apparatus increase, the achievement of a good quality earth for signal integrity and EMC reasons becomes more difficult. And as the integrated circuits within electronics get ever-smaller and hence more vulnerable to damage from lower levels of transients, the achievement of a good quality earth which will protect modern electronics from damage also gets more difficult.

When electrical signals and power entering or leaving a building, and lightning, are considered, it becomes necessary to connect the building's earth-bonding system to the local earth mass with electrodes which penetrate the soil, usually (at least) around the perimeter of the building.

There will also be a lightning protection structure (LPS) required for the external surface of the building, and this will be connected to the earth electrodes. The details of the construction of the LPS are considered further in Chapter 9.

Earthing is thus required for a large number of reasons, and a good earth-bonding system will deal with them all.

5.1.3.1   Independent earthing

Where the approach for historical reasons employed independent earthing (as the example in Figure 5.3(a)) it is now understood to create a safety hazard [97]. When heavy earth currents flow (usually due to nearby lightning strikes, or earth-faults in HV or MV distribution) the independent soil electrodes can take on very different voltages – perhaps exceeding 10,000 volts in the transient case.

This can cause electrical shock, as well as showering arcs, toxic fumes and smoke, equipment damage, personnel burns, and serious fire. This type of earthing scheme is also very poor for EMC purposes.

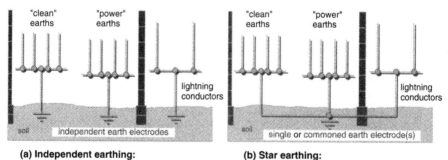

(a) Independent earthing:
not suitable for EMC, creates safety hazards

(b) Star earthing:
good for safety, poor for EMC at any frequency

Figure 5.3 Independent and star earth-bonding

5.1.3.2   Single-point or "star" earthing

Still found these days in older buildings is the "star earth" system, as shown in Figure 5.3(b). This eliminates the gross safety hazards of the independent earthing system, but was developed in the days when electronics was in its infancy.

When carefully installed and the topology maintained, star earthing systems are capable of providing safety at power frequencies. But for general use in modern installations of any size they are no longer considered adequate to control interference at the frequencies used by today's instrumentation, control, computing, and telecomm's technologies [97].

The traditional concepts of providing separate "clean" earthing systems for communication equipment has largely been dispelled with the code of requirements (BS7671 in the UK) of providing single point earthing to buildings, and the increasing use of earth as a reference potential for "single-ended" signalling in information technology and telecommunications apparatus [59]. Although earth-referenced signalling is not recommended [74], it seems that equipment manufacturers are increasingly using this technique to reduce the cost of their products.

Individual cables, wires, and structural steelwork of any practical length must all be considered to be effective antennae at the frequencies in common use, which severely limits the possibility that a star earthing system can provide any EMC benefits whatsoever. Without a great deal of attention to detail and rigorous checking and maintenance, star earthing leads to large amounts of interference with signal conductors between equipment [73].

It is very easy to compromise star earthing systems accidentally [61]. For example, a data cable might be connected between two items of apparatus that are on different limbs of the star, during a modification or refurbishment many years after the initial installation. Such cables may be installed on a temporary or permanent basis by personnel with no skills in earthing or lightning protection, indeed they may not even realize that such issues even exist. During a surge event, however, such unapproved modifications can cause damage to the apparatus at either end, or even fires or electric shock, due to the impression of large voltages and currents across the ends of the new data cable.

A similarly hazardous situation can arise where an item of equipment is moved to a new location (maybe only a few feet away), or when walls or boundaries are changed (e.g. partitions removed to make several small offices into one large one). Touch or step voltages which had been acceptable could now be unacceptable, resulting in serious hazards to personnel.

According to BS 6701, the UK voice telephony system requires a "star" functional earth (FE) from the customer's main earthing terminal (CMET), however a guaranteed isolation of the functional earth of a generic cabling system from the protective earth would tend to be unworkable and could in some cases be dangerous [28]. Time break recall, which does not need an earth reference, is used in the US (ANSI/TIA/EIA-607 [142] refers) and is becoming a standard function on most PABXs and therefore the need for dedicated FE networks (as per BS6701) in commercial building applications will gradually fall. Furthermore, recall signalling in modern digital telephone systems does not require functional earthing [59].

5.1.3.3   Three-dimensional meshed equipotential earth-bonding

The earth-bonding method required for safety in buildings is now the Common-Bonded Network, or CBN, shown by Figure 5.4.

Today's complex installations require their CBN to be enhanced to create a three-dimensionally meshed equipotential system, often referred to as a MESH-BN (for mesh Bonding Network). This bonds every piece of structural and non-structural metalwork together, including concrete reinforcement bars, girders, cable trays, ducts, deck plates, gratings, frameworks, raised-flooring stringers, conduits, lift (elevator) structures, window- and door-frames, and the metallic carriers of services such as gas, water, smoke, air, etc., to make a very highly interconnected system, which is then connected to the lightning protection system (LPS) at ground level (at least, see Chapter 9).

This highly-meshed three-dimensional system is then very highly interconnected to the screens and armouring of all electrical cables, and the frames or chassis of every piece of electronic equipment. Where existing metalwork or conductors do not already exist, large cross-sectional-area conductors are added to complete the mesh either

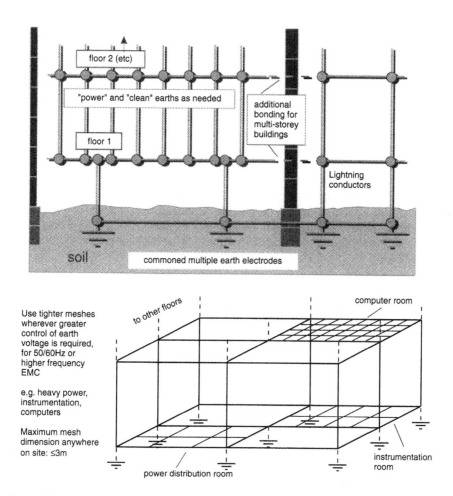

Figure 5.4 Common bonded network

vertically or horizontally so that nowhere is the mesh size greater than about 3 or 4 metres. The main earthing terminal (MET) for the incoming power supply to the building requires a number of bonds to the MESH-BN.

The MESH-BN is used for safety, functional, and EMC earthing, all at the same time – an integrated earth-bonding system. By creating such a MESH-BN it is possible to achieve the various aims of safety, signal integrity, equipment reliability, and EMC, which are often seen as being in conflict, at reasonable cost in a reasonable time, without compromises, and without any requirements for burdensome on-going management or restrictions on future modifications. Refer to [61], [97], [81], [74], [77], [129], [132], [63], [59], [73], and [11] for full details on MESH-BNs.

Nothing in this section should be taken to mean that safety considerations may be neglected. The requirements of the state-of-the-art for lightning protection and electrical safety for both the installation and the apparatus in it should always be followed. Due to the meshing of the earth structure some of the necessary calculations will be made more difficult, and some may well be made easier, nevertheless they should all be performed and the necessary actions taken.

Screened cabling systems require full equipotential bonding systems with defined current carrying capacity. CBNs implemented as MESH-BNs are ideal, making screened cable systems straightforward to install with no additional cable management problems, anything else is less than perfect [28].

Some IT and telecommunication system blocks will require an even higher quality earth reference than can be provided by the MESH-BN described above, this is sometimes called a System Reference Potential Plane (or SRPP) and usually requires a "bonding mat". The bonding mat is ideally a seamless metal plate under the entire system block of interconnected equipment and its internal cables, but it is more usual in commercial or industrial buildings to use a grid of copper conductors, or the metal framework of the raised flooring, or a hybrid of both. Refer to [74], [77], [129], [52], [59], and [73]. SRPPs and bonding mats are described in more detail in section 5.3.1.

Although each of the elements of the MESH-BN will have their own resonances and behave like antennas and/or poor earth-bonds at those frequencies, its highly interconnected nature will ensure that there are usually alternative current paths that are not resonating and so provide a high degree of equipotentiality over a wide frequency range [62], [59].

One consequence of this is that very regular bonding structures should be avoided, since all their elements will exhibit similar resonances at the same frequencies. This is usually not a problem for most buildings. Although the concrete reinforcing bars ("re-bars") and the structural steelwork in a building can have a very regular structure, the use of three-dimensional earth bonding between all metalwork at every possible point, and the bonding of cable screens and armour of every type at both ends and even along their length will randomize the mesh size and shape. This will ensure that there are always many possible earth paths for a current, all resonating at different frequencies.

As well as providing an equipotential earth, a MESH-BN with 3 to 4 metre spacing provides a degree of shielding, especially against the intense low frequency fields (<1MHz) created by nearby lightning strokes [74], [77], [129], [63]. Where the 3-D earth mesh size has to be larger than 3 or 4 metres (e.g. shipping bays, display windows) electronic equipment should not be fitted unless it is especially "hardened" to withstand higher levels of interference, particularly lightning surge (see Chapter 9).

The best electrical bonds are direct metal-to-metal, preferably seam-welded, but other types of bonding may be used. Practical techniques for three-dimensional earth bonding are discussed in more detail in section 5.2.2.

For safety from electrical hazards, all earthing and bonding conductors need to provide sufficiently high current conducting capability and low impedance according to the relevant safety standards. This is to avoid electric shock, fire, or damage to the equipment under normal or faulty operating conditions within an equipment or within the power distribution network, or due to the impact of induced voltage and current, e.g. by lightning [74], [77].

The return conductor for 48Vdc telecommunications power, and similar dc supplies feeding more than one item of equipment, are recommended to be made part of the MESH-BN, by connecting to it at multiple locations [74], [77], [129], [73]. This improves fault-clearance time and personnel safety as well as reducing the possibilities for interference.

For the control of voltage differences in the earth structure at higher frequencies for the same level of power, or at higher powers for the same frequency, the mesh size of the MESH-BN needs to be smaller. With the different types of apparatus having been segregated according to whether they are "noisy" or "sensitive" (see section 7.4.1), the building should then be partitioned into areas with different earth mesh sizes, depending on the earthing needs of each [97], as shown by Figure 5.4.

Note that each segregated apparatus area with its individual meshing or bonding is surrounded by a complete conductor. These are known as bonding ring conductors (BRCs, see section 5.2.2.3) and are important for protecting the apparatus in an area against lightning transients, earth faults, and other low-frequency surges originating outside of their area.

5.1.3.4   What do we mean by "equipotential"?

Electricians tend to refer to any earth-bonding structure which achieves safe potential differences (50Vac or less according to BS 7671) during normal use as "equipotential". But such earth-bonding networks can easily produce 160V AC for several seconds during an earth fault [26] and this may destroy inter-equipment I/O electronics [43]. During a transient surge such as caused by a lightning stroke such a weak earth-bonding system could expose electronic equipment signal and power cables to voltages of several kV.

A number of information technology (IT) and telecommunication standards have recommended for years (since the 8th edition of IBM's Cable Planning and Installation Guide in 1987, at least) that the maximum potential difference for the meshed earth-bonding within a system block must not exceed 1V AC RMS [28], [59], and [17]. In addition to maximum voltages of 1V, EN 50173, ISO 11801, and ANSI/TIA/EIA 568A all recommend that the earthing network that cable screens are connected to has a low impedance at the frequencies of concern [59].

The rationale behind the choice of 1V maximum for equipotentiality is governed as much by the need to control electro-galvanic corrosion in the earth-bonding system of DC powered telecommunication systems as it is for signal integrity. A bonding network that corrodes creates unreliable operation.

It seems that some manufacturers of IT and telecommunication equipment are increasingly using single-ended signalling, using the earth as the signal return. This is considered bad practice by [74] and [77], but it does make for low selling prices. Their instruction manuals apparently state that their equipment must be connected to a "clean earth", but it appears that "clean" in this case means equipotential *to no more than 15 milliVolts* over the frequency range of the signals employed (this might not be so clearly stated in the manuals). The installer (and user) is thus faced with the difficulty and expense of achieving a remarkably high quality of SRPP, usually using a bonding mat, and one can only hope that they fully understand this onerous requirement before beginning to install the equipment, and preferably before selecting it over possibly more expensive equipment which is more forgiving of installation type.

5.1.3.5   The bogey of ground loops

A common objection to the meshed earthing system just described is that it creates "ground loops" (equally known as "earth loops", but the word ground is used here to distinguish between incorrect and correct practice). Historically, currents flowing in ground loops, and their associated driving potential differences across different parts of the earth network, have been found to be particularly serious contributors to interference problems, and therefore a practice has developed of trying to eliminate all such loops. This practice, although often superficially successful, is unfortunately misguided.

In a situation where high earth potential differences exist, closing a loop between two such earth points will allow a high current to flow in the structure. If the conductors in that loop include a segment which either forms part of, or is closely coupled to, a signal or low-level power cable, then substantial interference can be induced in the circuits of that cable. If the loop is opened, the current no longer flows, and the interference disappears – although the high potential differences remain, ready to create problems again when another loop is closed somewhere else. This is the principle which is formalized in the star or single-point earth regime: remove all ground loops and live with the resultant high voltages between different parts of the earthing system.

Such an approach is fairly easy to implement and quite successful in simple low-frequency systems, but it represents a retreat from best practice. Now that interference frequencies are typically measured in MHz rather than Hz, it is untenable. This is because the star earthing conductors present a high impedance to these frequencies and therefore *de*-couple a system from earth, rather than couple to it. Also unfortunately, larger star systems tend to degenerate into accidentally ground-looped systems as time passes and systems and buildings are modified and added to, requiring a heavy management and control burden if their efficacy is to be maintained and safety and equipment reliability is to be ensured [62], [129].

In these circumstances the only reliable earthing system is a mesh. The mesh does indeed provide a multiplicity of ground loops, but they are small and controlled: voltage differences between parts of the structure are minimized, resulting currents are low and the interference consequences, if any, are negligible. The safety, reliability, EMC, and ease of use advantages of properly implemented MESH-BN systems far outweigh their possible disadvantages.

5.2   The impedance of the earth connection

5.2.1   Impedance of wires

Earthing wires do not offer a low impedance end-to-end at radio frequencies. Even at quite moderate frequencies, their impedance is significant. By way of illustration, Figure 5.7 shows the measured attenuation in a 50Ω system of a 60cm length of 24/0.2mm wire that is well bonded at one end to a reference plane. If the wire were a perfect short circuit, attenuation would be infinite; its actual impedance ensures finite attenuation. This measurement can be used to illustrate a number of factors.

5.2.1.1   DC – low frequency impedance

At DC, the impedance of this length of wire is purely resistive. Standard wire tables reveal that 60cm of 24/0.2mm should have a resistance of around 15.6mΩ. This will give an attenuation of over 60dB in the 50Ω system.

5.2.1.2   Low – medium frequency impedance

Any length of wire of length l and diameter d in free space has self-inductance. This can be determined from equation (5.1):

$$L = 0.002 \cdot l \cdot (\ln (4l/d) - 0.75) \; \mu H \text{ for } l, d \text{ in cm} \tag{5.1}$$

The inductive impedance part exceeds the resistive part at no more than a few kHz. For the 60cm of 24/0.2 (which has a diameter of about 1.1mm) the DC resistance of 0.0156Ω is equalled at 3kHz. At 2MHz, the impedance of 0.832μH is 10.5Ω ($Z = 2\pi \cdot f \cdot L$), which will give an attenuation in the 50Ω system of just 17dB – as we see in the plot. The impedance of any length of wire rises linearly with frequency for as long

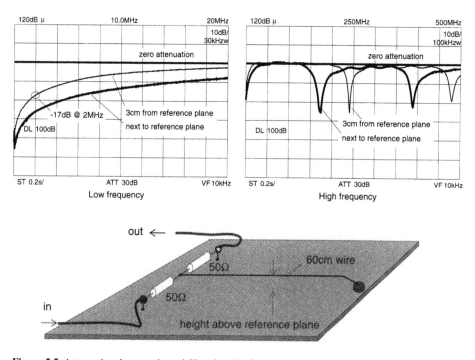

Figure 5.5 Attenuation due to a shorted 60cm length of wire

as the inductance dominates. The low frequency plot in Figure 5.7 also shows that when the wire is laid next to the reference plane, the impedance reduces – although it remains inductive, as can be seen from the slope of the plot. In this case equation (5.1) does not correctly predict the value of the inductance, which is reduced by the proximity of the plane.

5.2.1.3 High frequency impedance

As the frequency rises to a point at which the wire length becomes an appreciable fraction of a wavelength – which is another way of saying that it becomes resonant – then inductance alone is not sufficient to describe the impedance characteristic. In the high frequency plot of Figure 5.7, we can see that above about 50MHz the attenuation offered by the length of wire is negligible. This means that its impedance has exceeded 50Ω by a large margin. But at some higher frequencies there are some evident notches in the attenuation. When the wire is held 3cm away from the reference plane, the first of these occurs at 240MHz. At this frequency the impedance seen looking into the end of the wire has dropped to somewhat less than 5Ω. This is an example of the transmission line resonance of the wire; it is in fact the first half-wave resonance $\lambda/2$, that is, the end of the wire is exactly half a wavelength away from the earthing point (theory would predict 250MHz; the discrepancy is due to slight imperfections in the layout). Further resonances appear at multiples of the first, the next one visible on the plot being $2\lambda/2$ or the full wave resonance at 480MHz.

When the wire is placed next to the plane, the first resonant notch drops to 170MHz; the second one appears at 380MHz. The change in geometry has affected the resonant frequency but not the basic property of the line. Proximity to the plane and the presence of the wire's pvc sheath which now fills the gap has increased the capacitance of the wire, hence the lower frequency; this is another way of saying that the transmission line's phase constant, and hence its electrical length, has been altered. Discontinuities at each end mean that the geometry no longer approximates an ideal line and the second resonance is not an integer multiple of the first.

When the line is exactly a quarter wavelength long, its impedance is a maximum; at this frequency (which would be around 85MHz with the wire close to the plane, or 120MHz with it held away) the supposedly "earthed" connection is in fact almost completely decoupled, with impedances of thousands of ohms being typical.

5.2.1.4   Generalizations

A number of points can be drawn from this illustration, as follows:

- any length of wire becomes predominantly inductive above a few kHz; short fat wires have a higher transition frequency than long thin ones

- the inductive impedance of typical lengths of wire reaches ohms at around a MHz, and tens of ohms in the tens of MHz range

- the impedance of a length of wire connected at one end to an earth reference plane reaches a resonant maximum when its length is a multiple of a quarter wavelength, and falls to a resonant minimum at multiples of a half wavelength

- the exact frequencies at which these resonant peaks and nulls occur are strongly affected by layout; if any of them coincides with a susceptible or emissive frequency of the equipment, surprising and unpredictable variations in equipment performance will be brought about simply by moving such a wire by a few centimetres.

The general rule with earth wires is: short fat straps have the lowest impedance, as suggested in Figure 5.6. But even short straps are not perfect. Figure 5.7 shows the attenuation of a tinned copper braid 10cm long by 9mm wide by 2mm thick, in the same system as Figure 5.7. Clearly, although it is much better than a length of wire, the braid still has substantial impedance in the hundreds of MHz. Its merit is that those resonances which still exist are pushed much higher in frequency and exhibit a much lower Q, thus reducing their impact usually to negligible proportions.

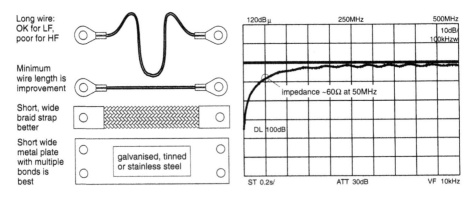

Figure 5.6 Hierarchy of earth conductors   **Figure 5.7** Attenuation due to a short earth strap

5.2.2   Effective bonding of joints

When multiple chassis or structural members are joined, the RF impedance across the joint determines the effectiveness of the whole of the metalwork as an RF reference. A uniform, infinite plane has no inductance associated with current flow from one point to another on its surface. Concentration of current at points of contact between structural members generates inductance across the joint, causing voltage differentials and external fields. To minimize this inductance, the <u>area of contact</u> and the <u>number of contact points</u> (Figure 5.8) must both be large. Wide short metal plates with multiple metal-to-metal bonds to each of the items to be bonded are better for high frequencies and/or high powers than either wires or braid straps. For best performance up to 200MHz or so, their fixings should be spaced at most every 100mm apart along the whole length of their mating surface by bolts or spot welds. For higher frequencies continuous conductive gaskets or seam welding is required.

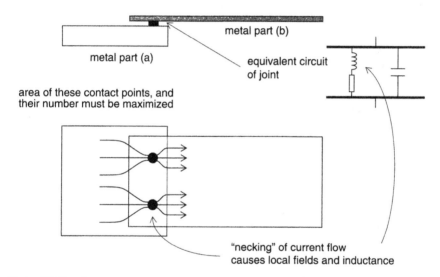

Figure 5.8 Bonding across metal joints

A large contact area has the secondary effect of reducing contact resistance, and this parameter can easily be measured by installation technicians using proper test instrumentation. A faulty high impedance bond can be identified from its contact resistance, a standard of 2mΩ across each bond and 25mΩ between any two points in the earth system being accepted in military situations.

Corrosion of the bond between dissimilar metals in the presence of moisture must be prevented by appropriate combination of metals with respect to the electro-chemical series (see Table 6.2 on page 148), by protective coating, and/or by the use of sacrificial washers. It is important that earth bonds remain effective over the life of the apparatus, even though subjected to dirt, vibration, atmospheric pollution, temperature cycling, and even condensation and spray in some environments [73].

Better control of high frequencies and/or high powers may be achieved when using wires, straps, or narrow plates, by using several such bonds. If using two, they should be positioned one on each side, but even better control will be obtained if several are used, spread all along the length of the mating edges. Where high-power RF equipment (such as induction furnaces) are involved, or powerful VHF broadcast transmitters are nearby, a short fat braid every 100mm must be regarded as the minimum for electrical bonding. Where UHF and higher frequencies are concerned, only wide metal areas in continuous contact will do, wires or braids are no use at all.

5.2.2.1   Use of gaskets

Closer spacing of fixing points gives better control of higher frequencies and/or higher powers, although if more frequent bonds are required it may be easier to slip a length of a suitable EMC gasket material between the mating surfaces (see also section 6.3.2.1 on page 145).

The pressure of the fixings should be sufficient to squeeze the gasket over the entire distance between the fixings with a pressure in excess of its minimum specification. Where the metalwork is less than sturdy, the pressure of the gasket material may make it bow between the fixing and look untidy. Worse than this, the metal may be bowed so much that a gap is opened up and the purpose of the gasket defeated. Additional strengthening plates may be found necessary to prevent this bowing, or else a lower-pressure gasket material could be used.

A very wide variety of gasket materials and products are now available, some of them with low cost, and some with additional chemical or environmental properties. Beryllium-copper or stainless-steel spring fingers may be used instead of foam or mesh type gaskets, to achieve good contact without excessive pressure. Care must be taken to ensure that the gasket materials are compatible with the metal types being bonded, taking account of the environment (condensation, spray, corrosive gases, etc.) to ensure that the electrical bonds last the life of the installation.

5.2.2.2   Bonding of equipotential mesh structures

As described above (section 5.1.3.3), a three-dimensional earth structure is required to provide equipotentiality over a wide range of frequencies.

All structural metalwork and cable supports should be RF bonded across all their joints, and RF bonded between each other whenever they are close enough, to make a 3-dimensional earth mesh. Plumbing, pipework, air ducts, chimneys, re-bars, I-beams, cable trays, conduits, walkways, ladders, ceiling supports, etc., should all be RF bonded, as shown in Figure 5.9. Building steel and reinforcing rods are recommended by [74] and [77] to have welded joints and a sufficient number of access points to them for frequent bonds to the earthing network to create the appropriate mesh size for the MESH-BN.

[73] suggests that bonding between metal elements can be by short thick round wires up to 1 metre in length, although these will only be adequate for the control of the effects of DC and 50/60Hz power. Adequate control of the effects of lightning surges requires the use of heavy-gauge round bonding wires under 500mm in length, or two or more wires in parallel each of up to 1 metre in length and spaced well apart (around 500mm). Flat wide copper strip or braid 500mm long provides improved bonding, possibly at frequencies of up to 30MHz. Braid or wide flat conductors are always better for RF than round wires of the same length and csa [73], but increasing the diameter or width of a bonding conductor is not an acceptable substitute for shortening its length.

The length of the connection between a structural item and the CBN should not be more than 500mm, and an additional connection should be added in parallel some distance away. Connecting the earthing bus of the electrical switchboard of an equipment block, or the earth bonding bar of a local AC power distribution cabinet, to

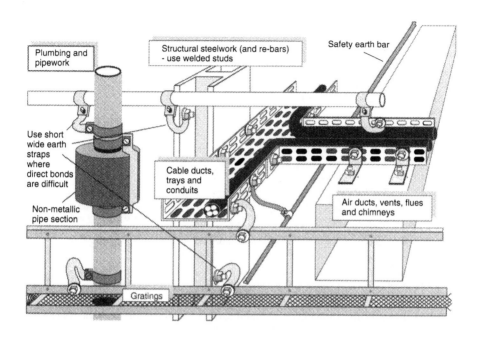

Figure 5.9 Bonding structural components of the equipotential earth mesh

the bonding network, should use conductors of under 1 metre length and preferably under 500mm, or two < 1 metre cables in parallel at least 500mm apart [74], [73].

[129] recommends that mechanical bonds of questionable continuity are bypassed by wire jumpers that are visible to inspectors. The jumpers should have low impedance at high frequencies. Achieving good signal integrity and EMC performance at frequencies of 100MHz and above requires direct metal-to-metal bonds at multiple points, preferably seam-welding, for each joint.

Note the these guidelines for the lengths of bonding conductors do not apply to termination of the screens of screened cables (described in section 7.2.3). "Pigtail" terminations of cable screens can render a screen totally ineffective at frequencies as low as 50MHz even for pigtails as short as 25mm.

5.2.2.3   Constructing a Bonding Ring Conductor (BRC)

Each segregated area is surrounded by a continuous conductor known as the Bonding Ring Conductor or BRC (Figure 5.10). This is generally a copper conductor of round or flat section with a large cross sectional area (csa). Other terms used for this include earthing bus conductor and interior ring bonding-bus. At the very least any building containing any significant amount of electronic equipment in it (including remote unmanned cubicles) requires at least one BRC around its internal periphery on each floor.

Where there are segregated areas within a building (including segregated system blocks connected only by LANs and power cables), each should have its own BRC [74], [77], [132], [63], [59], [73]. The use of BRCs is also a requirement for the creation of lightning protection zones, described in Chapter 9.

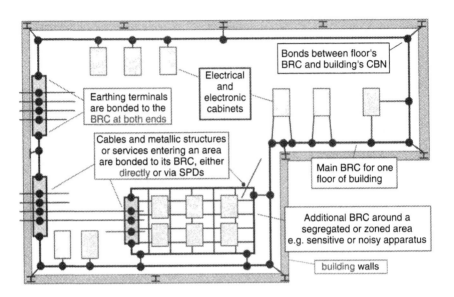

Figure 5.10 Application of Bonding Ring Conductors

A BRC is really an extension of the main earthing terminal for the floor, and should be bonded to the building's CBN at least at its four corners, preferably at many more points. The provision of a BRC allows equipment to be bonded by the shortest length of conductor, directly to the nearest point on the BRC. Where equipment is not near to the BRC, the BRC may be bisected by an additional conductor which bonds to the BRC at both its ends and "picks up" the equipment concerned along its path. Alternatively, equipment far from a BRC may use two or more widely-separated conductors to connect its chassis to the BRC.

In a metal room or building it may be possible to use the metal walls as the BRC, depending upon the construction of the walls and their equivalent csa.

Cables and metallic services entering a segregated area or zone must be earth-bonded to its BRC at the point where they cross it (see also section 9.3.3.2). This includes all cable armour and screens (with the exception of LAN cables guaranteed to be galvanically isolated for their entire path). Where surge protection devices (SPDs) and/or filters are fitted to power or signal cables to help protect a zone, they should be fitted at the BRC boundary [63].

A transient suppression plate (see Figure 5.18) is sometimes recommended to ease these boundary connections. This is a metal plate at least 1 metre square bonded to the BRC and to the CBN or MESH-BN, which provides an area suitable for the bonding of cable armour, cable screens, SPDs, filters. Not only does such a plate provide a handy area for fixings, it improves the high-frequency and high-current performance of the bonds made to it [129], [73]. Transient suppression plates may be horizontal (with cables passing across them), or vertical (with cables passing through them).

Electricians working in the hazardous atmospheres industries should be familiar with the BRC concept, as every metallic cable or service that enters a hazardous zone must be earth-bonded to the BRC for that zone (often a very sturdy conductor indeed), and standard explosion-proof cabinets are made with fixing points for ring-tags fitted to the cut ends of the BRC.

5.2.2.4   Bonding cable trays and ducts

Galvanized cable trays and rectangular conduits are best jointed by seam-welding, but it is often acceptable to use U-brackets with screw fixings every 100mm or less around the periphery of the U instead, as the top drawing of Figure 5.11 shows. Using lengths of wire will only control low frequencies (such as 50/60Hz). Shorter wires, or short fat braid straps, or multiples of each, all help increase the frequencies (or power levels) at which interference can be controlled.

Cable trays, ducts, and conduits will be required to act as "parallel earth conductors" (PECs), as described in section 7.4.3. The bonding methods at their joints and end terminations should relate to the frequencies it is important to control for the sake of the cables they are carrying (for both their emissions and immunity).

Where a rectangular cable tray or duct terminates at the wall of an equipment cabinet (or similar) a short wire or strap may be used for bonding which is effective at controlling DC and 50/60Hz disturbances. Two or more wires or straps will give better control of higher frequencies. An alternative is to cut away a few inches from the sides of the duct or tray, bend remaining floor section over and bolt it to the cabinet wall in at least two places, just below the aperture where the cables enter the cabinet. A U-bracket may also be used (with at least one metal-to-metal fixing every 100mm) and this can give very good control of electromagnetic disturbances up to very high frequencies. Figure 5.11 shows these practices.

Circular conduits are best jointed (either inline or at corners or junctions), using standard screwed couplings which make a 360° electrical bond. Similar 360° bonding glands should be used wherever a round conduit is terminated at cabinet walls, other types of cable ducts, or similar metal surfaces. These will generally employ some type of EMC gasket in their internal construction.

Choice of materials for trays, ducts, cabinets and other structural metalwork

Galvanized, tinned, or stainless steel are preferred, as they make it easy to achieve corrosion-free metal-to-metal bonds between surfaces when they are pressed together (for example by a bolt).

Aluminium is used instead of steel in some chemical industries. This should either have an unfinished surface, or be alochromed (a conductive finish). Anodising should be avoided if possible as it produces a very good *insulating* finish and is difficult to remove when preparing an electrical bond. Similar materials should be used for both sides of each bond, to reduce the risk that galvanic potentials will create corrosion which will limit the lifetime of the bond – this is most important for environments in which water or corrosive materials are present.

Where there is any paint or similar insulating protective finishes on one or both of the surfaces to be RF bonded, these should first be removed from the areas concerned, and a suitable corrosion protection will then need to be applied. (Even if the fixing pressure prevents corrosion at the actual contact, this will be no help when the material around the contact rusts completely away.) Corrosion can be a problem where metal finishes have been removed, or where holes have been drilled in galvanized steel. This should be dealt with in a way which does not degrade the quality and lifetime of the electrical bond. Petroleum jelly and copper grease are common coatings used to prevent corrosion at such places, although they can be rather messy and need to be applied with care so as to cover all exposed metal. Paint may also be used to protect a bond, but it must not have such low viscosity that it insinuates itself between the metal contacting parts and increases the bonding resistance.

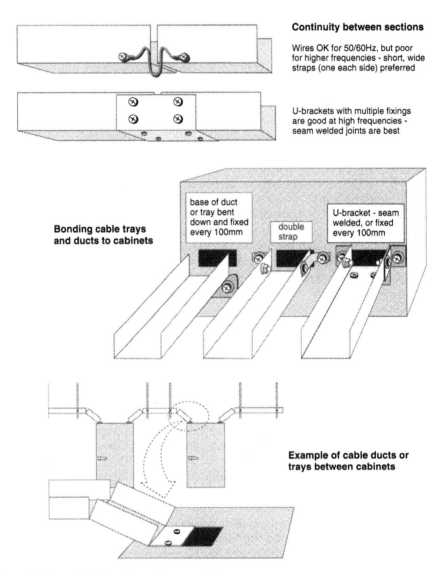

Continuity between sections

Wires OK for 50/60Hz, but poor for higher frequencies - short, wide straps (one each side) preferred

U-brackets with multiple fixings are good at high frequencies - seam welded joints are best

Bonding cable trays and ducts to cabinets

base of duct or tray bent down and fixed every 100mm

double strap

U-bracket - seam welded, or fixed every 100mm

Example of cable ducts or trays between cabinets

Figure 5.11 Methods of bonding cable trays and ducts

Electro-galvanic corrosion is a particular problem, especially where the earth currents contain a DC component. Protect against this at all bonds by choosing the metals to be in contact for a low-enough galvanic potential, given the physical ambient. Where this is not possible interpose a different metal which is mid-way in the galvanic series. 300mV has been found to be adequate for maintaining a low galvanic effect in a moderately corrosive atmosphere. Zinc chromate inhibitor or paste can be very useful, for example when fixing to aluminium with steel screws. Take great care with the electro-galvanic corrosion effects on the bonding of filters, which may have to maintain about 0.1mΩ to the enclosures of the susceptible equipment they protect over its operational life. Refer to prEN 50174-2 [73] section 5.14 and its Table 2 for more detailed guidance on this important issue. See also section 6.3.2.2 later.

As pointed out at the beginning of this section, the important issue for good control of high frequency and/or high power disturbances is the metal-to-metal bonding of the two mating metal surfaces, with a good flat area around the bonding point. Point-contact electrical connections are inferior to the pressing together of two flat areas.

5.2.2.5   Assembly methods for earth bonds in cabinets

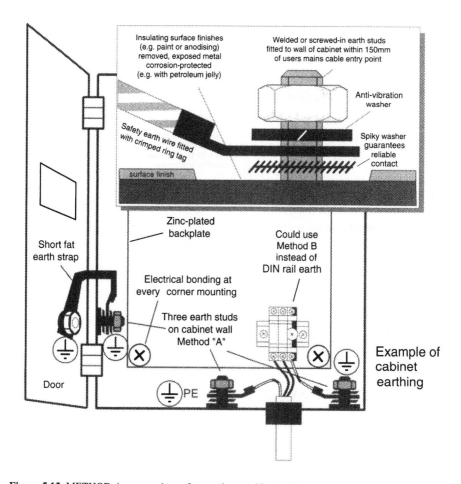

Figure 5.12 METHOD A: connecting safety earths to cabinet walls

Figure 5.12 shows an example bonding method for safety earths in cabinets. This uses welded or screwed-in earth studs fitted to the cabinet wall within 150mm of the mains cable entry point. For safety reasons, the user must connect the incoming protective earth directly to a method A stud on the cabinet wall. The standard for safety of electrical machinery, EN 60204-1, specifies that this connection must be the *only* one labelled "PE". A separate method A stud, near the PE stud, is used to connect the earth to the backplate and is labelled   .

Doors and removable panels should be earthed to the cabinet walls using their own earth studs and short wide earth straps or braids.

Method B (Figure 5.13) is used for internal earth wire bonding to a captive nut in the backplate, within cabinets. The captive nut is positioned so that the shortest length of earth wire may be used. For a plastic-bodied electronic unit the earth terminal should be bonded to the nearest possible backplate captive nut with a short, thick earth wire, as shown.

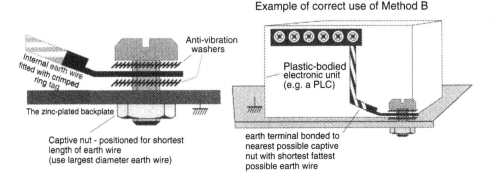

Figure 5.13 METHOD B: internal earth wire connections

Electronic metal-housed units (especially filters) must be bonded directly to the backplate at every mounting point. Figure 5.14 shows this.

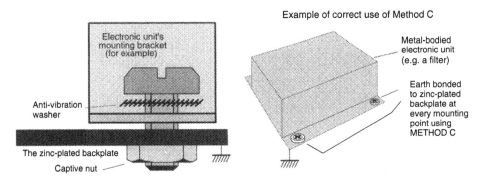

Figure 5.14 METHOD C: fixing metal enclosures of electronic units

5.2.2.6   Using fasteners in RF bonding

The body of a bolt (or other mechanical fixing method) is used ideally to apply pressure to mating metal surfaces to create an electrical bond. The metal mating surfaces must be conductive, clean, free from contaminants such as the remains of roughly scratched-off paint, and be protected from corrosion, all as described elsewhere. Such bonds are often acceptable for safety purposes as well as being good for EMC at RF frequencies.

Relying *solely* on the body of a bolt or other mechanical fixing (e.g. a rivet) to provide a bond may not be acceptable for safety purposes, and will not be as good at RF as directly mating metal surfaces.

The pressure of a fixing should be enough to ensure a gas-tight joint between the two mating surfaces, to help prevent corrosion. Larger bolt sizes are preferred because greater torque can be applied, improving the "gas-tightness" and longevity of the bond, and also because the increased surface area improves the bond's RF performance.

Washers with multiple anti-vibration "spikes" around their circumference are often used to help ensure gas-tight joints by concentrating the available mating pressure at their spikes. These may apply so much pressure that they cold-weld to the metal surfaces. Such washers may be inserted (with some difficulty) between metal mating surfaces to help overcome the lack of a good surface preparation, although the spacing they create between the mating surfaces will reduce the effectiveness of the bond at RF.

Such "spiky" washers may be used on both external sides of a bonding bolt, so that the body of the bolt also acts as a reliable bond between the two metal parts. Such external spiky washers can be very useful in *remedial* work when improving the bonding of existing painted or oxidized metalwork – providing that the spikes and pressure applied are both aggressive enough to guarantee biting through the insulating layers and achieving gas-tight points of contact to the underlying metalwork.

Shakeproof washers without circumferential spikes, e.g. spring washers, are not generally acceptable for bonds which depend upon the body of the bolt. Military equipment often uses wavy shakeproof washers, so as to prevent metal swarf from being created in high-vibration environments, but these only provide electrical bonds themselves when the metal surfaces they bear against have been carefully prepared and corrosion protected, as they may not generate sufficient pressures to guarantee gas-tight contacts.

5.3   Creating the meshed facility earth

Having explored the techniques of bonding between structural components, this section looks at the implementation of meshed earth systems in particular instances.

5.3.1   Constructing SRPPs and bonding mats for system blocks

Best-practice for typical commercial and industrial computer and telecommunication rooms (and the like) is to create a System Reference Potential Plane (SRPP) for each system block from a local bonding mat (Figure 5.15), as described by [74], [77], [129], [52], [59], and [73]. Bonding mat meshes are often made from wire of 6mm diameter or larger, or 25 × 3mm copper strip (of the sort used for external lightning protection). The advantage of using the "lightning down conductor" copper strip is that standard jointing clamps are available to ease assembly.

The bonding mat's mesh size must be related to the frequencies it is desired to control, whether signal or interference frequencies [77], [52]. The rule of thumb for general applications is that the mesh size should be no more than one-twentieth of the shortest wavelength of concern (i.e. the highest frequency), so for example: to have some control over signals and interference at 30MHz requires the mesh size of the bonding mat to be no greater than 500mm square. Higher frequencies and/or better control (i.e. more attenuation of interfering frequencies, better signal integrity) will require smaller mesh sizes.

The frame of each equipment cabinet should then be bonded to its nearest earth mesh conductor using the shortest heavy-gauge wire possible, better still a short wide braid strap. Round wires should not be used to bond the cabinet frames to the bonding mat where the frequencies of concern (either for signals or interference) exceed 10MHz. Where the wavelengths of concern are shorter than the longest cabinet dimension (e.g. >150MHz for a 2 metre cabinet) each cabinet may need multiple connections to the bonding mat from points on its frame separated by no more than one-tenth of the relevant wavelength, with a minimum spacing of 300mm [97], [59].

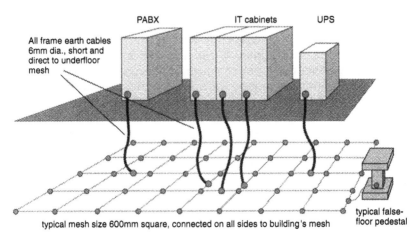

All frame earth cables
6mm dia., short and
direct to underfloor
mesh

PABX   IT cabinets   UPS

typical false-
floor pedestal

typical mesh size 600mm square, connected on all sides to building's mesh

Figure 5.15 Earth bonding mat for IT room

As discussed earlier, it is important that the mesh is not made too regular. If it is made very regular, then at some frequencies every mesh element will resonate together, and the earthing provided by the mesh could effectively "disappear" at those frequencies. A random allocation of ±20% dimensional differences will randomize the resonances and help maintain earthing integrity to higher frequencies. Even so, in some high-technology installations solid metal flooring may be required (described in the following section).

The bonding mat for a system block is nowadays recommended to be an integral part of the overall building's common-bonding network (CBN), and be bonded to it at as many points as possible. A continuous bonding ring conductor (BRC) should surround the bonding mat (which itself surrounds the system block it is the SRPP for) and be bonded to all the mat's conductors. This BRC then bonds to the building's CBN. In larger installations a system block may comprise a whole floor, in which case its BRC follows the perimeter of the building. Smaller system blocks may only take up part of a floor, so their BRCs will only enclose that part of that floor. Ideally the building's CBN vertical and horizontal meshes should be no more than 3 metres square. Figure 5.15 shows two IT/telecomms floors with meshed SRPPs, but the principle of bonding the CBN to each floor's BRC has universal application.

In computer/telecomm rooms with raised metal-coated flooring panels (usually used with conductive carpet materials for control of static build-up) an alternative is to use the metal-covered floor panels as the earth mesh, instead of creating one from copper wire or strip. The metal computer-flooring panels are already bonded together via the metal bridging contacts on the tops of their support pedestals. Equipment frame earthing leads are then merely fitted with a ring tag and screwed directly to the nearest floor panel (underneath the carpet). It is important that all the bridging contacts on the floor pedestals are in good condition and make a good electrical bond to each panel.

Where galvanic bonds cannot be guaranteed at every corner of the metal-covered floor panels, or at the crossing point of the framework, a grid of >10mm² csa copper conductors may be strung underneath the floor, bonding to the floor panels or stringers at every upright, although every second or third upright may be adequate [73].

Because the metal floor panels are only bonded by pressure contacts at the corner pedestals, this does not count as a solid metal floor system, but it should give equal or

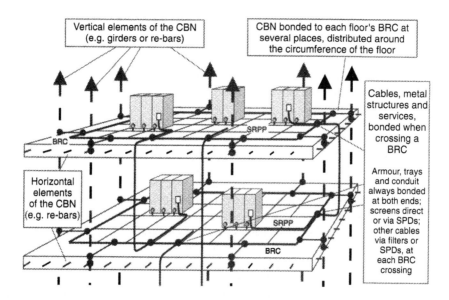

Figure 5.16 Application of vertical bonding (MESH-BN technique)

slightly better performance than a copper wire or strip mesh system (providing the pedestal bonds remain good). A possible disadvantage of using the raised metal flooring system is its intrinsic regularity, but the effects of bonding the floor to the various equipment cabinets, each with their own protective earth lead and cable screen bonding should help to randomize the bonding mat's frequency response.

Even where the raised-floor panels are not metal covered, it is often possible to employ floor stringers instead as a bonding mat. These are often used to add stability to higher floors, and may be used to provide a bonding mat instead or as well. When stringers are used as a bonding mesh they should not be used for protective earthing as well [59].

Bonding mats do not have to be underneath the system blocks they serve – they can be installed above them instead. As we will see later on, all the cables to and from the equipment cabinets must be run in close proximity to conductors forming part of their MESH-BN or SRPP bonding mat, and this may make an overhead bonding mat preferable. Overhead bonding mats may also be the best way to upgrade existing installations with minimal upheaval.

Despite the ideal of creating an integrated earth-bonding system, it is generally a good idea not to connect air conditioning, heaters, motors, and other electrical apparatus to a bonding mat used for an IT or telecommunication or similar electronic system block. They should be bonded instead (depending on the size of the building) to the building's main earth terminal, or the floor's main earthing terminal, or to the bonding ring conductor (BRC) that surrounds the system block, taking account of touch voltages during fault conditions and other safety issues [59].

5.3.1.1   Earth meshing computer/telecomm rooms for critical requirements

Where the electromagnetic environment is extreme[†], and/or when absolutely the best reliability is required from the apparatus in the room, the mesh earth described above may not be adequate. The bonding mat type of SRPP should instead be made from metal sheets of substantial thickness, all of them seam-welded to each other along all their mating edges. The equipment cabinets should be stood directly on this floor with any insulating feet removed, and RF bonded to the metal floor at each corner of their frame.

The solid earthing system needs to be complemented by providing a Single Point of Connection (SPC) to the rest of the building's structure. This is described again in section 5.3.2.1. Here, the SPC will generally be a region of the floor, or a vertical metal plate seam-welded to the floor, and *all* the cables and services entering will be electrically bonded all around their circumference at this point. Cables and services may *not* enter or leave the protected area at any other point. All unscreened cables will be filtered at this point, with the filter bodies bonded metal-to-metal to the SPC. For such rooms, metal-free fibre-optic cables will be preferred for all data and signals.

For installations required to survive the nuclear electromagnetic pulse, or which are so crucial to national security that they must emit no detectable electromagnetic disturbances, nothing but rooms made entirely of seam-welded metal, with specially screened doors, windows, and ventilation apertures will do. These rooms are essentially the same as the metal screened rooms used for some EMC tests, and what began as an earthing technique has turned into a shielding technique. For data fibre-optic cables must be used exclusively, and any metallic conductors entering the room must be as few as possible and all fitted with very substantial filters.

5.3.1.2   Applying the mesh-floor technique to other types of apparatus

Although the above refers to rooms intended for computers and/or telecommunications, the technique is also applicable to rooms where sensitive instrumentation is to be used.

A good example might be a scientific laboratory where experiments require high powers and/or high frequencies as well as sensitive detectors, for instance work with particle accelerators and cyclotrons. Segregation of the high power and/or high frequency equipment from the detectors is required (preferably by putting them in different all-metal shielded rooms), and each apparatus area should employ the techniques described above.

5.3.2   Improving the earth-bonding of older buildings

The improvement of the earthing and bonding of an older building which does not have a common-bonding network (CBN) requires consultation on-site. It is recommended to move towards an adequate CBN by enhancements to the earthing and bonding network to reduce the mesh apertures. It will also often be necessary to improve the outdoor lightning protection system (see Chapter 9).

The existing BN should be augmented with dedicated bonding ring conductors per

---

† These extreme techniques of interference control are included in a programme known as TEMPEST. Once restricted to military and government, TEMPEST is now available to the private sector to counter the increasing amount of commercial, financial, and industrial blackmail, terrorism, and espionage which relies upon the vulnerability of modern computers and computer networks to electromagnetic disturbances, and on their propensity to broadcast their data over a large geographical area where it may be picked up by sensitive receivers. In the USA these practices are becoming known as "information warfare".

room and per floor, and the bonding of cable ducts, troughs, racks, supporting metalwork, etc., at every possible point to create a CBN, and eventually a MESH-BN, although for practical reasons this may need to be accomplished over a period of time. In contrast to the traditional practice of using a small number of heavy-duty conductors, it is recommended to aim for a large conductive surface, e.g. by providing bonding at both side bars for joints in the run of a ladder type cable rack. Refer to [74], and [77] for more details.

Where independent earthing systems exist (they even used to be recommended by some older standards for specific purposes) these should be recognized as potentially hazardous and brought into the full CBN. Even if not fully meshed, they should at least have a single connection to the building's main earthing terminal.

TN-C type supplies (which use PEN conductors) within a building are very bad indeed for EMC and signal integrity, and it is recommended that such buildings are converted to TN-S supply systems (although TT and IT systems may be acceptable alternatives) with individual conductors for earth and neutral throughout the building (refer to [62], [81], [74], [77], [129], [132], and [73]). Where for some reason PEN conductors are practically unavoidable within a building, improving the BN to a CBN and then to a MESH-BN may still not create favourable conditions for modern electronic technology. Such buildings may need to use the insulated bonding network approach described below.

5.3.2.1   Application and construction of insulated bonding networks (IBNs)

In existing buildings which don't (or can't) have a full 3-dimensional equipotential earth mesh, earth currents (especially during an earth-fault or lightning event) can be very localized within the building. As mentioned above, TN-C supply systems (which use combined PEN conductors instead of dedicated earth and neutral) can also cause problems for modern electronics due to their powerful earth currents. In any of these three situations, bonding the earth mesh for an electronic system block at numerous points to the building's existing earth structure could encourage these undesirable continuous or transient earth currents to flow via the new mesh and give rise to very powerful electromagnetic disturbances in the vicinity of the system block, upsetting operation and even causing actual damage. These considerations have, in the past, led to the belief that ground loops are bad for signal integrity and EMC (5.1.3.5).

The recommended approach to this problem when installing modern electronic equipment is to modify the existing buildings as quickly as possible by:

- improving the lightning protection system to provide adequate protection for modern electronic equipment;
- using TN-S (or IT or TT) supply systems everywhere internally;
- creating a highly interconnected MESH-BN type of CBN, with a multiplicity of small ground loops (ideally no larger than 3 or 4 metres on a side).

However, sometimes one or more of these objectives are not possible, or other factors prevent their achievement, and an *insulated bonding network* (IBN) needs to be used instead.

IBN's are never totally isolated from the CBN of the main building, but will have just a single point of connection (the SPC) between their "insulated" bonding network and the building's CBN (actual isolation may be required in truly exceptional circumstances, but only after the approval of a comprehensive safety case by expert safety engineers).

The net current flowing through an IBN's SPC is ideally zero, and achieving this requires special measures to prevent the flow of common-mode currents (described later). When an IBN is completely disconnected at its SPC from the building's main bonding network it should be able to withstand at least 10kV of potential difference without arcing or leaking to the main building, so that it will not allow surge currents from the rest of the building to flow into the IBN during foreseeable lightning events.

An IBN is an example of a single-point (or star) connected bonding network, and a building might contain a number of them, of a number of different configurations as shown in Figure 5.17. A two or three-dimensionally meshed IBN is known as a MESH-IBN, and any IBN that contains system blocks may include bonding mats to provide SRPPs for these blocks. The construction of these mats is as described earlier, including their bonding ring conductor, except for the fact that they are not connected to the building's CBN other than at the SPC.

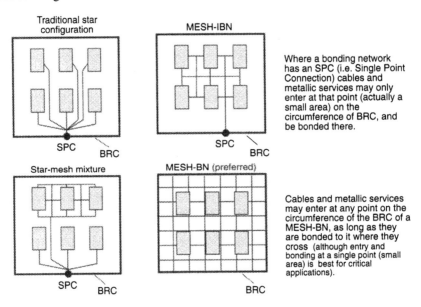

Figure 5.17 Examples of various star IBNs and star-MESH IBNs

The combination of one or more IBNs with the CBN of a building is called a hybrid bonding arrangement in IEC 61000-5-2 [97].

Increasingly, modern EMC standards are taking a tougher line than IEC 61000-5-2 towards the use of IBNs [77], [73]. Partly this is because the proper maintenance of an IBN is increasingly difficult, and partly because their utility is steadily being eroded as the electromagnetic environment becomes more polluted with ever-higher frequencies, and as the operating frequencies of electronic equipment continues to increase. IBNs can never be truly insulated at all frequencies: the inevitable parasitic capacitances to the surrounding equipment and its building CBN will cause increasing currents to flow into and out of the IBN at higher frequencies [129], destroying its effectiveness.

Insulated mesh-bonded systems are described by a number of standards, including ANSI/TIA/EIA-607 [142], but are probably most fully defined by ITU Recommendation K.27 [129], which is referenced by many other more recent standards. [129] considers that IBNs are only appropriate for telecomms buildings in exceptional circumstances (as outlined above) or due to some unavoidable incompatibility with pre-existing equipment.

When implementing an IBN, [77] says that co-ordination is needed regarding the routing of cables and the bonding of their screens, and also says that inspection, monitoring, and maintenance procedures are needed throughout its operational life to protect it from degradation.

The single-point connection (SPC) is the unique location in a MESH-IBN where a connection is made to the CBN of the rest of the building. The SPC is also used for bonding all the metallic services to the insulated area (water pipes, air ducts, etc.) as well as all cable armour, cable screens, the filters of unscreened cables, surge protection devices. Power cables should also have their protective earth conductors bonded to the mesh at the SPC, as should any other cables containing a protective or functional earth conductor [129]. In reality, the SPC is not usually a point, but must have sufficient size for the connection of conductors. The single-point connection window (SPCW) is the interface or transition region between an IBN and the CBN (for the building). All metallic conductors and services (including any fibre-optics which are not metal-free) must enter the IBN via this window. Maximum dimension of an SPCW is typically 2 metres [129].

The SPC can be a copper bus-bar, such as a normal earthing bar. However, where a number of cable shields or coaxial outer conductors (e.g. armour), or filters and surge protection devices, are to be connected to the SPC, it would be better realized as a frame with a grid or sheet-metal structure [129]. Reference [73] suggests using a "transient suppression plate" at the SPCW. This plate should have an area $>1m^2$ and be used as a potential reference onto which all EMC-related components are bonded.

The ideal connection to building earth for the SPC (SERP) of an insulated mesh is the metalwork of the low-impedance cable tray or duct that carries the cables for the new room. Where such a duct does not exist, a bond with as low an impedance as possible should be made to the nearest part of the building's earth structure. Consider also fitting the cables to the room with wired parallel earth conductors (PECs) at least for each different cable route (PECs are described in section 7.4.3).

There is a lot more to the successful realisation of an IBN than its earthing and bonding. Insulated systems attempt to interrupt the common-mode current by creating high impedances to its flow. Methods include unearthed systems, isolation transformers, opto-couplers, fibre-optics, power conditioners, and Class II equipment (i.e. double insulated and unearthed) [81]. But such attempts at isolation are fraught with problems and great care must be exercised if they have to be used in the absence of any alternative [62]. The most trouble-free way to prevent common-mode currents from flowing into IBNs is to use metal-free fibre optics.

5.3.2.2   Problems and special concerns with IBNs

Metalwork which is connected to the building's BN, but is near to an IBN, may need to be earthed to the IBN's SPC to prevent electric shock or flashover in the event of a lightning strike [129]. Other safety considerations, including insulation breakdown voltage and common-mode withstand voltage, and disconnectability, all need to be taken into account when designing the insulation means for an IBN and design guidelines and installation instructions are given in [73].

There is a concern that craft-persons and other non-expert personnel may accidentally compromise the insulation of an IBN. Violation of an IBN's insulation due

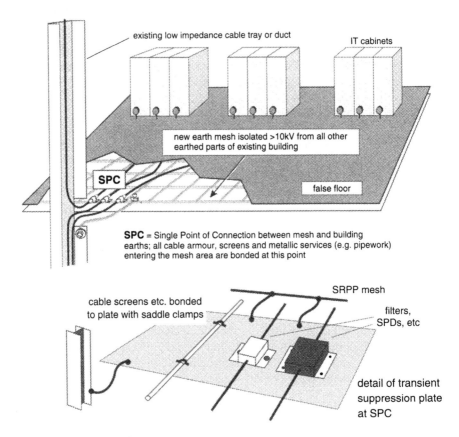

existing low impedance cable tray or duct

IT cabinets

new earth mesh isolated >10kV from all other
earthed parts of existing building

SPC

false floor

SPC = Single Point of Connection between mesh and building
earths; all cable armour, screens and metallic services (e.g. pipework)
entering the mesh area are bonded at this point

cable screens etc. bonded
to plate with saddle clamps

SRPP mesh

filters,
SPDs, etc

detail of transient
suppression plate
at SPC

Figure 5.18 The insulated bonding network in an unmeshed building

to maintenance (or any other) work may lead to failures in system operation or even physical damage during lightning or power faults. Hence the successful use of IBNs relies upon implementing appropriate control procedures, including regular maintenance and inspections, these are described in more detail in [62], [129], and [73].

Looking back at successful star earthing installations (which we now recognize as IBNs) that have helped to create the traditional view that the creation of ground loops is a bad thing, we notice that they were often broadcasting facilities (such as the BBC or ITA) or other enterprises with a high proportion of very skilled electrical engineers. The skilled engineers were needed on a daily basis to ensure continued operation of the facility, and in many cases design and/or maintain the electronic equipment, and this used to be considered an acceptable overhead. Comparable modern facilities contain a very great deal more electronic technology operating at vastly higher frequencies, the detailed internal operation of which is mostly unknown even to the now very small proportion of skilled electrical engineers who may be employed. Some modern installations employ electrical engineers with almost no knowledge of modern electronics technology at all, this being left to the IT department whose computer science courses left them completely unaware about electrical safety or any of the issues covered by this book. Most programming or IT support staff, and quite a few operational staff, would feel equal to the task of stringing a long data cable between two computers in different parts of their building (or even between buildings), or running an extension lead from an available mains socket in a nearby room, and would not even conceive that there were vital electrical safety and reliability issues involved. This is a serious worry, and one of the main reasons why IBNs and star earthing techniques are increasingly not recommended for general use.

5.3.3   Maintaining earth-bonding networks

Regular inspection and testing of earth-bonding networks should include the continuity testing of:

- cable supports
- bonding straps
- cable screen terminations
- connections

and also the checking and testing of :

- cable entry points to segregated areas
- corrosion effects (especially on earthing systems)
- integrity of equipment zoning (segregation)
- surge protection devices and filters
- the insulation of any IBNs.

In addition, stringent monitoring of all electrical installation extensions and alterations must be implemented, and records kept.

5.3.4   Non-IT installations

So far these guidelines on earthing and bonding have tended to reference IT (i.e. information technology, including telecommunications) standards and conference papers. This is because these are very large industries in their own right with a strong commitment to standards making.

However, the best practice techniques described here are not limited to areas intended for computers and/or telecommunications, they are of general applicability, especially where electrically sensitive equipment (such as some type of instrumentation) or electrically noisy apparatus (e.g. metal welding, induction heating, plastic welding) is involved.

Many industrial installations consist of machines with associated dedicated control panels. As long as these are individually mesh-bonded to the metalwork of their machines (assuming the machines use metal construction) by taking advantage of all the cable trays, conduits and armour and adding additional bonding conductors where necessary, there should be no EMC problems as long as the panel only communicates signals to and from its own machine. Increasingly, machines are networked together, and/or to some central SCADA or other managerial system. As long as the interconnecting cables use a galvanically-isolated communications method specifically intended for the electromagnetically harsh industrial environment (Fieldbus, CAN, etc.) the vast majority of installations will not need to go any further (e.g. by using an SRPP, such as a bonding mat). However, some industrial processes (such as the manufacture of DRAMs and similar semiconductors) are extremely sensitive to electromagnetic disturbances, and require the full metal flooring and mesh-bonding technique, even sometimes to the extent of screening entire rooms or even whole buildings.

A good example of another difficult installation might be a scientific laboratory where experiments require high powers and/or high frequencies as well as sensitive detectors, such as work with particle accelerators and cyclotrons. Segregation of the high power and/or high frequency equipment from the detectors is required (preferably by putting them in different all-metal shielded rooms), and each apparatus area should employ the earthing techniques described above for "demanding requirements".

Hospitals sometimes have similar requirements, e.g. the use of sensitive and life-critical anaesthetic machines on a patient simultaneous with the use of high-powered RF diathermic "knives". Clearly in such a situation the option of placing each item of equipment in its own metal room is not feasible, and the only recourse is to specifying sufficiently robust equipment and checking that the final system works as intended, taking remedial actions as necessary.

Professional audio or video installations can also be difficult installations because of their use of low-level signals and the very high signal-to-noise ratios demanded. Indeed, it is commonplace for complex audio and video systems not to function to specification when first installed. Installers of such systems should take heart from the fact that the best-practices described in these guidelines are known to work very well in pro-audio applications. Unfortunately it appears that many manufacturers of pro-audio/video equipment do not design them in such a way as to allow the benefits of best-practice earthing and bonding techniques, especially when it comes to I/O design and cable screen bonding. Also unfortunately, many professional audio and video installers have a traditional horror of earth loops, and as a consequence spend many days, or even weeks, "debugging" their traditionally designed installations before they even begin to function adequately, sometimes even compromising safety by removing protective earth conductors to try to break a ground loop.

Embracing these guidelines in full, replacing "traditional" pro-audio/video techniques where necessary, and using equipment designed to take advantage of these modern techniques will not only save time and money and reduce commissioning times, but will also provide improved signal quality, higher reliability, and a lower risk that future extensions and modifications will have a negative effect on signal quality. Overall, the cost of ownership will be reduced by using these EMC best practices.

Chapter 6

Cabinets, cubicles and chambers

6.1 The purpose of a metallic enclosure

In the context of systems EMC, a metal enclosure can have a number of purposes as well as its primary one of providing physical protection and mounting:

- to provide a local earth reference for the internal equipment
- to provide and demarcate a zone of increased EM protection
- to prevent radiated field coupling to and from the internal equipment.

You will notice that the conventionally understood function of a metal cabinet – to provide shielding – is placed last in the above list. This is deliberate. As we will see, many examples of metal housings are likely to function very poorly as shields, because of their surfeit of inadequately treated apertures and seams. But this doesn't mean they have no EMC function. In fact, the first purpose – to provide a local earth reference – is nearly always the most important. This can be achieved by an enclosure which is effectively *un*shielded, provided that care is taken to use the bulk metal of the cabinet in the right manner. Such an approach is very cost effective, since many of the assembly, installation and maintenance implications of a fully shielded enclosure can be dispensed with.

Simply constructing a metal cabinet or cubicle to house the equipment in the physical sense, is not adequate for electromagnetic purposes. The constructional and assembly methods of the individual structural parts, and the provisions for cable entry/exit, are at least as important as the mere fact of a conductive enclosure. Interference currents in the structure are expected and have to be controlled. So, electrical interconnections between the parts, and any discontinuities in the form of apertures, seams and cable penetrations, must be fully specified.

This chapter will first discuss the merits of right use of the structural components for local earthing, and then consider the important aspects of a classically shielded enclosure, before going on to look at architectural shielding.

6.1.1 Transfer impedance of the earth reference

The function of providing a local earth reference is critical to the effectiveness of a metal enclosure, whether this enclosure is deliberately intended as a shield or not. Chapter 5 has already looked at the principles of earthing and Chapter 7 will look at cable layout and termination in detail. In between is the area where the cables and equipment are terminated. This area must provide the lowest possible transfer impedance to the internal circuits and equipment, so that interference currents do not couple between the enclosure and the sensitive or noisy circuits within it.

Transfer impedance was discussed in section 5.1.2.3. Figure 5.2 on page 106 shows the hierarchy of structural shapes for improving transfer impedance, from which you might deduce that if all cabinets were cylindrical with the connections taken out from each end, all EMC problems would be over. Life is not quite this simple, but the principle remains that the cabinet structure has an important part to play in providing the best overall EMC performance.

6.1.1.1   Cabinet backplate earths

That figure showed that, after a cylindrical or rectangular enclosure, a large flat plate over which the relevant circuits are mounted gives the best transfer impedance performance. Industrial control cabinets in general include a backplate for physical mounting, whilst smaller enclosures provide some sort of metal chassis. Telecomms and IT cabinets have an internal support structure. Mounting all electronic modules on, and terminating all cable screens and parallel earthing conductors to, a backplate or similar chassis provides the lowest achievable transfer impedance in practical terms. But since using the backplate in this way means that it must carry interference currents, the way in which contacts are made to it, and its conductivity, become significant. Zinc plating, and clamp-style cable screen connections, are both recommended.

At high frequencies only a metal area (mesh or plate) can give a reliable low-transfer-impedance earth, so you are best advised to use a solid backplate or chassis of an enclosure as the earth for all internal electronic equipment *instead of* using green/yellow wires to a star point. This calls for heavy zinc-plated metalwork, not painted; if any passivation is applied it should be the yellow type.

At the cost of some inconvenience, you can continue to use painted metalwork as long as the following precautions are taken for all the earth connections:

- remove the paint
- use star washers to bite into the metal
- apply suitable corrosion protection after the joint is made.

Terminations of screened cables to the backplate or chassis should be carefully planned and implemented so that all common mode interference currents flow directly through it, and not into the circuits mounted on it. The methods of section 7.2.3, particularly the use of clamps rather than pigtails, should be followed rigorously.

6.1.2   Layout and placement within the enclosure

6.1.2.1   Cable runs

Coupling of external fields to internal cables, and of local electric and magnetic fields between cables, is greatly affected by the route a cable follows around a system. To minimize coupling of cables with external fields, run the cables close to a well-bonded metal structure which can act as a low impedance earth reference – this is usually the backplate or chassis. Where a cable leaves the backplate/chassis, ensure that it follows a conductive structure which is electrically bonded to the backplate/chassis. Avoid running cables near to apertures in the structure or enclosure or near to breaks in the bond continuity (Figure 6.1), as the localized fields around these points are high. (This advice is really the same as saying that the internal construction of the enclosure acts as a continuation of the PEC for the cables, as described in section 7.4.3.)

To minimize coupling of cables with each other, segregate different classes of cable and run them with at least 150mm (see section 7.4.2) of separation. Do not allow long runs of closely spaced cable of different classes.

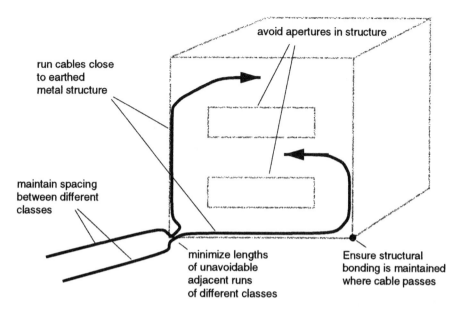

run cables close
to earthed
metal structure

avoid apertures in structure

maintain spacing
between different
classes

minimize lengths
of unavoidable
adjacent runs
of different classes

Ensure structural
bonding is maintained
where cable passes

Figure 6.1 Cable runs within the enclosure

6.1.2.2   Module placement

Carefully position the various electronic, electrical, pneumatic, hydraulic, etc., units on the backplate/chassis to keep sensitive units such as PLCs, computers or analogue instrumentation away from electrical noise sources (e.g. switches, relays or contactors), and to help achieve segregation of the different cable classes (Figure 6.2). The

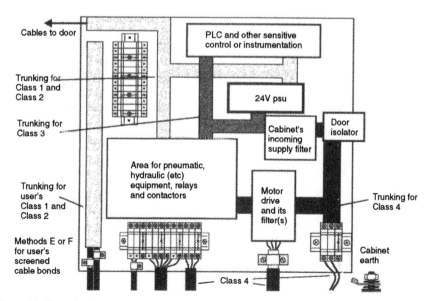

Cables to door

Trunking for
Class 1 and
Class 2

Trunking for
Class 3

Trunking for
user's
Class 1 and
Class 2

Methods E or F
for user's
screened
cable bonds

PLC and other sensitive
control or instrumentation

24V psu

Cabinet's
incoming
supply filter

Door
isolator

Area for pneumatic,
hydraulic (etc)
equipment, relays
and contactors

Motor
drive
and its
filter(s)

Trunking for
Class 4

Cabinet
earth

Class 4

Figure 6.2 Example of backplate layout for a simple industrial control panel

important principle here is to assess each item for its interference potential, and to specify the internal layout accordingly.

Have no internal trunking at all for Class 4 cables by minimizing their internal lengths as much as possible. This means fitting Class 4 associated units such as inverter drives near to the edge of the enclosure and to their cable entries, and/or filtering any Class 4 incoming supplies at their point of entry to make them Class 3 or even 2.

Figure 6.3 gives two sketches of an industrial enclosure, showing the cable route to door mounted equipment, with cables strapped along the short earthing braid between door and cabinet wall; and an example of backplate layout in a motor drive area. The purpose of the cable following the earth strap across the door opening is to minimize coupling of the cable with the door aperture, which would compromise the transfer impedance of the cabinet for this cable.

Because of their extremely aggressive emissions, inverter drive motor connections are always important. The example of Figure 6.3 (b) should be regarded as a generic approach, and suppliers' instructions (e.g., [5]) should always be followed if they are available. The important issue here is to provide a local return path to the filter for the switching noise currents which are flowing back down the earthing conductor and/or the screen of the cable to the motor. These currents can easily pollute the rest of the cabinet and even other equipment in the immediate area of the drive-to-motor circuit, if proper high-frequency earth bonding between the cable screen and the filter is not provided. Similar considerations apply for other "noisy" transducer drivers, such as RF-stabilized welding, spark erosion, and Class D audio amplifiers.

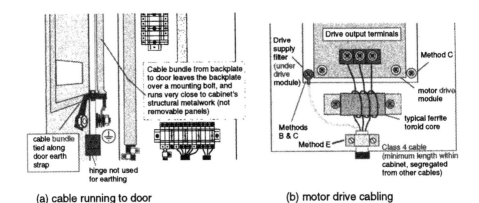

(a) cable running to door   (b) motor drive cabling

Figure 6.3 Detail of cable routing internally in cabinet

6.1.2.3   The clean/dirty box approach

A frequent and effective approach for industrial and other enclosures is to segregate the enclosure into a "clean" compartment and a "dirty" compartment. Either the cabinet can have a partition welded into it (Figure 6.4), or an additional "dirty" enclosure can be bolted or welded to the side of the main "clean" cabinet instead of a dividing plate.

The "clean" compartment is then used for all the electronics which must be shielded from the external environment. All apertures in this part of the enclosure are rigorously controlled. Connections through to the clean volume must be made via 360° screen bonds to the partition plate, or via effectively earthed filters – no untreated cables are allowed. Through-bulkhead filters present no problem, but chassis-mounting filters must keep the leads passing through the partition plate as short as possible and possibly treated with ferrite sleeves to minimize HF propagation across the partition.

The great advantage of this approach is that the dirty volume can be used for many or all of the field-installed connections. The interface through the partition can be pre-wired and checked before the system is shipped from the factory. Then, all of the strictures in this book about ensuring correct installation practices are taken out of the hands of the installation technician and given to the system assembler and designer. This is of particular benefit if the system supplier has no control over the installation methods at all – a regrettably common problem.

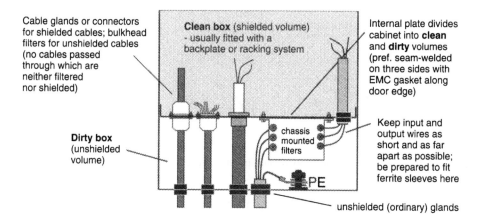

Figure 6.4 The "clean/dirty" segregated shielded cabinet

6.2   Shielding theory

6.2.1   Shielding effectiveness

Shielding effectiveness is defined as the ratio of the field strength impinging on a shielding barrier to the field strength propagated away from the other side of the barrier. It is calculated from measurements of the coupling between two antennas set at a fixed distance from each other, over a range of frequencies, both with and without the shielding in place around one of them.

This simple description belies the complexities both of real shielding effectiveness measurements, and of the variables that affect shielding performance even in an ideal geometrical situation.

One variant of classical shielding theory uses the "transmission line analogy" to describe how a shield barrier works. The impinging wave with its characteristic wave impedance is analogous to a wave propagating along a transmission line: it encounters a barrier of a different characteristic impedance, some of the wave is reflected and some is absorbed in the body of the barrier. The attenuated wave is further reflected at the far side of the barrier, and the remainder propagates into free space beyond the barrier. This process is shown conceptually in Figure 6.5, and it results in the equation

$$SE(dB) \quad = \quad S_R(dB) + S_A(dB) + S_{MR}(dB) \qquad\qquad (6.1)$$

where S_R is the total contribution from reflection at each surface

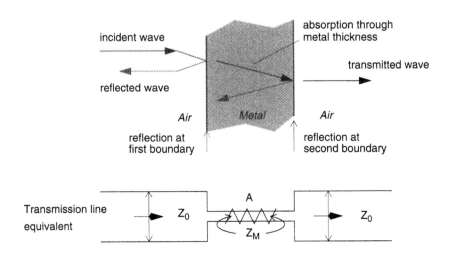

Figure 6.5 The transmission line theory of shielding effectiveness

Absorption S_A is a function of frequency and material and is described in terms of skin depth through the barrier (see 6.2.1.2 shortly). It is independent of the nature of the field source. S_{MR} is negligible unless S_A is low, since multiple reflections are attenuated on each pass through the barrier.

6.2.1.1   Reflection

This leaves S_R, which can be given by

$$S_R = \frac{(Z_0 + Z_M)^2}{4Z_0 Z_M} \approx \frac{Z_0}{4Z_M} \text{ if } Z_0 >> Z_M \tag{6.2}$$

Both Z_0 and Z_M are frequency dependent. Z_M is a function of the material but will always have a low value, for a conductive barrier. Z_0 depends on the nature of the source; we can discuss it for three extreme conditions, but real shielding problems fall somewhere within the extremes, and are rarely calculable.

Near field of an electric dipole

In this case the wave impedance Z_0 is high but falling with frequency and distance from the source. Very high reflective attenuation (>200dB) is theoretically possible at low frequencies and close to the source. This is described as S_E.

Near field of a magnetic dipole

Here, the wave impedance is low but increasing with frequency and distance from the source. For attenuation (S_M), the reverse of the previous situation holds: less than 10dB is possible at low frequencies (e.g. 50Hz) and close to the source.

Far field, plane wave

The far field condition implies that the source is at a distance greater than $\lambda/2\pi$ from the barrier. In this case, Z_0 is constant and the attenuation is independent of distance or frequency.

6.2.1.2   Skin depth and absorption

Reflection at the boundaries of the barrier is only one part of the story: absorption is the other. The important characteristic of the barrier material is its "skin depth" δ. This is the distance into the material at which the current density has reduced to 1/, (0.37 or 8.7dB) due to the skin effect. For every distance δ into the material the current density drops by 8.7dB. Put the other way, the absorption attenuation within a barrier will be 8.7dB per skin depth thickness. The skin depth is inversely proportional to √f and depends on the material parameters. Figure 6.6 shows the skin depths for copper, aluminium, mild steel and mu-metal.

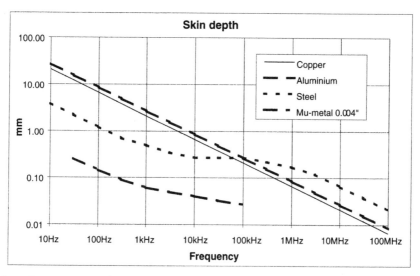

NB figures for mu-metal derived from Telcon Metals data for 0.004" thick sheet

For a conductor with relative permeability μ_r and relative conductivity σ_r at frequency F Hz:

$$\text{Skin depth } \delta = \frac{0.0661}{\sqrt{(F \cdot \mu_r \cdot \sigma_r)}} \text{ metres}$$

Figure 6.6 Skin depth for copper, aluminium and mild steel

6.2.1.3   Total shielding effect

It is now possible to combine the reflective and absorptive attenuations and show the total shielding effectiveness of a conductive barrier, as in Figure 6.7. Some conclusions can be drawn from this:

- at low frequencies, screening a magnetic field with normal metallic barriers is difficult; because of its higher permeability, steel is more effective below 100kHz, but because of its higher resistivity, not above this frequency;

- at low frequencies, screening an electric field with normal metallic materials is very easy;

- at high frequencies, absorption ensures that normal metallic materials are effective whatever the field source;

- the thicker the barrier the greater its absorption, but this is less important for electric field shielding.

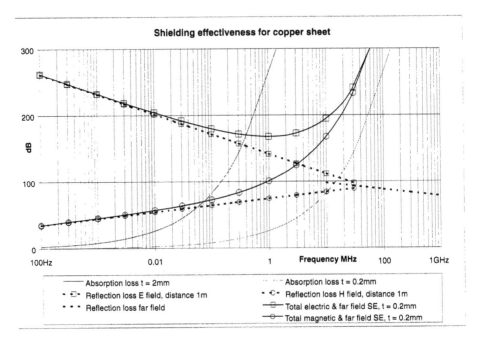

Figure 6.7 Combined shielding effectiveness versus frequency

The figures of upwards of 100dB look very impressive. But it is important to appreciate that Figure 6.7 is based on a theoretical concept which is not realized in practice: that of a uniform conducting barrier of infinite extent. Real enclosures are finite in size and, just as importantly, they have apertures and penetrations. For all applications except low frequency magnetic shielding, these are the aspects which actually determine the shielding performance.

6.2.1.4   Shielding effectiveness specifications and IEC 61000-5-7

In an attempt to systematize the description of shielding effectiveness a number of test methods have been used. The most common of these, historically, has been MIL-STD-285 (the German military standard VG 95373 part 15, and an adaptation of MIL STD 285, MIL G 83528B, are also sometimes used). This standard is useful for the characterization of screened rooms, but less so for enclosures, since it stipulates antenna separation and a few predetermined measurement frequencies. This makes it harder to apply meaningfully to small cabinet-sized items.

At the time of writing, a draft standard is in progress in the IEC [98] which will define screening effectiveness tests and classes for enclosures. Its eventual publication target date, as IEC 61000-5-7, is spring 2000. The approach taken in this document is to make swept or stepped frequency measurements over each frequency decade for which performance data is required, of the loss between two antennas. One of these is either open (unshielded) or it is located inside the enclosure to be tested (shielded). The shielding effectiveness is then the difference between these two measurements. The frequency range covered by the standard extends from 10kHz to 40GHz, split into decade sub-ranges. To cover the whole range, different types of antenna (magnetic loops at low frequencies, dipoles at high frequencies and horns at microwave frequencies) are specified.

A systematic designation of shielding effectiveness (the EM code) similar to the IP environmental classification system operated in IEC 60529 is proposed. This will show minimum shielding performance according to the above tests in 10dB steps over the appropriate decades of frequency across which the minimum performance is maintained. Whilst this may appear to be a simplistic way of describing shielding performance, the proposed system does have the merit of allowing quick comparison between different enclosures, and could integrate easily with other evaluation techniques such as the root-sum-of-squares (RSS) method of summation of emissions from several different units [60].

Because of the dynamic range required, shielding effectiveness tests are not trivial. Antennas need to be small in order to fit inside the enclosure to be measured without significantly affecting their properties, but this means they are insensitive. Therefore to obtain a good dynamic range the transmitted signal must be at a high level, but this means the test must be conducted inside a screened chamber. Reflections from the chamber walls must not affect the measurement, so the chamber must be anechoic[†], over the frequency range required for the test. This means that only a fully-equipped EMC test house can carry out such tests.

6.2.2   The effect of apertures, seams and penetrations

The implication of the previous description of a screening barrier is that it relies on current flow within the body of the shielding barrier, or, at higher frequencies, near to its surface. This current provides the mechanism for absorption losses and is also a necessary consequence of the reflected wave. Therefore, anything which interrupts this current flow will affect the shielding effectiveness.

Any discontinuity in the conductive material will interrupt the current flow. Discontinuities are due to a number of factors:

- joints between structural members and/or panels
- large openings: windows, access doors
- small openings: ventilation holes, connectors, controls.

Any enclosure for electrical or electronic apparatus will have most or all of these features. Proper treatment of them will allow the enclosure to achieve some degree of shielding effectiveness, but it will rarely approach the theoretical values given in Figure 6.7.

6.2.2.1   Degradation of shielding effectiveness

Much academic effort has been expended in trying to quantify the general effect of apertures and discontinuities in shielding. Apart from the special case of a sphere, a simple analytical expression based on the fundamental field equations has been unobtainable, leaving us until recently with recourse either to numerical computer modelling of a particular situation, or to various empirical rules of thumb. The difficulties are compounded by the possible number of variables:

† Mode-stirred techniques can also be used.

- the relative positions and distances between the field source, the shield barrier and the apertures in it, and the victim
- the frequency and source impedance of the field
- the shape, size, position and number of apertures
- the geometry of the shielding enclosure

all of which have to be taken into account by the model, even disregarding the possible variations in the properties of the shielding material itself.

Some developments are still occurring in the formulation of the shielding problem. Recent studies [48] have looked at modelling an aperture in a rectangular shielded box as if it were a transmission line discontinuity, in effect extending the scope of basic shielding theory, and these show reasonable agreement with experiment. If adequate computing resources and skilled modelling staff are available, some shielding problems can be successfully modelled by numerical methods and substantial academic work has gone into improving these. Results have yet to be made available in a useable form to wider industry applications (but see 6.2.2.3).

This leaves us with some general guidelines and a few simplistic equations. The most widely quoted of these equations (e.g. , [8]) is

$$SE = 20 \log (\lambda/2l) \text{ dB (below resonance)} \tag{6.3}$$
where λ is wavelength and l is the maximum aperture dimension

which suggests that the shielding effectiveness SE degrades proportional to frequency and inversely proportional to aperture size until the aperture size is a half wavelength, at which point shielding is zero. This has the merit of highlighting the general dependence on frequency and size of aperture: that is, the larger the aperture the greater its effect, and the higher the frequency the less is the available shielding. But as a means of predicting actual shielding effectiveness the equation is quite inaccurate, since SE also depends on the dimensions of the enclosure and the point at which SE is measured.

Work at the Universities of York and Nottingham [48] does show good correlation with measurement, and more especially with position of the point of measurement with respect to the aperture, as well as predicting enclosure resonances. It treats the enclosure as a length of rectangular waveguide shorted at the far end, with the aperture appearing as a shorted transmission line at the entrance to the waveguide. An example result is shown in Figure 6.8 for two different positions of the measuring probe.

This shows the general form of frequency dependence as predicted by equation (6.3), but by no means the same value. It also shows the worsening of shielding effect as the measurement position is moved closer to the aperture; and most significantly it includes the confounding effect of the enclosure resonance.

6.2.2.2   Enclosure resonance

Resonance is evidenced by the sharp worsening in SE in Figure 6.8, even leading to negative SE or *enhancement* of radiated coupling as the frequency approaches the first resonance of 700MHz. This is a feature of all rectangular box-type shielded enclosures. At any frequency at which the box forms a resonant cavity, the maximum current flow occurs in the structure, and this magnifies the degrading effect of any discontinuity. For an *empty* box, the frequencies at which resonance occurs are given by

$$F = 150 \cdot \sqrt{\{(k/l)^2 + (m/h)^2 + (n/w)^2\}} \text{ MHz} \tag{6.4}$$

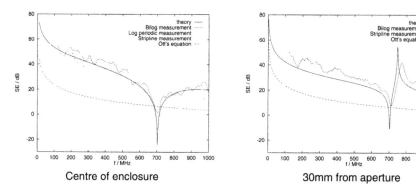

Centre of enclosure   30mm from aperture

Calculated and measured SE for 300 x 120 x 300mm rectangular enclosure with 100 x 5mm aperture
NB line referenced as "Ott's equation" refers to equation 6.3

Figure 6.8 Shielding effectiveness degradation due to an aperture (source: [48])

where l, h, m are enclosure dimensions in metres and k, m and n are positive integers, only one of which may be zero

The lowest resonant frequency will occur for the two larger dimensions of the box; for instance, for the $0.3 \times 0.3 \times 0.12$m box shown in Figure 6.8, the lowest frequency is calculated as

$$F = 150 \cdot \sqrt{\{(1/0.3)^2 + (0/0.12)^2 + (1/0.3)^2\}}\text{ MHz} = 707\text{MHz}$$

Many higher frequency resonances occur above 1GHz for this box, at the higher order modes given by further values of k, m and n. For larger enclosures, of course the resonant modes will occur well below 1GHz; for a typical 19" 37-U height rack cabinet (0.8m depth, 0.6m width, 1.8m high) the lowest frequency will be around 200MHz.

Accurately predicting the resonant frequencies of an enclosure *with a typical load of internally-mounted modules* is not normally feasible; the presence of the internal components and wiring detunes the resonances to a relatively unpredictable degree, although it is usual to find that the frequency is shifted downwards, and the depth of the resonant notch is reduced, sometimes substantially [57]. The components themselves, and any significant apertures, also add their own subsidiary resonances. The best use to which equation (6.4) can be put is to gain an idea of the frequency range above which screening is likely to be degraded and variable.

6.2.2.3 *Multiple apertures and the shape of the aperture*

The shielding degradation is principally due to the aperture's longest dimension. The theory given in [48] includes a term for the height of the aperture but this has a second-order effect on the result. Shielding degradation over the majority of the frequency range (resonances excepted) is approximately proportional to the square of the length.

A square-shaped aperture has been found to have an identical performance to a round one of the same area [58].

Practical enclosures will usually have several apertures, some of which may be combined into groups, as in ventilation louvres or meshes. Adding more apertures *of the same size* can be predicted to worsen the shielding effectiveness roughly in proportion to the number added, e.g. doubling the number will give a 6dB reduction in shielding. However, because of the impact of the maximum dimension of a given aperture, for a specified total open *area* – for a given ventilation air-flow, for instance – it is better to have more, smaller holes than fewer, larger ones. The work reported in [48] and [58] leads to the conclusion that for a constant area the shielding effectiveness improves proportional to √(number of holes). It was also shown that the difference between three 160 × 4mm slots and 20 12mm diameter holes was up to 30dB, the smaller holes being better over the whole frequency range. Dividing a long slot into two shorter ones increases shielding by about 6dB.

Other theoretical work reported in [50] which uses the same approach as [48] has resulted in analytical expressions for the shielding effectiveness of a perfect rectangular enclosure with a mesh or grid as one face. It is likely that the near future will see CAD application software becoming available which will implement these new theoretical approaches.

6.2.2.4   Guidelines for unavoidable apertures

Despite the advances in theory described above, proper calculation of the effect of apertures is still difficult, and it is more straightforward in engineering terms to apply a set of design rules to any enclosure with the intent of maximising its shielding effect and minimising its transfer impedance to internal parts. Section 6.3 looks at the various techniques and hardware that are available, but the basic guidelines can be summarised here:

- Any discontinuity (aperture or seam) should be treated with the techniques of 6.3.2 if its longest dimension is greater than a tenth of a wavelength at the maximum frequency of concern. For 1GHz, this converts to 30mm; for 200MHz, 150mm.
- Internal placement of wiring and modules should avoid proximity to apertures and seams in the enclosure. Where compromises must be made, classify the wiring as per section 7.4.1 and make a similar assessment for the noisiness or sensitivity of the modules, and keep the most critical items the furthest away from the largest apertures.

6.3   Shielding techniques

6.3.1   Bonding structural components

Enclosures are normally built out of a number of components. Typical larger enclosures will use structural supports – pillar and frame members – and panels which bolt to them. Various standards for enclosures, such as the IEC 60297 series give the mechanical dimensions and other requirements to enable such enclosures from different manufacturers to integrate successfully. Such standards do not have anything to say about the electrical integrity of these enclosures, beyond the need to ensure that parts are galvanically connected for safety purposes.

The safety bond – a length of green-and-yellow wire interconnecting panel and frame, or different panels – is familiar to system builders. It is vital to realise that *this is not adequate* for EMC bonding. The purpose of a wired safety bond is to prevent different parts of the structure from assuming different potentials and hence presenting an electric shock hazard at power frequencies: it has no other purpose. It cannot give a low impedance connection at RF.

This is not to say that the safety bond is forbidden for EMC purposes: it can coexist quite happily with a proper EMC bond. But the one is no substitute for the other. If you are intent on building an adequate enclosure for RF purposes, then full metal-to-metal connection at all joints is required. Chapter 5 (section 5.2.2) gives greater detail on bonding and you are referred there for a proper discussion. In the context of enclosure design and construction, the following points should be noted:

- any un-bonded joints between panels and structural members present discontinuities in current flow and as such will degrade shielding effectiveness and transfer impedance – see section 6.2.2

- bonds are best made by surface-to-surface conductive contact at frequent intervals, or preferably continuously along a seam; "bonding" straps, although necessary in many circumstances and preferable to wire connections, are a second best option

- bonding between parts requires removal of insulating layers, for instance paint or anodising, and often the treatment of mating surfaces to ensure conductivity, for instance zinc plating or chromate conversion

- positive pressure is required to make a bond; fasteners will provide this but usually the gap between two fasteners does not allow pressure to be maintained, hence the use of conductive gaskets (see 6.3.2.1 shortly)

- once a bond is made between two surfaces, it should be protected from corrosion by being made gas-tight or by applying some type of overall coating.

6.3.2   Shielding hardware

To improve the bonding of various aspects of a conductive enclosure, many types of hardware are available. These can be classified either as conductive gaskets which are intended to be applied between mating surfaces, and other conductive hardware products which are for application at different sorts of intentional aperture.

6.3.2.1   Conductive gaskets

When two conductive surfaces are mated, electrical contact is unlikely to exist along the whole length of the seam or join. However accurately machined the components may be, some areas will be under greater pressure than others, and these will carry the current across the joint in preference to the poorly-mated areas. The majority of commercial enclosures are not machined at all, and contact across the joint in between those points of pressure provided by fasteners is uncertain and unlikely. Thus current will be diverted or channelled through the assured contact points and not elsewhere; this in turn creates discontinuities in the current flow and hence degradation in the EMC effectiveness of the structure. Even if the two surfaces are very close together and so have an appreciable capacitance between them, the impedance of this capacitance is orders of magnitude greater than that which results from actual contact, and it does not significantly improve the effectiveness of the joint. If no contact can be made at all, then some capacitance due to overlap of two non-contacting surfaces is better than none, but this situation is rare in conventional enclosures.

The purpose of conducting gaskets is to remedy the situation of occasional contact at pressure points, by providing a continuous contact path across a joint in a conformal manner (Figure 6.9). There is then much less variation in joint impedance along the length of the joint and hence less diversion of current paths. The optimum situation is where the gasket provides the same impedance as the bulk impedance of the metal; different types of gasket approach this ideal more or less closely. If no pressure were applied between two surfaces, they would only make contact at the three highest points; as pressure is applied by fasteners or other means, the irregular high spots are flattened resulting in a greater number of contact points and more surface area. A gasket provides this effect more widely at lower closing pressure. The ideal gasket will bridge all irregularities within its designed deflection without losing its resilience, conductivity or stability.

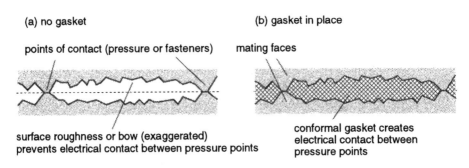

Figure 6.9 The effect of a conductive gasket

Table 6.1 surveys the various types of conductive gasket available and compares their properties and applications. Figure 6.10 shows methods of using these gaskets on typical doors and panels.

Mechanical properties are at least as important as the electrical shielding performance; the latter can usually be adjusted by choice of conductive material, silver, copper and monel being the preferred variations. Compression set (the gasket takes up a particular shape after repeated compression/relaxation cycles, leading to reduced contact force) is an important consideration for gaskets in applications such as doors and panels which will be subjected to repeated mating and opening. Compression force (the mating force needed to achieve the specified electrical performance) and required deflection to achieve it is also more important in these applications, and interacts with the mechanical design of the housing. These factors will dictate the primary choice of type of gasket.

6.3.2.2   Electrochemical compatibility

Corrosion caused by galvanic action is a major problem, both between mating surfaces directly and between gasket material and its adjacent surfaces. As a rule of thumb, electrochemical potential difference should not exceed 300mV to prevent excessive corrosion, although this is not the only criterion: surface resistivity is also important. Galvanized sheet metal develops a coating of high resistance oxidization, and cadmium or chromate passivation also results in a fine surface protective film. For these surfaces you will need a gasket construction that has hard points, such as wire mesh or oriented wires, which penetrate these films to contact the underlying conductive surface.

Tin and solder coatings are compatible with most metals, but are soft and easily corroded by vibration and abrasion. Silver oxide is as conductive as the metal itself, but may be affected by sulphur emissions under some atmospheric conditions; silver is more expensive than other finishes. Untreated aluminium is wholly unsuitable since aluminium oxide is highly insulating, and anodising the aluminium is worse since this deliberately grows a thick layer of surface oxide. Alochroming (chromate conversion) is a technique much used by the military for providing a conductive and non-corroding

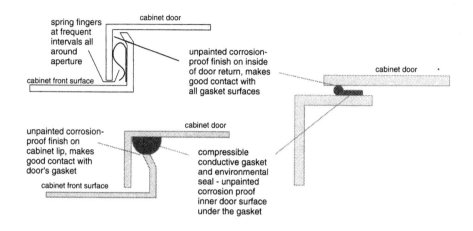

Figure 6.10 Gaskets on doors and panels of shielded cabinets

Table 6.1 Types of conductive gasket

Material	Shielding	Mechanical	Environ-mental seal	Variations/application
Conductive elastomer	Variable depending on filler, poss. up to 120dB	High compression forces, possible problems with compression set, brittleness and ageing	Yes	Wide choice of fillers, elastomers, mouldings or extrusions
Oriented or woven wire in elastomer	Medium 40–80dB, wires will puncture surface films		Partial	Suitable for thin, flat-formed gaskets e.g. connectors
Be-Cu spring finger	Good, up to 130dB, poorer above 1GHz	Fragile, limited range of compression forces	No	Best used for wiping contacts
Knitted wire mesh	Up to 100dB, poor above 1GHz, wires will puncture surface film	Problems with compression set, significant compression force	No	Can be stand alone or over elastomer core to improve compression properties
Conductive fabric over foam core	Good, 60–110dB broadband depending on fabric material	Low compression force, very little compression set	Partial	General purpose

finish for aluminium enclosures. It costs about the same as anodising, but does not provide such a hard-wearing and scratch-resistant finish so is less suitable for aluminium front panels and the like.

The optimum metals for gaskets and for surface treatment are nickel, monel, or stainless steel. Table 6.2 shows the common grouping of metal types by electrochemical compatibility. Selecting adjacent materials (gasket and mating surface) from a single group in the table will give the least galvanic corrosion. Materials in adjacent groups can also be used together, with appropriate protection. Materials from separated groups (e.g., aluminium and brass) should not be used together.

Table 6.2 Electrochemical grouping

Group I	Group II	Group III	Group IV
Magnesium and alloys	Aluminium and alloys	Cadmium plate	Brass
Aluminium and alloys	Beryllium	Carbon steel	Stainless steel
Beryllium	Zinc, Zinc plate	Iron	Copper, copper alloys
Zinc, Zinc plate	Chromium plate	Nickel, nickel plate	Nickel/copper alloys
Chromium plate	Cadmium plate	Tin, tin plate, solder	Monel
	Carbon steel	Brass	Silver
	Iron	Stainless steel	
	Nickel, nickel plate	Copper, copper alloys	
	Tin, tin plate, solder	Monel	

6.3.2.3 Other conductive hardware

Whilst conductive gaskets form a large class of products for mating two surfaces, there is also a growing range of more specialized products for particular shielding applications.

Cable penetrations

As is described elsewhere (section 7.2.3.2 on page 170), cables entering or leaving an enclosure should have their screens properly terminated to the enclosure wall. (*Un*screened cables should enter or leave via a suitable filter if the shielding effectiveness is to be maintained.) This means that full 360° contact should be maintained around the outer surface of the cable screen.

Mechanisms for ensuring this are similar to conventional cable glands for environmental sealing, except that the appropriate parts are fully conductive. Most of the traditional manufacturers of cable accessories are now aware of the importance of EMC aspects, and provide EMC-specific cable glands as part of their stock range. A typical construction using tapered washers to compress an iris-type spring against the cable screen outer surface is shown in Figure 6.11. This is one of the most common methods of clamping to the screen, but others are possible, including collet or other clamping mechanisms, elastomeric compression modules, or folding the screening braid back over a conductive tube.

Aspects which you need to consider when specifying a screening cable gland system are

- mechanical compatibility: the cable screen outer diameter must match the gland's construction, often within quite tight tolerances;
- electrochemical compatibility: the materials used for screen, gland and enclosure panel should discourage corrosion;
- ease of assembly, especially if unskilled or poorly trained technicians are expected, or the working area is restricted;
- conductivity across the joints, which directly affects shielding effectiveness;
- least disturbance of the screen;
- whether or not an environmental seal is also required.

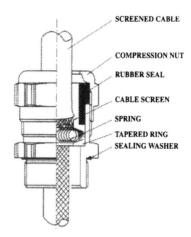

Figure 6.11 Construction of a typical shielding
cable gland assembly (Elkay Electrical)

Waveguide tubes

When a hole in a thin panel is extended so that the depth of the panel is greater than the
hole diameter, it becomes a tube. A tube acts as a waveguide for electromagnetic
energy. A waveguide will allow propagation of an electromagnetic wave along its
length only if its diameter (or maximum cross-section dimension, if rectangular) is
above the minimum required to support the dominant electromagnetic mode. This is
directly related to wavelength, and therefore there is a "cut-off" frequency below which
energy is attenuated rather than propagated down the waveguide. For a square cross-
section tube with air as the medium, the cut-off frequency is [9]

$$f_c \quad = \quad 150/d \text{ MHz} \tag{6.5}$$

where d is the side dimension of the tube in metres

So for instance a 3cm tube will have f_c = 5GHz. If the frequency is substantially lower
than f_c, then the attenuation or shielding effectiveness due to one such tube is given by
the simple expression

$$SE \quad = \quad 27.3 \cdot l/d \text{ dB} \tag{6.6}$$

where l is the length of the tube and d is the side dimension as above

Thus a tube of length-to-width ratio of 4 will give an attenuation exceeding 100dB,
more than enough for most purposes.

Waveguide tubes are widely used in shielded enclosures to carry non-electrical
services and metal-free fibre optic cables through the walls of the enclosure without the
penetration compromising the shielding effectiveness. The usual form of construction
is a simple metal tube welded into the wall or fixed to it in the same manner as a cable
gland. It rarely matters whether the tube projects into or out of the enclosure, or equally
on both sides of the wall.

What is vital is that *no conductors are passed through the tube*. It is emphatically
not suitable for passing metallic cables of any sort, or metal pipes. These destroy the
waveguide properties of the tube and render it as useless for shielding integrity as would
passing a cable through an ordinary hole.

Ventilation panels

One application for waveguide tubes is to provide ready-made ventilation panels for shielded enclosures. Since the walls of the waveguide can be made quite thin and the tubes can be stacked together with little effect on the electromagnetic attenuation, an assembly of such tubes can be built to provide a high degree of through airflow at the same time as a high level of attenuation. These are available as pre-packaged units (Figure 6.12) known as "honeycomb panels" which can simply be fitted into the wall of an enclosure (observing the proper precautions regarding bonding all around the periphery of the assembly) to give any reasonable level of ventilation. These are much more effective at screening than a mesh of holes of the same open area in a thin panel, but of course are more costly and require some thickness in addition to the panel. The wire mesh approach may be adequate for many low-performance applications.

To avoid variations in attenuation due to wave polarization, it is common to stack two layers of honeycomb panel against each other, with different orientations of the waveguide tubes in each layer.

Figure 6.12 Honeycomb ventilation panel (TBA ECP Ltd)

Shielded windows

Viewing apertures can represent the largest size hole in the apparatus. If the display behind the window is a serious source or victim of disturbances, then the entire aperture needs to be shielded. Special conductively treated windows are available for this purpose; they need to be installed, as always, with great care to ensure that they are bonded to the surrounding panel all the way around the edge. The conductive treatment must be brought out to the edge of the window in such a way that good contact can be made to it with no breaks – a metal or metallized frame is often the best way of ensuring this.

The material itself tends to be available in two varieties:

- glass, polycarbonate or acrylic which has been coated with a very thin coating of a conductive film, usually indium tin oxide (ITO), which is substantially transparent. The coating is usually quite fragile and needs some form of protection, and its poor conductivity prevents it achieving high levels of shielding effectiveness. Its visual properties are good, though, and it does not degrade image resolution noticeably;

- glass or polycarbonate laminated with a metallic mesh. These are more rugged, and can achieve better shielding than the coating type, but the mesh can substantially degrade image resolution due to Moiré fringing effects, although these can be avoided with careful design.

In both cases there is a trade-off to be made between electromagnetic protection and transparency or light transmission; generally, the more transparent a window, the less shielding effect it can give.

Because shielded windows are expensive, especially for large areas, wherever possible it is better design practice to find other ways than shielding to protect equipment (such as monitors) with large displays. Fortunately, most types of display are not serious victims or emitters of RF fields (active matrix LCDs can be an exception), at least at the levels required from commercial standards.

6.3.3   Installation and maintenance of screened enclosures

As apertures and seams in screened enclosures such as racks and cabinets can affect the screening performance drastically, it is very important to ensure that measures which are taken to control their effects at the system design stage are not degraded by installation and maintenance procedures (Figure 6.13).

Bonding integrity must be maintained continuously. Any surfaces which are intended to mate must not be allowed to corrode and must never be painted until after they have been assembled. Normally a robust conductive finish such as zinc, alodine or alochrome is used, but if the environment is corrosive (such as in a naval installation) then more specialized measures, and more frequent maintenance, will be needed. Where fastenings provide a conductive path they must all be kept in place and at the correct tightness. Replacement of short, wide bonding straps by loops of wire is unacceptable as their inductance is excessive.

Doors, panels and hatches which make contact via gaskets or spring finger stock must be installed and treated with care so as not to damage or distort the contact surfaces, which should be regularly checked and cleaned if necessary. Filtered inlets and shield penetrations must make assured 360° contact to their host panels; a DC continuity check is rarely adequate to confirm that this is present.

It should be clear by now that *requiring* a cabinet or other enclosure to exhibit good shielding is not a simple or inexpensive option. The requirement affects all aspects of the installation throughout its intended life cycle. Also, maintenance and installation personnel need to be trained in the principles and techniques involved, since they may otherwise unwittingly compromise shielding integrity just by following their own established practices. For these reasons, shielding is often best regarded as a means of last resort if other, lower-cost EMC options are unavailable.

6.4   Architectural shielding

Shielding at the level of an installation, usually called architectural shielding, is not easy to do at low-cost. In general it is much better to purchase apparatus that meets its specifications for electromagnetic performance, i.e. it withstands its electromagnetic environment with acceptable performance degradation, and it does not emit electrical disturbances that cause problems for other apparatus. New apparatus could use any of a number of techniques to achieve this, including if necessary shielded cabinets and shielded cables, and so would not need a shielded room.

Nevertheless, architectural shielding is sometimes required (e.g. for X-ray rooms, to help protect other apparatus). A number of shielding techniques exist, but all are based upon attempting to encase the entire room volume inside a seamless conductive wall.

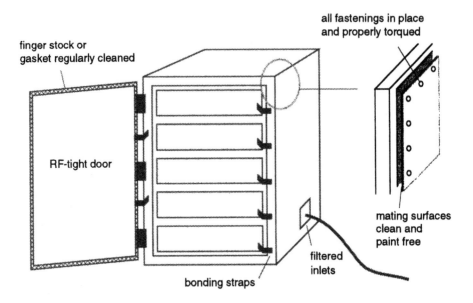

Figure 6.13 Screened enclosure maintenance

Shielding entire rooms against very low frequencies (say, 50 or 60Hz) is generally regarded as impractical except for high-value military or scientific projects where cost is no object (see the discussion in section 6.2.1.1).

At frequencies exceeding 10kHz, practical shielding is becoming possible, and by 1MHz good results may be had with quite thin metal shielding materials. "EMC wallpaper" is a thin copper foil with a building paper backing, which it is relatively easy to glue to existing surfaces. Metallized fabrics and expanded metal have also been successfully used as a form of shielding wallpaper, with the advantage that they allow air to pass through them to some degree.

Making a framework for a room from timber studding, and cladding it on both sides with metallized fabric or expanded metal, can provide quite acceptable values of shielding effectiveness provided that the seams between the metallized sheets are properly dealt with. However, for the best shielding performance there is no substitute for two layers of seam-bonded metal plates, and this is how the basic structure of most EMC test chambers is made.

6.4.1   Apertures

The real problem with shielding a large volume, instead of an electronic unit or equipment cabinet, is that its shielding effectiveness is always compromised to a great extent by all the necessary apertures: e.g. windows, doors, cable entries, ventilation and air conditioning. Accidental apertures, especially imperfections in the joints of the supposedly seamless metal skin, also compromise shielding.

A metal room that ought to achieve 80dB shielding effectiveness at 200MHz can be reduced to 10dB or less by the penetration of the cable that powers the lights, or can be reduced to 40dB by a single hole not much wider than one could put a finger through.

All the real expertise in achieving a screened room is in the details of how to make very long metal joints effectively, and how to get people, air, light, services, goods, (and sometimes vehicles), in and out of the room. This requires a fair degree of specialization.

Most of the best practices described in this book do not need special expertise to implement, and typical installation engineers can easily cope with them. But constructing a shielded room is not something that should be attempted by the non-expert. The best approach is to contact one of the many companies that offers to design and build shielded rooms and subcontract them to do it all for you.

There are a number of such companies, offering a range of services at a range of prices. The most important thing for the installation's project manager to do is to make sure that the contract specifies:

- The shielding performance target for the frequency range of interest, and how it is to be measured and specified;

- The air and light quality within the room, and any movement of people, goods, services, and vehicles in and out of the shielded volume;

- The verification methods to be employed and the minimum results they must achieve (for instance, using MIL-STD-285);

- That it is the contractor's responsibility to achieve those results for the agreed price and in the agreed timescale. Holding a substantial part (or all) of the price of the shielded room back, until verification has been satisfactorily completed, is a powerful way to get the desired performance from an experienced contractor. (Don't even think of using a contractor with no proven track record for this work.)

Remember that a shielded room is not shielded when any of its doors are open.

6.4.1.1   Interconnecting shielded enclosures or rooms

Figure 6.14 attempts to show how two shielded enclosures are connected together and to the rest of the installation, without compromising their shielding effectiveness. These enclosures could be any size: equipment cabinets, or large rooms. The screened cables that enter or leave them have their screens terminated by 360° electrical contacts in shielding glands (or connectors) at the enclosure walls.

The unscreened cables that enter or leave them only enter the cabinets via filters mounted to the cabinet walls. Bulkhead-mounted filters built into the wall of the enclosure are the best for this purpose. A number of companies make "room filters" for EMC test chambers that are also suitable for use in architecturally shielded rooms or buildings.

Metal-free fibre optic cables would be better than metallic conductors for such rooms, because they may be passed through waveguide tubes (page 149) so that the small aperture they make in the metal wall does not cause any leakage. Unfortunately it is still impossible to transfer substantial amounts of electrical power into a room by the use of fibre optics (although it has been done at levels of a few tens of watts), so we are still left with metallic power cables and their filters.

A PEC (e.g. at least one cable tray, duct, conduit, or item of structural steelwork) is bonded to each cabinet and runs between them, carrying all the cables between the cabinets whilst maintaining their segregation by Class of cable.

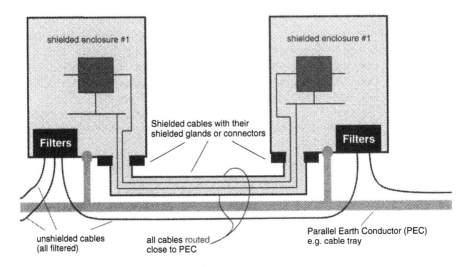

Figure 6.14 Installing connections between shielded enclosures

Chapter 7

Cabling

Cabling between equipment is a principal route for interference coupling, both into and out of equipment. Cables themselves are of course passive, and therefore do not come within the scope of any EMC regulations. Although there are occasional sales claims to the effect that this or that cable type "complies with" the EMC Directive or FCC rules, such claims are entirely spurious. However, use of a particular type of cable may *enable* a system or apparatus to comply with EMC requirements. Choice of cable is a crucial aspect of system design for EMC.

Just as important in a systems context is what you do with the cable in order to install it. Layout, routing and terminations will all affect coupling of interference to or from the cable and can therefore have an impact on the overall system performance. The equipment designer may need to specify a particular cable type for a particular interface in order to meet EMC requirements for the equipment; the system designer will have to ensure that this is used, and will need to implement the best installation practice.

This chapter will look first of all at cable coupling, and then explore issues of use of screened and unscreened cables, before finally discussing the questions of classification, segregation, routing and parallel earth conductors.

7.1 Coupling to, from and within cables

The distinction between differential and common mode coupling has been discussed in section 4.2.4. Nowhere is this distinction more important than with cables.

7.1.1 Differential mode

In virtually all cases the wanted or intended signal current(s) within the cable are flowing in differential mode. That is, two conductors in the cable are devoted to each signal: one carries the "flow" or "go" current and the other the "return" current. The designers of the equipment at each end have allocated connector pins for each conductor and the source and load circuits are connected to these pins. If the circuit could be considered entirely in isolation, the only current that would be flowing in these conductors would be that due to the intended signal.

This description applies equally to power or signal circuits. Examples of such circuit pairs might be mains live and neutral, DC supply plus and minus, each signal and its 0V in an RS-232 data line, audio or data signals in a telecomms network, RF or wideband data signals in a coaxial cable where the inner conductor is paired with the outer sheath. Whether or not the circuit is "balanced", i.e. the voltages are symmetrical about earth or some other fixed reference, is not relevant to the differential mode classification – although it is important in other respects. The important point is that the current in one half of the conductor pair matches and reflects the current in the other half. In mains filter parlance, this mode is sometimes known as "symmetric" mode because the current is symmetrical across the two halves (in the live line and in the neutral line) of the filter. In a three-phase power circuit, the differential mode is not related to the question of whether the load is balanced across the phases; in the three-phase supply cable, the differential mode currents (including the neutral in the star configuration, or not in the delta configuration) always sum to zero.

It is also the case that in a multi-conductor cable, differential mode can exist quite happily when only one conductor takes the return current for all the other conductors. This is the case, for instance, in the RS-232 serial data specification [141]. The cable may be carrying anything from two (TXD and RXD) to twenty data signals, depending on the complexity of the application; but there is only one connection for the 0V reference, which takes all of the return currents for each of the signal circuits. As far as the whole cable is concerned though, the differential mode condition is satisfied as long as the net total of all the currents in the cable is zero (see Figure 7.1(c)). Because of the shared return, this is in many respects an EMC-hostile type of connection, which is one of the reasons for the limitation of the RS-232 specification to low data rates and short distances; but its advantage of low cost often outweighs these disadvantages, hence its huge popularity.

7.1.1.1   Magnetic coupling in differential mode

It was seen in section 4.2.2.1 on page 81 that coupling between the magnetic field around a circuit and the current flow within it is related to the enclosed area of the total loop, provided that the current of interest stays within that loop. This is directly relevant to the above description of differential mode currents within a cable. The loop area for a total circuit is dominated by the distance between its cable go and return conductors, multiplied by the cable length. The larger this area, the greater the resulting magnetic coupling.

However, we are usually more interested in the coupling with a small segment of the cable, especially if this runs near to a severe emitter or a sensitive potential victim. In this case, it is not a question of coupling with the whole loop, but with two closely-spaced conductors carrying equal and opposite currents. If the two conductors were co-located (a physical impossibility, but approached by the coaxial cable configuration, and to a lesser extent by the twisted pair) the net current would be zero and therefore the net emitted magnetic field would be zero. In practice, each conductor generates an equal and opposite magnetic field which is separated slightly in space. As the observation point moves away from the conductors, this separation becomes proportionally less significant and the combined field decays rapidly towards zero. The impact of this geometrical result is easily calculable (see Figure 7.2) for any point in a plane normal to the two conductors, at frequencies for which the distances involved are small compared with a wavelength. No magnetic field is coupled in the longitudinal direction, i.e. the direction of current flow.

This characteristic of magnetic coupling results in the advice to keep all cable pairs as closely coupled within a cable bundle as possible.

7.1.1.2   Electric coupling in differential mode

Electric field coupling is related to the voltages on the conductors rather than to the current flowing in them. Here, we are interested in the voltage developed between a conductor pair as a result of an external electric field applied across them, or conversely

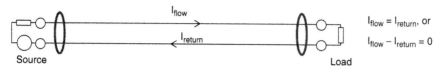

(a) single pair, e.g. mains live/neutral, dc +/-, signal/0V

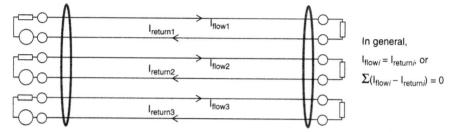

(b) multiple pair, e.g.audio or data, balanced or unbalanced

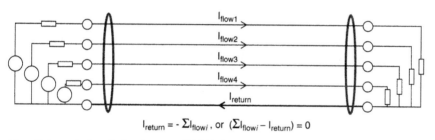

$I_{return} = -\Sigma I_{flowi}$, or $(\Sigma I_{flowi} - I_{return}) = 0$

(c) multiple conductor, single return. e.g. RS-232 data

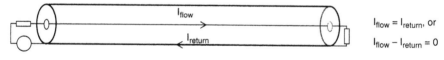

(d) co-ax, e.g. RF or LAN

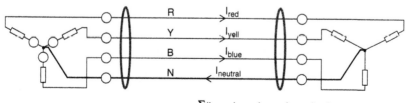

$\Sigma(I_{red} + I_{yell} + I_{blue} - I_{neutral}) = 0$

NB this condition holds in the cable even

(e) three-phase star   when the phases are unbalanced, i.e.
have different loads

Figure 7.1 Differential mode current in cables

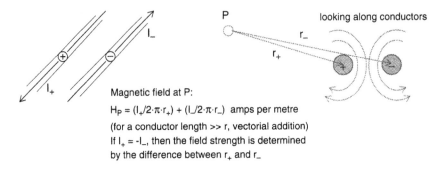

Magnetic field at P:

$H_P = (I_+/2 \cdot \pi \cdot r_+) + (I_-/2 \cdot \pi \cdot r_-)$ amps per metre

(for a conductor length >> r, vectorial addition)

If $I_+ = -I_-$, then the field strength is determined

by the difference between r_+ and r_-

Figure 7.2 Magnetic field at a distance from a conductor pair

the electric field which appears as a result of the voltage applied between them. The coupling is sensitive to the polarization, that is no coupling exists for fields normal to the plane of the wires; maximum coupling exists within this plane (Figure 7.3). For a constant field strength, the voltage induced is proportional to the conductor spacing, hence again the advice to minimize this separation. It is also dependent on the exposed surface area of the two conductors and on the impedance of the circuit.

The external field geometry and distribution is heavily affected by all other external conductors, whether they are part of an intended circuit or not, and therefore it is far less practical to calculate and predict electric field coupling than it is magnetic. Fortunately, the electric field is far easier to screen against and in practice it only becomes important for unscreened cables or for those whose screens are imperfectly implemented.

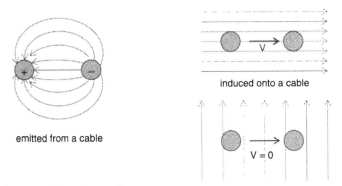

emitted from a cable

induced onto a cable

V = 0

Figure 7.3 Electric field cable coupling

7.1.1.3   High frequency coupling in differential mode

The above descriptions, separating electric and magnetic coupling, can only be applied accurately for low frequencies where the cable length is a small fraction (of the order of a tenth or less) of a wavelength. When the dimensions become equivalent to or greater than a wavelength, the phase variation of the fields with spatial position is important. The resulting phases of the induced currents and voltages vary with their position along the cable. At the point of interest – usually the end of the cable where it connects to the equipment – the quantities can be anywhere from zero to a maximum, depending on the relative phases at each incremental point along the cable, and vary periodically with frequency and the spatial distribution of the fields (Figure 7.4). In principle the induced interference at any given frequency can be calculated if the geometry and electrical parameters of the cable and fields are known [9], [10], [12]. This is of no practical use for the systems installer though, since these parameters are either unknown, uncontrolled or subject to frequent change. In most circumstances the "envelope" of maximum possible coupling is the most useful aspect that can be predicted.

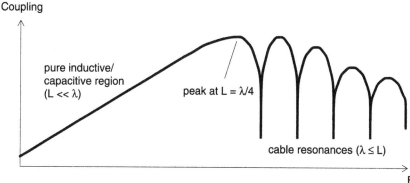

Figure 7.4 Typical variation of cable coupling with frequency

7.1.2   Common mode

Cable conductors will also be carrying currents in common mode, simultaneously with the intended and desired signals in differential mode. The common mode currents are almost invariably unintentional or undesired side effects of the required signals, or they are wholly a result of external disturbances. A well-designed cable installation is very largely one which is successful at keeping the common mode disturbances separate from the differential mode desired signals.

Put simply, the common mode currents are those which are flowing on all conductors of the cable equally in the same direction, and therefore returning via some external path. They can be seen as due to two principal causes:

- $I_{flow} \neq I_{return}$ in a conductor pair, so that $(I_{flow} - I_{return}) \neq 0$; the difference between these differential mode currents is the common mode current and is directly related to the signal carried on the cable;

- external influences, unrelated to the power or signal being carried, couple with the cable equally on all conductors and induce current equally on all conductors – the cable appears as a single conducting structure.

These two causes can be modelled separately but both are usually present in a typical EMC situation.

7.1.2.1   Differential to common mode conversion

The first cause can be seen as a mechanism by which the intended differential currents are partially converted to common mode. The equivalent circuit for this can be represented as in Figure 7.5, in which the real-world parasitic components are shown along with the designer's expected circuit. If we ignore the distributed parameters of the cable – that is, the frequency is low enough that the cable length is much less than a wavelength – then each of the three parts of the circuit is modelled as having lumped stray reactances to some external reference. This is often assumed to be "earth", but in reality it can be any indeterminate potential, due to the aggregate of conducting structures that are coupled to the cable and the apparatus at each end.

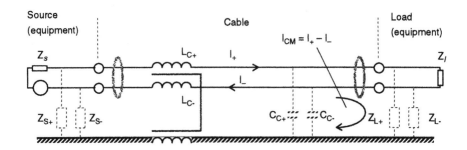

Figure 7.5 Stray impedances converting differential to common mode

The plus and minus terminals at the source equipment end will have Z_{S+} and Z_{S-}; the terminals at the load end will have Z_{L+} and Z_{L-}. These impedances will be dominated by stray capacitance if the circuit is galvanically isolated, but will be a mixture of resistive, inductive and capacitive if the circuit is earth-referenced at some point. The cable conductors will have stray capacitance C_{C+} and C_{C-} along their length, and also will be inductively coupled to the external reference conductors, as shown by L_{C+} and L_{C-}. Strictly speaking, the cable resistance should be included as well, but it rarely contributes significantly to the common mode conversion mechanism.

If all of these stray components were balanced – that is, if all those with the subscript + equalled their opposite numbers with the subscript - – then no conversion from one mode to the other would take place. This is never the case in reality. Imbalances between Z_{S+} and Z_{S-}, and Z_{L+} and Z_{L-}, are present at each end because of differences in PCB track length and area, connector and circuit asymmetry. Even if the cable were perfect, these imbalances would result in some difference between I+ and I-. This difference is related to the ratio of the desired circuit impedances to the undesired impedances; that is, the lower the value of $Z_{S+/-}$ and $Z_{L+/-}$ by comparison to the circuit impedances Z_s and Z_l, the greater the effect of their imbalance. If these impedances are mostly capacitive, the imbalance effect becomes more significant at high frequencies.

The difference between I+ and I- is of course the common mode current I_{CM}, which flows through the stray impedances at each end and whose return path is via the nominally "earth" reference. One of the functions of the Parallel Earth Conductor (PEC) discussed later (section 7.4.3) is to provide a controlled path for this return current which is closely coupled to the cable, in contrast to the uncontrolled and chaotic paths that would otherwise be taken without it. These uncontrolled paths still exist even with a PEC in place, but the PEC is designed and terminated to ensure that it provides a preferential, i.e. low-impedance, route for the common mode return currents.

The cable also contributes to the circuit imbalance, through differences in C_{C+} and C_{C-}, and to a lesser extent in L_{C+} and L_{C-}. These differences will be partly due to imperfections in the cable construction, but mostly to variations in the environment through which the cable passes and with which it must couple – typically, the proximity of other metal objects. If, for instance, a cable passes edge-on near to some conducting structure then one conductor is nearer to the structure, and hence has a higher capacitance and/or partial inductance with respect to it, than does the other conductor. If the cable geometry could be maintained perfectly flat against all possible nearby conductors then balance could be maintained, but instructions to this effect would not be popular with installation technicians.

Imbalances in cable strays can be negated to a large extent by screening and by twisting the conductor pairs, and these techniques allow a parameter to be developed for any cable type which is called "longitudinal conversion loss" (LCL) [128]. This is a measure of the degree of conversion from differential to common mode of the circuit and can be seen as a quality factor for a particular cable – the higher the value, the better. Some cable manufacturers are able to quote LCL figures for some of their range, especially those which are intended as wideband data cables, for which the impact of common mode conversion in controlling RF emissions is important. The LCL can be used to estimate common mode current emissions for a given circuit, as follows ([68], Annex E):

$$I_{CM} \text{ (dB}\mu\text{A)} \quad = \quad V_T \text{(dB}\mu\text{V)} - LCL \text{ (dB)} - 20\log_{10} \cdot |(2Z_0 \cdot (Z_{cm} + Z_{ct})/(Z_0 + 4Z_{cm})| \qquad (7.1)$$

where V_T is the differential signal voltage,
Z_{cm} is the common mode impedance of the item having the worst LCL,
Z_{ct} is the common mode impedance of the item with the higher LCL,
Z_0 is the differential mode characteristic impedance

7.1.2.2   Separately induced common mode currents

Common mode currents which are related to the signals carried by the cable are usually generated by the conversion mechanism described above. In contrast, there are also common mode currents on any cable which have no relation to the intended signals carried by the cable. These may be generated by the equipment at either end, and then radiate from the cable acting as a disturbance emitter (Figure 7.6); or they may be induced on the cable by external sources, in which case the cable is acting as the receptor of the disturbance, and conducting it into the equipment at either end (Figure 7.7).

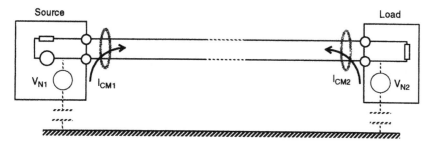

Figure 7.6 Separately generated common mode emissions

Common mode emissions which have nothing to do with the signals carried by the cable are a common cause of failure to meet compliance requirements [31], [47], and are often very hard for non-EMC engineers to understand. They are also becoming more prevalent, with the introduction of high-speed digital circuits (microprocessors) into many originally analogue applications. Such emissions can be seen to occur even when a single wire is connected to a supposedly fixed circuit node, such as circuit 0V. Figure 7.6 shows the equivalent circuit for this mode of emissions generation. The crucial aspect is the presence of an aggregate noise source V_N which appears between the point of connection of the cable, and the external reference. This noise source is the result of the normal operation of the circuit, developing potential differences across the mesh of stray impedances that are formed by the PCB tracks and wires within the equipment. Aggregate V_N values of only a few millivolts are enough to cause a breach of emissions limits, if connected directly to an external cable. Equipment designers minimize the problem by careful PCB layout to reduce V_N, and by careful interface design to prevent it from being directly coupled to an external cable.

Notice that even the apparatus which forms the load for the wanted signal, if it also contains other operating circuits, can cause emissions of this sort (V_{N2} in Figure 7.6). Power cables, which do not carry any intentional high frequencies, can still offer a coupling path for these emissions.

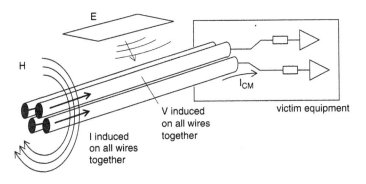

Figure 7.7 Cable acting as a disturbance receptor

The effect of the second mechanism (Figure 7.7) is that the common mode voltages and currents which are presented at the connectors of the apparatus, may experience conversion to differential mode and then appear at internal circuit nodes differentially and cause malfunctions. This is the most common coupling route in cases of susceptibility. The conversion to differential mode acts in the inverse manner to that discussed earlier, as a result of differential circuit and stray impedances. Good equipment design is aimed, firstly at preventing the common mode signals from penetrating into the circuits, and secondly at laying out the circuits so as to minimize the common mode to differential mode conversion. Because coupling is reciprocal, these are the same techniques as are used to control common mode emissions.

As with the case of differential mode, the coupling can be classified as either via magnetic fields, electric fields or at high frequencies, a combination of the two. However, the crucial difference is that now the relevant separation distance is that between the cable as a whole, and the reference structure that is carrying the common mode return current. In this respect the internal cable construction is irrelevant; the separation between the individual conductors has no effect on the coupling. The outer of a screened cable will carry common mode currents just as do the conductors of an unscreened cable. In fact the screen is intended to do this. Segregation and routing of cables is more of a vital factor in controlling common mode coupling, since the inverse square law with separation distance that is evident in Figure 7.2 does not operate – the magnetic field around a cable carrying common mode currents decays only proportionally to distance, as does the electric field. For both emissions and immunity, then, the cable is acting exactly like an antenna, and its coupling efficiency is very much greater than in differential mode.

7.1.3   Crosstalk

The issue of crosstalk occupies the never-never land between EMC and signal integrity. Crosstalk within a cable is not a problem for external (inter-system) EMC in the sense proposed in section 1.1, but it does have a bearing on *intra*-system EMC, that is, the ability of a system not to interfere with itself. The problem is essentially one of coupling between separate circuits in a cable loom.

Visualize two circuit pairs in a single cable (Figure 7.8). Along the length of the cable there is distributed capacitance between every conductor and each of the other three conductors. Similarly, there is mutual inductance linking every conductor to each of the others. At frequencies where the cable length is much shorter than a wavelength, the L and C can be simplified to a matrix of discrete reactances, but at higher frequencies it is necessary to assign elemental reactances to an infinitesimally short length and then integrate these over the length of the cable.

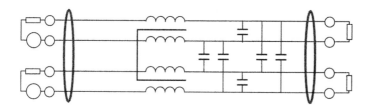

Figure 7.8 Intra-cable crosstalk

The mutual capacitance and inductance between the two conductors that form each circuit pair are benign, and determine the characteristic impedance of that pair ($Z_0 = \sqrt{L/C}$, assuming no losses). In conjunction with the circuit driving and load impedances they will determine the bandwidth capability of the cable/equipment system. But the mutual impedances *between the pairs* are undesirable. These result in crosstalk interference between the two circuits.

7.1.3.1   *Capacitive crosstalk*

The voltages appearing on + and – of pair 1 are coupled by the mutual capacitances onto + and – of pair 2. The amplitude of the induced voltage is determined by the values of the capacitances and the circuit impedances, and the rate of change of source voltage (dv/dt). Balanced circuits, and a balanced cable construction which equalizes the capacitances, will minimize the effective crosstalk since voltages induced on or from the + conductor will be nearly cancelled by those induced on or from the – conductor. Unbalanced circuits with high dv/dt and high impedances will be the most susceptible to capacitive crosstalk. Screening each pair individually (Figure 7.9) will remove the capacitive crosstalk almost entirely, since the mutual capacitances between pairs are eliminated, to be replaced by mutual capacitance from each pair to its screen and mutual capacitance between screens. (Screens without 100% optical coverage, such as braids, will still allow a small amount of capacitance directly between conductors, through the gaps in the screen.) The screens must of course be connected to a fixed potential (which may be system earth, or sometimes circuit 0V); voltages will still be developed longitudinally along the screens as a result of their resistance, and this along with other screen imperfections is then the limiting factor in capacitive crosstalk suppression.

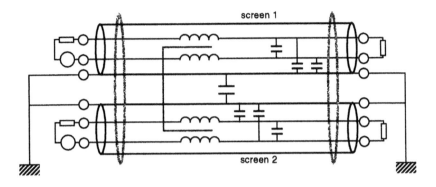

Figure 7.9 Screening against capacitive crosstalk

7.1.3.2   Inductive crosstalk

Currents flowing in each conductor will induce a longitudinal voltage in all other conductors as a result of mutual inductance within the cable. The amplitude of this voltage is proportional to the di/dt of the source current and the mutual inductance linking the conductors; circuit impedances do not affect it. If the mutual inductance from one source conductor to each conductor of the other circuit pair were to be equal, then the same voltage is induced in each and the net effect on that circuit is nil. By itself this is not usually the case; but when the opposite sense contribution from the other source conductor is included, in a cable with good symmetry the total contributions can cancel each other. For this reason inductive crosstalk by itself is rarely as serious a difficulty as capacitive.

7.1.3.3   Distributed crosstalk

At high frequencies the cable must be considered as a distributed structure and the mutual impedances of elemental lengths have to be integrated over the whole length. The phase differences along the length become significant, and the contributions of inductive and capacitive crosstalk result in constructive interference at each end at some frequencies, and destructive interference (nulls in the coupling) at each end at others. What is more, the sense of the wave travelling down the cable becomes significant, and it is necessary to talk of "near-end" crosstalk (NEXT) as distinct and different from "far-end" crosstalk (FEXT). This book does not attempt to go into the detail of distributed crosstalk analysis – see Chapter 10 of Paul [9] or section 4.3 of Tsaliovich [12] for an in-depth treatment.

7.2   Cable screening techniques

7.2.1   Options for cable screening

Any quick glance through a cable catalogue will reveal a huge range of types of screened cable. This implies that they are all imperfect in some way (counting high cost as an imperfection), and that it is going to be quite difficult to choose the most affordable type with the fewest imperfections for the job.

Manufacturers of apparatus should provide information on the cable and connector types to be used in their installation instructions. They should have had their equipment EMC tested in some way, and they will have discovered which cable and connector types were at least adequate when they did so. Quite often they will specify a cable type as (for example) "Belden 9829 or equivalent". It goes without saying that you need to be fairly knowledgeable to determine what is an equivalent in EMC terms, so it is safest to stick with the specified type, provided of course that it is readily available.

If the manufacturer does not specify the cable type, they should be asked to. It is important to discuss cable and connector details with possible suppliers of apparatus well before making the decision to purchase from one of them. The cost of installing expensive made-to-order cable and/or exotic connectors can often swing the purchasing decision to a manufacturer whose quote is higher because his product is designed to achieve the required EMC performance with standard low-cost cables and connectors.

Where screened cables have to be chosen without the benefit of instructions from the manufacturer, a few guidelines may help:

- Twist the send and return current conductors together for every single load. Do not rely on any other wire or cable (or cable screen) to carry the return currents, even if they appear capable of doing so. This means using twisted pairs at least (sometimes twisted triples or quads are needed).

- Despite their prevalence for RF signals, co-axial cables are not very good for low-to-mid frequency EMC, because the screen is used to carry both the signal return current and the interference currents, and they get mixed up. Only use them when the manufacturer has specified them as being suitable for the intended operational electromagnetic environment.

- The drain wire in a foil-screened cable is not there to provide a neat termination for the screen. The foil screen should be bonded 360°, and the drain wire can either be bonded with it or fixed to the body of the connector by some other means.

- Overall braid screens are electrically superior to overall spiral-wrapped foil screens.

- Braid is generally better the greater its optical coverage (the less you can see through it).

- Braid and foil are generally better the heavier their gauge, since this gives greater conductivity; but the skin effect limits the improvement at frequencies over a few MHz.

- Longtitudinally-wrapped foil (i.e. wrapped lengthwise) is better than spiral foil, but is hard to come by and not as flexible.

- Metal foil, the thicker the better, gives better shielding performance than metallized plastic film in a full shielded cable. Unfortunately, it also makes the cable less flexible.

- Overall braid and foil screen, or double braid, is much better than a single screen, even when the two screening layers are not isolated from each other. The best configuration for braid and foil cables is when the braid lies against the conductive side of the spiral foil so as to "short out its turns".

- Individual twisted pairs (triples, quads, etc.) in an overall screened cable may need individual screens to prevent capacitive crosstalk between signals as in Figure 7.9 (usually individual spiral foils are adequate) – this will be dictated by the circuit parameters.

- Multiple isolated screens are better than non-isolated screens, in general (although resonances can make them no better than single screened wires at some frequencies).

- Better cables for screening are generally thicker, heavier, have a larger minimum bend radius, and cost more.

- Almost the best possible double-screened cable is one which is run in its own circular metal conduit (pipe) with 360° bonding at all joints and to the equipment cabinets at both ends.

- "Low capacitance" cables, and/or "high velocity factor" cables, will generally be better at carrying data for longer distances.

- For high-rate data (or long distances) the characteristic impedance of the cable and its connectors must match the specification for the "physical layer" of the data standard, which may also specify the detailed build of the cable. Where the datacomms specification used is proprietary to a manufacturer, he must be able to specify cable and connectors precisely. Where it is a published data communication standard, read the standard or obtain one of the many industry guides.

- Screened cables subjected to tight bend radii, or to repeated flexing, can find gaps opening up in their screening layers and reducing their effectiveness. This is a particular problem for screened cables to the moving parts of robotic systems. Careful control of bend radii and choice of flexible cable is required.

7.2.2   Cable transfer impedance

The screening performance is formally specified in the "transfer impedance" parameter Z_T of the cable. This is the same concept as the transfer impedance of structures discussed in section 5.1.2.3 and 6.1.1, but specifically related to screened cables. It offers a convenient way to describe and compare the quality of different types and make of cable. So far, it is rarely found in catalogue descriptions except for high-performance types, but manufacturers can often supply details of a particular type on request.

7.2.2.1   Single coaxial cable

For a single conductor coaxial screened cable the transfer impedance is simple to define. It is the voltage V_t which appears between the inner and outer of the coax at an open circuited end, divided by the interference current I_S flowing on the screen which produces this voltage (V/I has dimensions of ohms). Figure 7.10 shows the equivalent circuit. The far end of the coax is shorted so that no voltage appears at this end. The transfer impedance is specified as a differential parameter with respect to the length of the cable, and is quoted per unit length in units of ohms per metre. If the screening were perfect, then no matter how much current flowed in the screen, no voltage would appear on the inner, and Z_T would be zero Ω/m.

The screening imperfections can be modelled as resistance in the screen conductor, and leakage inductance between the inner and outer. At very low frequencies and DC, only the resistance is significant, and the transfer impedance actually equals this

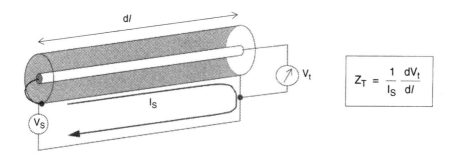

Figure 7.10 Transfer impedance of a single-conductor coaxial cable

resistance. This can be seen intuitively: no current flows in the inner conductor, so the voltage to earth at the measurement end equals the voltage to earth at the shorted end, which must be $R_{screen} \times I_S$.

At higher frequencies the leakage inductance between inner and outer becomes dominant. This is the difference between the self-inductance of the cable as a whole, and the mutual inductance of the two conductors. It can be written as $L_c(1 - k)$, L_c being the cable inductance and k being the coupling coefficient. As k approaches unity the leakage vanishes; this is the situation for a cable with a 100% solid outer sheath, such as rigid copper coax. In this case the transfer impedance actually improves (reduces) with frequency through another cause, which is the separation of inner and outer sheath currents through the skin effect. Typical braided coax has a k value of around 0.996 as a result of apertures and other effects in the braid, so that its transfer impedance starts to rise above 0.5 –1MHz.

Measurements of transfer impedance show these effects (Figure 7.11) for various types of cable. Once the frequency rises above that for which the cable can be considered electrically short (length < λ/10), Z_T measurement becomes complicated. This is because the Z_T definition above does not take into account standing waves on the inner conductor; it assumes that the measurement of the induced voltage is "current-free", i.e. terminated in a high impedance and not in a matched load. In reality, such a measurement set-up is untenable once the cable length approaches resonance. Measurement at higher frequencies therefore implies either shorter lengths of cable, causing sensitivity issues and contamination of the results by artifacts of the end terminations, or the use of matched triaxial systems. This has resulted in a "perplexingly large number" of Z_T measurement methods [56] which seek to overcome the problem. Although only one of these is quoted in IEC 96-1 [80], which defines screened cable performance parameters, many others remain in widespread use, defying proper comparison of cable data.

7.2.2.2   *Multiconductor screened cable*

Although Z_T is most usually quoted for single conductor coaxial cables, it is equally applicable to multi-way screened cables. In this case the voltage of interest can be developed either between individual internal circuit pairs, or between the screen and all the inner conductors connected together, as a result of the interfering screen current (Figure 7.12). Whilst the former is of more use for determining the performance of a system using that cable, it cannot be defined independently of the circuit in which it will be used. Screen currents here develop a common mode disturbance in the wanted

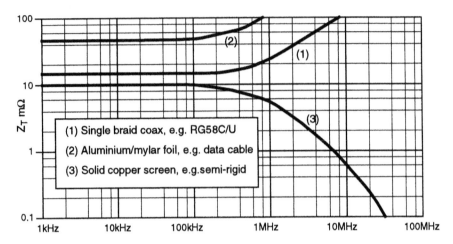

Figure 7.11 Typical transfer impedance curves for various cable types

circuit, and the overall impact depends on the common mode rejection of this circuit. This is not usually dominated by the cable. In other words, it is more sensible to define a transfer impedance for the system rather than a per-unit-length Z_T for the cable alone.

Z_T in the case of multi-conductor cables is still dominated by the leakage inductance $L_c(1 - k)$ at high frequencies. The common aluminium foil shields with drain wire are less effective than overall braid shields, partly because of their higher resistance, and partly because the drain wire distorts the magnetic coupling of the shield with the inner, giving a lower value of k. Figure 7.11 includes curve (2) for such a foil-shielded cable, configured as in Figure 7.12(a).

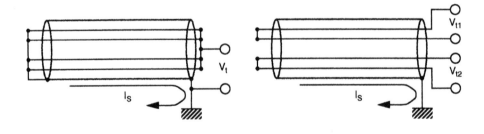

(a) screen to all conductors together   (b) screen to individual circuit pairs

Figure 7.12 Transfer impedance of multi-conductor cables

7.2.3   Terminating the screen

The previous section may have implied that the performance of a screened connection between two pieces of equipment could be described by the transfer impedance of the cable alone. In reality, the cable is only part of the whole; the end terminations also contribute to the *system* transfer impedance, and if poorly made, will dominate it.

Cable screen RF bonding requires glands (or connector) backshells which achieve 360° (full circle) electrical contact to the gland (or connector) right around the circumference of the screen. Saddle-clamps or P-clips may also be used, although they are not as good for very high frequencies or high powers. It is also important that connector backshells bond 360° to the body of their mating connector, and that the bodies of glands or connectors bond 360° to the metal surfaces they are mounted on.

The maintenance of the 360° coverage of the cable screen and connectors, right through any joints, connectors, or glands from one electronic circuit to another, is vital. The EMC performance of the cable type is wasted if 360° screen coverage is not continuous from end to end.

360° screen coverage for an entire interconnection helps ensure that the currents and voltages caused by external electromagnetic disturbances remain as far as possible on the outside of the cable and connector screens, and do not get mixed up with the internal (signal) currents. It also minimizes the amount of signal voltage or current that is coupled to the outside world, helping to prevent problems with signal integrity as well as the emission of disturbances.

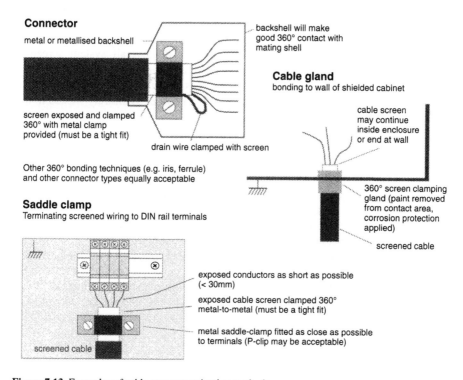

Figure 7.13 Examples of cable screen termination methods

7.2.3.1 Connectors

Connectors that are properly and intentionally designed for screened cable pay considerable attention to the series impedance presented by the screen termination. A good example is the N type RF connector, which can maintain system performance up to 18GHz. In this construction the braided screen is compressed by a metal ferrule against the connector body, which is mated through a threaded cylinder to the complementary female half of the connection. Not only is 360° contact maintained throughout, but the diameters of the various parts are closely controlled to minimize deviations in characteristic impedance through the connector.

General purpose screened electrical connectors do not normally need to maintain characteristic impedance, but they should ensure proper 360° contact through the screen to the body of the connectors and then to the mating screen or chassis. The effect of this, in transfer impedance terms, is to maintain a high value of coupling coefficient k through the connection, which minimizes the Z_T of the connection. An alternative way of seeing this is to say that a poor or inadequate 360° shroud increases the inductance of the screen connection.

Commercial screened connectors vary quite widely in their form of construction and hence in their Z_T. Both the method of terminating the cable screen and the method of mating the shells are important. For instance, the principal distinction between so-called "EMC" subminiature D-type connectors and their lesser brethren, is the dimples on the male connector's shell, and the tin plating on the shells of each half. These two features markedly reduce the contact impedance between the shells as compared to their non-EMC-dimpled, cadmium passivated relatives. But equally, whatever backshell is used needs a secure method of terminating to the connector shell and of clamping to the cable screen.

It should be clear that the system Z_T defines the overall system screening performance, and this is made up of the sum of the cable Z_T as discussed in 7.2.2, and the connector Z_{TS} at either end. If the connector Z_{TS} are substantially greater than the cable, they will dominate the whole. A cable with a good Z_T is entirely wasted in this situation.

7.2.3.2   Direct to metalwork

A variety of low-cost 360° cable screen termination methods which tie the screen direct to the chassis are becoming available.

Some new techniques use stainless-steel cable ties, and are very fast and low-cost indeed, although not as aesthetically pleasing as more traditional connectors and glands. Alternatively, elastomeric blocks are available, which grip all around the cable screen and transfer the connection to the metalwork via a housing frame.

The conventional cable gland is entirely acceptable provided that it intentionally makes contact all around the cable screen, and all around the edge of the hole in the metalwork. "EMC" cable glands are marketed for this purpose with internal iris or ferrule arrangements that ensure good contact is made with the screen. Any paint or other coating must be removed from the chassis contact area – a partial contact to the chassis, i.e. not 360°, is unacceptable.

7.2.3.3   The problem of the pigtail

"Pigtails", lengths of wire soldered to braid shields, or the drain wires of foil-shielded cables, are surprisingly ineffective at providing a good earth termination even when they are as short as 25 millimetres. The pigtail introduces inductance in series with the screen-to-earth connection [36], [46], which will dominate the transfer impedance of the complete assembly. The flux from the interference current in the pigtail wire links with the inner circuit(s), whereas the interference current on the braid generates flux which does not link with the internal circuit(s) except by leakage through the braid apertures or other inductive effects as discussed in 7.2.2. The mutual inductance of the pigtail section is proportional to the pigtail length; for a 25mm length the mutual inductance is a few nanohenries, which is substantially greater than the contribution from leakage inductance of a typical braided cable.

Pigtails cannot be recommended for EMC practice, except where the frequencies of concern for either emissions or immunity are very low indeed (say, less than 1MHz [36]) and/or the signals concerned are only slightly sensitive or slightly interfering (Classes 2 or 3 in section 7.4.1). In any case, keep all pigtails less than 30mm long – just long enough to assemble without too much difficulty.

Once suitable tools are available to assist with proper 360° screen bonding, pigtails may be found to be more expensive and time-consuming to assemble, and less attractive to installers.

7.2.4   Which end to earth?

This section describes why it is that cable screens should nowadays be bonded to earth *at both ends*, in general, and how to prevent the resulting earth loops from causing problems.

7.2.4.1   Single end versus both ends

Cable screens should generally be RF bonded to their local earths at both ends, but to do this without suffering cable damage due to heavy circulating currents (often called "earth loop" or "ground loop" currents) requires the installation of equipotential mesh earthing systems (MESH-BNs, see section 5.1.3.3 on page 108) and the routing of the cables along high-conductivity parallel earth conductors (PECs, see section 7.4.3) such as conduits, for their entire path [61], [97], [74], [77], [130], [59], [73].

Even though cable screens have a high resistance compared to other elements of the MESH-BN, and cannot carry heavy currents, when bonded to earth at both ends they improve the MESH-BN because there are many of them [74].

Traditionally, building earth structures have only been bonded to achieve personnel safety. It has been common practice (and is still recommended by some trade associations) for the earthing of a site to be improved only up to the point where the touch voltages do not exceed 50V rms. This is known as equipotential bonding, but it is clear that it is not equipotential enough to meet modern requirements for equipotentiality for EMC purposes (refer to section 5.1.3.4). This traditional practice also required cable screens to be bonded at one end only, because where large potential differences exist between earths, bonding cable screens at both ends has been known to cause problems with cable overheating.

Compliance with national wiring installation rules (e.g. BS 7671 in the UK) which limit touch voltages (e.g. to 50Vrms), can still allow higher voltages to be experienced during an earth-fault, when it is not uncommon for touch voltages to rise momentarily to 160V [26]. Electrical storms and other high-energy transient surges in the earth network can cause this voltage to rise momentarily to considerably more than 160Vrms. These high voltages can cause serious damage to electronic circuits not designed to cope with them, and could give a nasty shock to any personnel subjected to them.

Bonding screens to earth at only at one end provides no screening protection from certain orientations of magnetic fields, which require a current to be driven along the length of the screen to provide the screening effect. This applies both to emissions from the cable, and its susceptibility [73].

The final problem with bonding screens at one end only is that the unbonded end will provide a "window" in the overall high frequency shielding, through which external fields will couple with the inner conductors. Screening must be complete for the entire route of a signal [28], [73], [129]. Even small discontinuities in the screening, e.g. holes, and lack of 360° screen bonding (such as the use of pigtails) can degrade screening effectiveness unacceptably – as we have seen in 7.2.3. Traditional practices were not so concerned about screen bonding at both ends, or the use of pigtails instead of 360° bonding, because the frequencies in common use were not as high as they are today.

With the high data rates commonly used these days, and the worsening of the electromagnetic environment, it is no longer acceptable to waste the EMC performance of expensive screened cables by only bonding their screens at one end, or using pigtails instead of 360° screen-bonding techniques. Transient interference poses a further threat; in a paper describing the advantage of screened ribbon cable [45], the author describes in wry fashion some of the ESD susceptibility tests that were performed:

> First, the shield was brought very close to the connector, then cut off, as might be done to avoid ground loops. The voltage on the center conductor [during an applied 10kV ESD] was estimated to be far greater than 500 volts. After this test, our oscilloscope was returned to the manufacturer for repairs.

Present day best-EMC-practice for bonding cable screens is as follows:

- use equipotential 3-dimensional mesh earthing techniques (MESH-BN) to reduce the earth potential differences between electronic units (that need to communicate over metallic conductors) to < 1V rms;
- 360° bond cable screens to the local earth at both ends, preferably directly to the wall or backplate of equipment cabinets (see 7.2.3);
- run all cables (whether screened or not) very closely along elements of the MESH-BN earth system, using them as Parallel Earth Conductors (PECs), along their entire run.

PECs can also be constructed to provide significant assistance with emissions and immunity (and crosstalk between cables) at frequencies much higher than 50/60Hz, potentially allowing the use of lower-cost, lower-specification cables than would otherwise have been necessary.

7.2.4.2   Special cases

Co-axial RF cables must always have their screens bonded at both ends, and will always generate a high-frequency common mode (CM) current path between their outer conductor (braid) and earth [59]. This is because the braid of a co-axial cable carries the return current of the signal carried by the centre conductor, and is why twisted-pair or twin-axial cables (which do not use their braid to carry return signal currents) are preferred for LF and wideband EMC and signal integrity reasons [97]. To reduce the emissions and immunity problems caused by the CM currents of co-axial cables they should always be routed close to a PEC (see section 7.4.3), and it may also be necessary to bond their screens to the PEC at irregular intervals averaging every 5 to 10 metres for long cable runs to prevent the CM currents leaking from the braid from becoming too troublesome.

If a cable screen is not bonded to local earth at both ends, this should be for a very good technical reason, usually because the manufacturer of an EMC-compliant apparatus has specified this deviation from general good practice, and not because of some traditional proscription against earth loops. Where cables interconnect units of electronic apparatus it is normal to bond their screens to the cabinets at both ends. Screens bonded at one end only are usually only associated with certain models of transducers (often plastic-bodied ones) that contain no electronics.

LAN connections

When a network cable is run from one equipotential area to another (e.g. between IBNs), if the cable screens cannot be bonded at both ends because of the resulting high levels of circulating currents in the screens, then galvanic isolation must be used. Measures are also required to make sure no-one can touch the unearthed end of the screen (e.g. insulation to the maximum potential transient potential). Surge protection devices (SPDs) may also need to be fitted between the signal conductors themselves and their cable screen ("in-line" types), and from cable screen to the local earthing terminal (parallel types). Such SPDs will certainly be required where such a cable goes between buildings (even underground) [17]. For coaxial (BNC) screened cables, the RF earthing is maintained along with the isolation by capacitively coupling the screen (outer) terminal to the equipment case.

ECMA 97 recommends that LANs be isolated from earth [34], because they cannot be guaranteed to stay in one equipotential area, and recommends the use of galvanic isolation barriers rated at >1500V and preferably 2kV, but states that 500V rated isolation may be used in specified circumstances. Since ECMA 97 was written the best-practices have shifted away from the use of IBNs inside buildings, and now we can see that when we have a proper implementation of a MESH-BN throughout a building it is preferable to bond LAN cable screens 360° at both ends, and run LAN cables closely along elements of the MESH-BN at all times, using them as PECs.

7.2.4.3   Alternatives to bonding cable screens at both ends

In some instances it is impossible or undesirable to bond cable screens at both ends. There are a few alternatives.

Capacitive bonding at one end of a cable screen

Capacitor bonding to earth at one end of the cable screen, with direct RF bonding to earth at the other, is sometimes suggested. This avoids the need for equipotential bonding, MESH-BNs, or PECs, by providing RF bonding at both ends without creating troublesome "earth loops". Capacitor values of 1nF to 100nF are commonly used for this technique, with the lower values being better for higher frequencies but less effective at low frequencies.

Where no true equipotential bonding is available, capacitive screen bonding is an option [28], but you should be prepared to use adequately engineered application-specific components for this, rather than just connecting wired capacitors to earth. This is because it is difficult to assemble a wired capacitor between the screen of a cable and the earth and still have it operate effectively over a wide range of frequencies.

This technique also suffers from the fact that where the cable is not run over a well constructed MESH-BN over its whole route, it does not help to prevent the unbonded end of the screen from flashing over, or protect the electronics from overvoltage damage, during earth-faults or electrical storms. Also, the capacitors used will need to be rated for the full fault and surge voltages expected, which are likely to be high where the earth bonding is poor. Measures are also required (e.g. adequate insulation to cope with lightning surges) to make sure no-one can touch the unbonded end of the screen. Of course, creating a true equipotential MESH-BN or SRPP over the whole route of the cable, and running the cable close to its elements at all times, will remove the problems of flash-over and exposure of electronics and people to high voltages, but then the screen may as well be bonded to earth at both ends anyway, since the additional earth loop created is more likely to be an advantage overall than a problem.

So capacitor screen bonding cannot in general be recommended as a best practice for either EMC or safety. However, it may prove adequate in some situations when carefully implemented by an experienced engineer aware of all the issues. It may also be a helpful remedial technique for dealing with specific interference problems on a pre-existing "traditional" installation which does not have equipotential bonding or PECs, and where cable screens had been bonded at only one end.

Galvanic isolation

The only wholly commended alternative to the use of PECs and bonding cable screens both ends, is to use galvanic isolation barriers rated at full fault/surge voltage. Fibre-optics, wireless, microwave, laser, and infrared (e.g. IrDA) communication techniques are very suitable for communicating data without metallic connections at all. All of these techniques are becoming more readily available and their cost is falling.

In some instances, manufacturers of wireless data systems are claiming that the overall cost of installing their system on an industrial site is less than the cost of installing a single cable. Wireless systems should still be treated with caution for other EMC reasons, though. Any radio-based system is inherently more susceptible to blocking by many different sources of radio interference. Intelligent digital signal processing can mitigate but not remove this problem.

Fibre-optic communications are probably the most EMC-friendly of any of the available options, capable of supporting almost limitless common mode potential differences. A minor disadvantage is that the electronic receivers and transmitters at each end may add a weak point for interference susceptibility and unreliability.

When using fibre-optics, care must be taken with any cables that use metal elements, such as for strengthening or armouring. Their metal elements may accidentally compromise the creepage and clearance distances required for safety, and/or may arc over during surge events and cause intense (though localized) high-frequency disturbances and/or fire risks. Such metal elements should be cut, isolated, and protected where they cross the boundary of an equipotential zone.

Infrared communications tend to be line-of-sight, and microwave and laser communications are always line of sight, so if they are used thought must be given to the possibility that the communication path might be blocked, for example by a maintenance worker.

7.2.4.4   The problem of earth faults

Earth loop currents are mostly caused by voltage differences in the earth structure due to heavy power consumption, so are generally at 50 or 60Hz and related harmonics. The low impedances of PECs at low frequencies diverts most of earth current away from the cables' screens, allowing RF bonding of cable screens at both ends without fear of cable overheating.

Magnetic fields are another source of earth loop currents. These currents are reduced to negligible amounts by running the cables concerned very close to their PEC. "Close" means no more distant than 25mm from the base of a tray, duct, or conduit or no more than a few millimetres away from a dedicated green/yellow earth wire or copper strip being used as a PEC.

A problem frequently encountered in practice is the effect of earth loop currents caused by earth faults, that is, when a conductor that is earth-bonded at one point but is intended to be isolated from earth along the rest of its length, becomes linked to earth at some other point through an insulation failure (Figure 7.14). This can happen either to screened cables, passing near an earthed structure where the jacket is abraded or pierced, or to the neutrals of supply cables which are not RCD-protected. (It can also happen to live conductors, of course, but in that case the fuse or circuit breaker should operate, which alerts maintenance personnel to the problem immediately!)

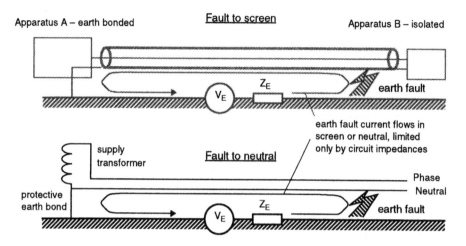

Figure 7.14 The earth fault

The effect of an undiscovered earth fault can be pernicious. The current that flows in the conductor which is faulted depends on its impedance and the earth potential difference at the position of the fault. This may be low enough that, although a fault current flows, its effect is negligible and the fault is never manifest. On the other hand, it can be several tens of amps, in which case it is usually discovered as a result of the cable damage. In between, though, there are several potential hazards:

- longer-term fire hazard: the cable heats up enough to present a fire hazard, but not enough to signal its presence quickly;

- magnetic field interference due to the high current flow over a large loop area – the "wobbly monitor" effect is a typical symptom of this sort of fault [49];

- induced interference in telecommunications systems if the affected cable is part of, or close to, such a system.

Some of these effects are quite difficult to diagnose on-site, and result in several man-days of wasted and expensive effort. In some cases, the potential for the problem is so acute that a system will be deliberately designed to be isolated from earth and incorporate an integral earth fault detector, so that maintenance engineers can be alerted to a fault occurring and trace and correct it quickly. The use of a properly installed MESH-BN will reduce the impact of such faults – by reducing or eliminating the earth potentials that are the root of the problem – and often will result in an installation being able to tolerate much higher levels of fault occurrence.

7.3   Unscreened cables

Many cable-coupling problems are better addressed without screening the cable, which is after all an expensive and inconvenient option from the installer's point of view.

Instead, the equipment designer implements filtering and other forms of interface protection so that, whatever interference is present on the signal conductors at the cable terminal, the equipment can deal with it. In this respect common mode protection is particularly important.

Two particular types of unscreened cabling have relevance to EMC solutions: these are twisted pair (often referred to as Unshielded Twisted Pair, UTP) and ribbon cable.

7.3.1   Twisted pair

7.3.1.1   The purpose of twisting

At various places in this book we have emphasized the importance of ensuring that a signal and its return are closely coupled in an adjacent conductor pair. The closer the coupling, the less interference is induced within the differential circuit. Twisting the pair together improves the situation in two ways:

- the magnetic field coupling with the pair is continually changing orientation. If the cable is viewed as a succession of half-twists, the coupling with each successive half-twist will cancel that from the previous one;
- electric field coupling is more balanced to each half of the pair than if the pair were untwisted.

The number of twists per unit length may be significant if the cable is passing through an area of rapidly varying field strength, since the reduction in coupling by twisting depends on the area of each half-twist being small by comparison to the field variation. In an unknown disturbing environment, a high rate of twists per unit length is desirable [24]. The overall performance of a twisted pair cable relates to its "balance", which in turn is determined by how tightly controlled the cable geometry is – uniformity of twist rate, cross-section and dielectric material. This implies that the handling during installation of such cables can noticeably affect their performance, particularly in respect of bending and crushing, which changes the geometry at particular points.

7.3.1.2   Coupling to and from twisted pair cables

The crosstalk coupling to a twisted pair has both capacitive and inductive components. The equivalent circuit (Figure 7.15) depicts the capacitive coupling as a current source I_C onto each conductor half-twist while the inductive coupling is a voltage source V_I in series with each conductor half-twist. Each elemental V_I pair cancels the next one, leaving only the induction at the ends of the circuit un-cancelled.

The effectiveness of twisting a signal/return pair is not achieved in isolation from the signal circuit carried by the pair. It depends on the impedance and the balance or unbalance of this circuit. For unbalanced circuits, capacitive coupling dominates at high impedances and there is little reduction in overall coupling by twisting. As the circuit impedance drops so differential capacitive coupling reduces and the inductive part becomes dominant, so that twisting becomes progressively more beneficial. Twisting together power conductors (circuit impedances of a few ohms) is therefore good practice.

Balancing the circuit relegates the capacitive coupling to a common mode effect, since the I_C induced onto one conductor appears in the same sense and with the same amplitude as that induced on the other, so that the circuit's common mode rejection works to maximum effect. The crosstalk is then determined mainly by residual inductive coupling. This will be sensitive to an odd or even number of half-twists, or

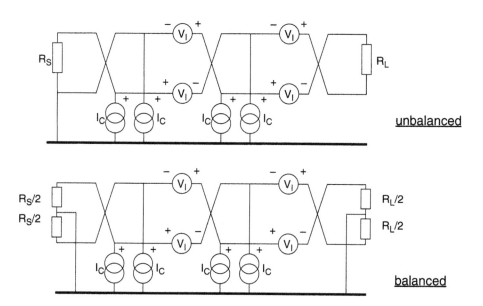

Figure 7.15 Coupling model for twisted pair

more properly to the differences in voltages induced in the enclosed area at each termination.

Twisting the pair does nothing directly to reduce common mode coupling with the cable; the common mode currents that are present will radiate just as effectively, and external fields will induce common mode currents just as effectively, as if the cable was untwisted. The key aspect which is improved by a good twisted pair is the conversion from common mode to differential mode or vice versa, i.e. the LCL (see 7.1.2.1).

7.3.1.3   Structured cabling and the UTP/STP debate

The proliferation of Local Area Network and other data communications applications in the last decade has resulted in a specification for cable performance, originally in EIA/TIA 568 and more recently in ISO/IEC 11801 [109] and EN 50173 [72]. These documents refer to "Category 3" and "Category 5" cables, with further categories in preparation. The categories define performance specifications for particular applications, Cat. 3 referring to operation up to 10Mbps and Cat. 5 up to 100Mbps. Key cable characteristics are near-end crosstalk, attenuation and impedance up to 100MHz. The purpose of these specifications has been to allow a building to be installed with "structured cabling" without a detailed knowledge of the applications that the cable system will support in future. These requirements can be met by UTP cables and this has allowed the implementation of low-cost LAN installations – in fact it could be said that the large installed base of UTP cabling has driven the design of successive generations of LAN architectures – but there is as yet no specific requirement for the EMC performance of the cable system.

Since installations nowadays do have EMC compliance issues, this situation has led to a rash of claim and counter-claim by cable suppliers, that a particular cabling system has proven EMC characteristics and is better than some other system in this respect. This has particularly surfaced in the argument over whether shielded twisted pair (STP) is better or worse than unshielded. In practice, the argument is irrelevant: an equipment designer has to ensure that his product is compliant when connected to a particular cabling system. If that system requires shielded cable, then it is unreasonable to expect a UTP installation to achieve the desired compliance status. If the equipment has been designed to use only unshielded cable, then it will be impossible to terminate the cable shield correctly and a shielded cable installation may well be worse than a proper unshielded one, especially if inadequate earthing arrangements are made ([28], and see Chapter 5). Although it might seem that screened cable will always be better for EMC purposes, this hinges on the methods used to install it. Apparatus that has been specifically designed for a UTP interface can be perfectly acceptable, provided that the cabling system itself is properly installed.

7.3.2   Ribbon cable

Multi-conductor ribbon cable is a very common and popular method of running multiple signals, particularly parallel data buses, between and within apparatus. Its popularity is mostly due to the ease with which it is terminated, using insulation displacement connector (IDC) techniques. Its correct use inside an equipment enclosure is important to the equipment designer but less so to the system designer. But when it is brought outside an enclosure, difficulties can arise.

The disposition of the conductors in a ribbon cable is rarely within the control of the system designer, but it has a serious impact on the interference coupling to and from the cable. The issue is, as always, the enclosed loop area for signal and return currents. All too often, the circuit 0V, which forms the return for all the wideband signals in the cable, is relegated to a single conductor, which in the worst case is located on one edge of the ribbon. Figure 7.16 shows why this is the worst case: for the example of a 50-way ribbon of 1m length, the loop area for the signal on the opposite side of the ribbon is 635cm^2 – which makes a mockery of other good EMC practices such as a well laid-out PCB!

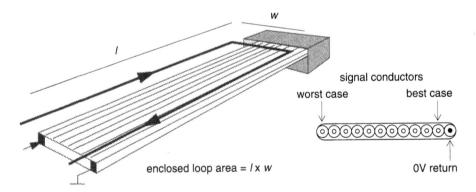

Figure 7.16 Loop area issues in ribbon cable

Leaving a large area between signal and return is inviting differential mode coupling into and out of the ribbon and hence immunity and emissions problems. Bringing such a ribbon cable outside a protected enclosure is not advisable. If it is going to be run within a cabinet, all the precautions advised in section 6.1.2 on routing within cabinets are essential.

Improvements in EMC performance of ribbon cables can be achieved by three methods:

- multiple 0V return conductors adjacent to each critical signal conductor (Figure 7.17(a)). This is easy to implement at the equipment design stage but not thereafter, and is costly (potentially doubling the number of conductors) and difficult in pin-limited applications;

- using ribbon cable with an integral ground plane layer (Figure 7.17(b)). This is a better electrical solution than (a), but is difficult to implement adequately in production since a mass-termination method of connection to the ground plane layer is needed. It is generally unpopular;

- fully shielded ribbon cable. If properly terminated this can give good EMC performance, but all the issues relating to screened cable become relevant. The manufacturers of shielded ribbon cable normally use the foil-and-drain wire method which forces a "pigtail"-style termination, compromising the available shielding performance. Also, shielding is least effective at the edges of the ribbon, so the most critical signals should be placed in the centre [45];

- "round and flat" cable rolls a ribbon cable up into a tube and sleeves it with an overall braid screen. Externally it looks just like any other multi-way screened cable, but it allows the use of mass-termination connectors with a proper 360° screen termination and so avoids the "pigtail" problems of flat screened ribbon cables.

If any of these precautions are implemented, external runs of ribbon cable are useable. But bearing in mind their difficulty, all in all, ribbon cable for wideband data transmission outside a protected enclosure should be avoided if at all possible.

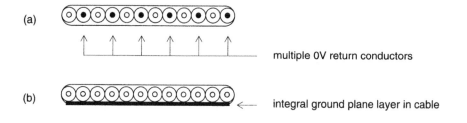

Figure 7.17 Ribbon configurations for improved EMC

7.4   Installing cable systems

7.4.1   Cable classification

Cables can be divided into four Classes with respect to their interference potential. These classes must be physically segregated and preferably run in different trunkings as described in section 7.4.2. Their trunking must keep the cables close to a parallel earthing conductor along their whole length, and if cable classes must cross they should only do so at right angles.

7.4.1.1   Cables CLASS 1 – for very sensitive signals

CLASS 1 generally applies to all low level signals such as thermocouples, thermistors, RTDs, strain gauges, load cells, microphones, etc.; also high-rate digital and analogue communications such as Ethernet, video, RF receiver cables from antennae and all other signals with full-scale range less than 1V, 1mA, or where their source has an impedance higher than 1kΩ, or their signal frequency or rate is greater than 1MHz.

CLASS 1 cables must always use good quality highly-screened twisted-pair (or similar) cables and shielded connectors with no breaks in 360° screening and no screw terminals. Cable screens may be terminated to backplates using clamps, and/or to the walls of shielded cabinets using glands, as long as the cable screen remains unbroken over the entire signal path. Where CLASS 1 cables are closer than 1 metre to medium and high voltage power cables (1kV and above), they should be segregated within a covered metal conduit.

Always ask the manufacturers of the electronic units using such sensitive signals for their recommendations for cable types, connectors, and wiring routes, and make sure that their recommendations are followed exactly.

7.4.1.2   Cables CLASS 2 – for slightly sensitive signals

This covers ordinary analogue (e.g. 4–20 mA, 0–10V, less than 1MHz) and low-rate digital bus communications (e.g. RS232, RS422, RS485, Centronics). These signals should use screened cables. Flat ribbon cables must be screened either with flat screening jackets or screened "round and flat" (a flat cable is rolled up in a round cable with an overall braid screen).

CLASS 2 also covers digital (i.e. on/off) inputs and outputs such as limit switches, encoders, non-data-bus low rate signals, etc., and the outputs of internal DC power supplies (but not the DC links of power convertors or motor drives, which are CLASS 4). For these, use screened cables, multiconductor cables, flat cables, or even single wires.

Externally supplied AC and DC power distribution (up to 230/415V) may be treated as CLASS 2 after it has been internally filtered in the cabinet. For such conductors single wires may be used, preferably twisted together, but in any case always run all the phases and the neutral associated with a load in the same bundle. (To allow a mains cable to be treated as CLASS 2 the supply filters used must be adequate for the full range of supply interfering phenomena and mounted properly. Refer to section 8.2 on the use of filters.)

Twisted pairs/triples/quads/etc. are always preferred for both screened and unscreened cables for CLASS 2, but be aware of the problems that can occur where an electronic unit only provides a single return terminal or connector pin when there are a large number of return wires. You may have to make special arrangements for connecting all the return conductors to the return terminal.

7.4.1.3   Cables CLASS 3 – for slightly interfering signals

CLASS 3 covers externally supplied AC (≤ 230/415V) or DC power which does not also supply noisy external equipment, e.g. welders, drives, power convertors, etc. (External equipment must be considered to be noisy unless it is known to have emissions in line with harmonized EMC emission standards. Such units will generally be fitted with internal supply filters.)

CLASS 3 also covers control circuits with resistive and inductive loads where all inductive loads (e.g. the coils of solenoids and contactors) are fitted with suppressors recommended by their manufacturers.

Induction motor (e.g. squirrel cage) cables are CLASS 3 providing they are on/off controlled or run from adjustable speed drives with sinusoidal output filters, and providing their AC supply is itself CLASS 3 or better.

CLASS 3 signals and power can use screened cables, multiconductor cables, or single wires. Screw terminals are fine as long as exposed conductors are less than 30mm long.

As for CLASS 2, twisted pairs (or triples, quads, etc.) are always preferred.

7.4.1.4 Cables CLASS 4 – for strongly interfering signals

CLASS 4 applies to the cables associated with the power inputs and outputs and DC links of adjustable speed motor drives (DC, AC, steppers, servos, etc.), welding equipment, and similar electrically noisy equipment.

Noisy equipment may be known to be noisy, or there may be some doubt which may be reinforced if it is clear that some or all of its power inputs and outputs are unfiltered. Always check with the drive (or other) manufacturers whether they specify or recommend special cables, routing, or other details such as ferrite toroids/clips or filters – and then always follow their advice exactly.

Cables to on/off controlled DC motors or slip-rings are also CLASS 4, although sparkless types such as pancake motors, or motors fitted with spark suppressed rotors or integral filters could be CLASS 3.

All the above CLASS 4 cables should use good quality braid screened types with properly made screen connections. Do not use pigtails for CLASS 4 screened cables.

The conductors in all CLASS 4 cables should preferably be twisted pairs, triples, quads, or as required, to twist together all the send and return current paths associated with one signal or load. Screw terminals are generally fine as long as no exposed conductors are more than 30mm long, but some manufacturers' instructions may provide more stringent requirements, so always check.

CLASS 4 also covers control cables to unsuppressed inductive loads (relays, contactors, solenoids, etc.), which may use unscreened twisted pairs but only as long as they operate infrequently (every few minutes or longer). If they operate more frequently than every few minutes they should either be suppressed, or use screened cables as above.

CLASS 4 can include all LV supplies which also power external noisy electrical equipment (e.g. drives, power convertors, welding, etc.) which may not meet their relevant harmonized EMC emissions standard, or which lack supply filters. These may use unscreened cables or bundles of single wires containing all associated phases and neutral. (The use of unscreened cables for CLASS 4 supply cables is forced on us by practicality and by historical installation practices, not because it is acceptable for EMC.) Unscreened Class 4 power cables should always run close to a PEC, which can be the cable armouring where this is available.

Cables to RF transmitting antennae are also CLASS 4, and must only be exactly as specified by the manufacturer of the electronic unit concerned, for cable type, termination, and routing. For such RF cables, no breaks in their 360° screening are allowed.

7.4.1.5   Cables CLASS 5 and 6 – for MV and HV

Cables carrying AC or DC supplies at voltages higher than 400/230V are commonly called High Voltage (HV), although more correctly these should be divided into Medium Voltage (MV) and HV according to the IEC's definitions. MV and HV cables are much more exposed to external disturbances such as lightning and powerful transients and surges caused by the operation of heavy-duty switchgear, than are LV cables.

MV power distribution cables are CLASS 5, and HV power distribution cables are CLASS 6. The types of cables used are defined by the power network engineers, and rarely by EMC engineers. However, running all the conductors associated with a given load in a single cable or bundle of cables, and using its armour as a PEC, is possible where load currents allow, and will help ensure that these cables don't cause unnecessary disturbances.

7.4.1.6   Some details common to all cable classes

All the conductors associated with a signal or a load must always be run together in a bundle over their whole length, even if this uses more copper (see section 7.4.2.3 on cable routing below).

Each multi-conductor cable, or bundle of single wires, if not armoured or screened, should include an earth conductor bonded to the local earths at the equipments to which the cables are terminated, to create a PEC. (Armoured cables can also use their armouring, although it is a good idea to include a separate earth conductor as it is less likely to be disconnected during future modifications.) This additional earthing conductor acts to reduce magnetic coupling at low frequencies, and also helps to reduce crosstalk.

When bundles are run over properly bonded MESH-BN PECs in a properly meshed site earthing system, this earth conductor probably need not be any greater cross-section than the other wires in the bundle. If the PEC or rest of the earth structure is lacking in some way, this earth conductor will provide a PEC (at least for the lower frequencies) but will need to be sized appropriately for the continuous and surge currents it may be called upon to handle.

As has been mentioned before, screened wires must use 360° RF bonding (described above) at all connectors and glands, unless their signals really are non-aggressive and insensitive, when very short "pigtail" screen terminations may possibly be acceptable.

7.4.1.7   Signal cables from PLCs and computers

PLC and computer digital control and analogue I/O signals may be thought to be benign because the signals they carry are slow or very infrequent. However, if poorly designed they may carry electrical common mode noise generated by the operation of their internal digital processing, and this has a much higher frequency content than would be expected from the signals themselves (see Figure 7.6 on page 161).

So, unless the manufacturer or supplier of the equipment or systems clearly states otherwise (and you have confidence in him), always treat them as Class 1 cables, using 360° screened cables along their entire length with 360° RF bonding at all connectors and glands.

7.4.1.8   External cables where powerful emitters are nearby

Where the electromagnetic environment suffers from a high-level electromagnetic field, for example from a nearby radio or TV broadcast transmitter, medical diathermy, plastic welding, induction heating or similar apparatus, additional measures may be required. These strongly polluted environments may only exist for a small area near to an item of equipment, and the best approach is segregation – avoiding placing apparatus or their cables in these areas altogether. Where this is not possible, the following methods may be used.

Using better quality screened cables

All signal or power cables that have to be present should use an additional screening layer which is 360° RF bonded at all joints, connectors, or glands. This is regardless of the cable Class based on the signals or power carried by the cable itself, so a Class 1 cable would require a double-screened cable. (Double screens do not need to be isolated from each other, except for the most demanding performance. A braid screen simply wrapped over a foil screen layer can make a large improvement in a cable's screening effectiveness.)

Using covered cable ducts or conduits

If the cables are only run through this environment in rectangular ducts (used as PECs) fitted with conduit with RF-bonded covers, or in circular conduits, all RF bonded using techniques appropriate to the frequencies concerned (see section 5.2.2.4 on page 119), then it may be possible to use the normal cables for their Classes. The covered duct or circular conduit acts as another screened layer to the cable.

Circular conduit with 360° bonded joints and terminations can provide the most excellent screening performance at almost any price, so this is a remarkably cost-effective technique for extreme electromagnetic environments.

Using additional bonds between cable screens and PECs

Another technique, where the strong fields concerned are below 20MHz, is to expose an inch or so of cable screen or armouring every 10 metres or so, and bond these exposed places to the PECs the cables are following. The distances between these bonds should not be regular, and should be randomly varied between (say) 6 and 12 metres.

Capacitors can be used instead of direct bonds, to help keep 50/60Hz currents out of the cable screens, but this should not be necessary where proper earth mesh equipotential bonding has been followed throughout PECs and the site.

Measures such as this have been used as a remedy to protect computer systems from electromagnetic disturbances created by nearby electric train services and/or marshalling yards. (Electrified railway systems usually produce their maximum radiated disturbances in the 1–20MHz region.)

Although time-consuming and costly, as a remedial measure 10-metre screen bonding may be much easier to apply than the above two techniques. It suffers from the problem that the exposed cable screens or armour may allow water ingress or corrosion, so steps may need to be taken to seal these bonding points. Where the electromagnetic environment contains high levels of higher frequencies, screen bonding every ten metres may not be frequent enough. But bonding more often than this starts to be less cost-effective and the other two techniques above become more appealing.

7.4.1.9   *What if you can't easily tell what Class a cable belongs to?*

Sometimes there may be some doubt about the "noisiness" of an item of apparatus or its cables. This often occurs when a new apparatus is being installed in a pre-existing site which already has apparatus and cables in place. A simple and quick test with a portable spectrum analyser and hand-held close-field probe is all that is required to identify electrically the "noisier" cables and apparatus. The probe (or probes) and analyser should cover the frequency range from 10kHz to 100MHz, at least.

The above four cable classes are relative, not absolute, and field-strength values cannot be given for the close-field probe results. However, a little time spent with the probe will soon show which cables are to be kept well away from the others. The problem with "noisy" cables is that their function may not clearly reflect their disturbance capability, especially power cables, and this is where this close-field probe method can save a lot of time and difficulty.

Unfortunately, it is not so easy to identify the *sensitivity* of cables with a portable probe (it can be done, but takes much more time and effort). Happily, most site electrical engineers will know what the cables are for, and cross-checking against the functional lists provided for Classes 1 and 2 should soon identify them.

It is quite likely that, in existing installations, different cable classes will be mixed up together and not segregated. This does not mean it is adequate to continue in this fashion. New cables should be segregated from each other, and from existing cables. The additional time and cost will be well spent when, once every few installations, it prevents unreliable operation.

7.4.2   Segregation and routing

7.4.2.1   *Physically segregating cables by their classes*

The purpose of determining which Class a cable belongs to is to be able to choose the correct cable type and terminations, but it is also so that different classes of cables may be run segregated from each other to prevent them from interfering with each other [21] (crosstalk, see 7.1.3). Figure 7.18 shows the minimum separation distances that should be maintained between the different cable classes.

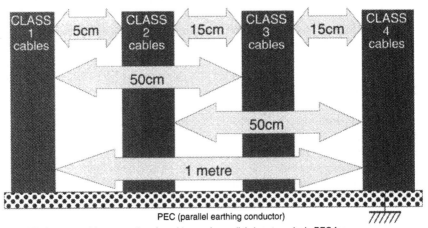

NB: these are minimum spacings for cables run in parallel close to a single PEC for runs of up to 30 metres; for longer runs multiply these spacings by L/30 (L in metres)

Figure 7.18 Minimum spacings between cable classes when run over a single PEC

This assumes a continuous flat metal PEC under them all. Where separate PECs are used, a larger gap between them is required, although this may be offset by the use of PECs with side-walls (e.g. trays, open ducts), the taller the side-wall, the better. In the absence of a nearby PEC (not advised), the spacings between two numerically adjacent classes of cable should be at least 10 times the diameter of the larger bundle [97].

Where the PECs used completely enclose the classes of cables (e.g. covered ducts, circular conduits) and the PECs are correctly RF bonded along their length and at both ends, there is no minimum spacing between PECs. Even so, it is still best practice to arrange the ducts or conduits in order of cable class so that Class 1 is always the furthest from Class 4.

When covered ducts are used for their EMC performance, it is important to ensure that all covers are replaced and properly RF bonded after any maintenance or modification work. Similar checks are needed for any part of the earthing structure or PECs (have all the multiple fixings at a joint been re-fitted? is the cable armour still 360° bonded at all joints?). This is because many engineers may not realize that these metal structures are doing a lot more than merely providing mechanical protection or support.

Remember that all cables must run close to their PECs at all times (see section 7.4.3 for PECs). "Close" means touching or within a few tens of millimetres – certainly not as much as 100mm away unless the PEC is a large circular or covered conduit containing only one class of cable.

If cable classes must cross, they should only do so at right angles with as much space between them as possible. It is best to try to avoid cable class crossings wherever possible, or else arrange for a PEC to run between them, this is especially important when classes 1 and 4 must unavoidably cross (in which case enclosing one in a circular conduit or covered duct is not unreasonable).

Narrow metal ducts or conduits (or circular conduits) should only contain one Class of cable, if they are too narrow to maintain the minimum spacing between classes inside them.

The above figure only shows the first four cable classes. Class 5 (for MV power distribution cables) should be segregated from Class 4 by 150mm at least (and from Class 1 by 750mm minimum), and Class 6 (HV power distribution) should be a further 150mm away from Class 5.

Remember, all these spacings are minimum values, for cables run for 30 metres over a common PEC. Where Class 5 or 6 cables are closer than 1 metre to Class 1, the Class 1 cables should be run in a covered metal duct.

7.4.2.2   Segregation within Classes

So far, this discussion of segregation has assumed that all the cables in a Class may be bundled together, but this may not always be advisable, especially for the more extreme Classes. Sensitive analogue Class 1a cables should not be bundled with high-rate digital signals in twisted-pairs (Class 1b) and neither of them should be bundled with high rate digital in co-axial cables (Class 1c). It is recommended that these sub-classes be bundled separately and not run too close to each other (separation of at least 10mm between each pair in the order: 1a, 1b, 1c), keeping each bundle as close as possible to the metal surface of the PEC at all times.

Different Class 4 cables may also require individual routings. The cables from adjustable speed inverter drives to their motors are often specified by the drive manufacturers to have 600mm or more spacing from any other parallel run of cable. Quite how this should be interpreted when running a dozen drive cables along the same general route is something that should be taken up with the drive manufacturers. It is difficult to make general rules for segregation within Class 4, because a cable may be very noisy in its own right, but still sensitive enough to pick up sufficient interference from neighbouring Class 4 cables to upset the electronics it is connected to. This is another reason why the use of sinusoidal output filters for inverter drives might be preferred: such motor cables become Class 3 and may be bundled together with each other and with other Class 3 cables with few worries, as long as the drives are compatible with the filters.

7.4.2.3   Routing

All cables between equipment units should follow the same route

All the cables between items of equipment should ideally follow a single route along a single PEC, whilst also maintaining their segregation by Class (see above for details of PECs and cable classes).

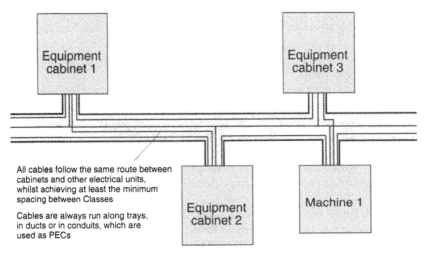

Figure 7.19 Installation cable routing

Figure 7.19 shows the two main principles of cable routing:

- the cables between two items of equipment must always follow the same route, and
- there should be a single interconnection panel for each item of equipment.

The single connection panel per unit helps prevent the inevitable circulating currents in the cables and metalwork from flowing through the apparatus and possibly causing interference with the internal electronics. Where there has to be a multiplicity of routes and/or connection panels, each should have its own PEC, and higher PEC earth currents should be expected.

Stacking cable trays along a route

Because of the minimum segregation distances required between cable classes, it is generally impossible to run cables of all four classes along one cable tray (they are usually not wide enough). Figure 7.20 shows how this is overcome by running a "stack" of cable trays. In this example each tray is carrying just two cable classes.

The cable trays are stacked vertically and electrically bonded together at all of their support pillars. They all follow the same route between two items of equipment. Where cables are segregated vertically in uncovered trays, as shown in Figure 7.20, their

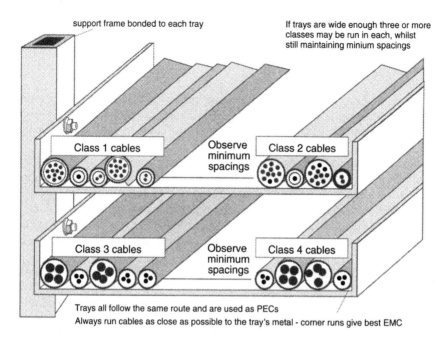

support frame bonded to each tray

If trays are wide enough three or more classes may be run in each, whilst still maintaining minium spacings

Class 1 cables

Observe minimum spacings

Class 2 cables

Class 3 cables

Observe minimum spacings

Class 4 cables

Trays all follow the same route and are used as PECs

Always run cables as close as possible to the tray's metal - corner runs give best EMC

Figure 7.20 Segregation of cable classes in trays

vertical separation according to their respective classes should equal or exceed the horizontal minimum spacings of Figure 7.18.

Running all send and return conductors close together at all times

Figure 7.21 shows the technique of running all the send and return current conductors involved with any signal or electrical load together along all the entire route (even where this uses more copper). This technique is important to reduce magnetic field coupling.

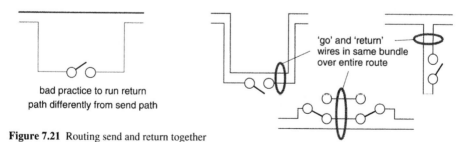

bad practice to run return path differently from send path

'go' and 'return' wires in same bundle over entire route

Figure 7.21 Routing send and return together

Often, this technique is not applied. For example, when the live lead to a load passes through a switch or relay, the neutral is usually routed directly or even "picked up locally" to the load. This is bad practice, as shown by the first example in the sketch. Send and return current paths must always be as close together as is possible, which is why twisted pairs are preferred for signal cables.

Of course, in this example the unswitched conductor cannot follow exactly the same route, since it does not pass through the switch, relay, contactor, or whatever. Even where double-pole switching is used (or triple-pole, for three-phase supplies) there will always have to be some physical path difference between conductors in the vicinity of the electromechanical elements. The idea is to minimize all path differences everywhere else.

Dealing with heavy power conductors

The send and return conductors of heavy power cables are usually not twisted together (as is preferred for signal cables) due to their size. Such cables are best laid side-by-side (over their common PEC) as close as they can be together, to achieve the maximum EMC benefits.

Unfortunately, where cables are carrying very heavy powers the mechanical forces on their copper or aluminium conductors can be strong enough to make them move enough to wear out their insulation. These mechanical forces are created by the very magnetic fields that placing the cables close together is intended to partially cancel. Because of this it may be necessary to place the cables further apart and forgo the EMC benefits. Extra segregation distances from sensitive equipment should then be provided.

A recent example of an 8000 amp DC drive (see section C.3 in Appendix C), where the send and return conductors were 30 metres long and 1 metre apart along their length, showed that standard VDU screen displays could suffer unacceptable distortion and movement at distances of nearly 30 metres from the cables. This could have had a strong impact on the siting of the control room, but in this case LCD display screens were used instead of VDUs because they are not sensitive to magnetic fields. Running the DC motor cables closer together would have significantly reduced their magnetic field emissions, but could have given the cables a reduced lifetime (and re-wiring costs were estimated at £100,000).

Connections to cabinets

There should only be a single connector panel for a cabinet. All external cables should enter a cabinet at only one side, rear, top, or bottom, and they should also enter the earthed backplate along one of its edges.

This is so that, in conjunction with the other techniques described here, the high-level circulating currents flowing in the long cables in many industrial situations will flow from cable-to-cable via the connector panel or backplate edge via the screen-terminations or filters mounted in that area, and will not flow through the rest of the cabinet or backplate structure where they might upset the electronic units.

7.4.3   Parallel Earth Conductor (PEC) techniques

Modern best practices for EMC in installations (according to IEC 61000-5-2:1998 [97] and prEN 50174-2 [73]) also require the use of cable trays, conduits, and even heavy-gauge earth conductors as Parallel Earth Conductors (PECs) to divert power currents away from cables and their screens. The cabinet and backplate should provide the means for the connection of the necessary PECs.

7.4.3.1   Constructing PECs

The first function of a Parallel Earth Conductor (or PEC) is to divert heavy earth loop currents from both screened and unscreened cables. Since earth currents are usually at 50/60Hz, and the surges from lightning events have most of their energy below 10kHz, it is enough for this purpose that the PEC has a very low resistance and a sufficient current-carrying capacity. Most cable support systems have enough metallic cross-sectional area to provide this low resistance and current capacity, especially on earth-meshed (MESH-BN) sites where earth currents in PECs are lower.

Any screen or earth conductor external to a cable should be treated as a PEC and bonded to earth at both ends. So, cable armouring should always be regarded as a PEC. It is vital to ensure there are no breaks in the electrical continuity of any armour used for this purpose. Cable installers traditionally regard armour merely as mechanical strengthening or protection, and may not be used to bonding it at joints and to the local earth at both ends.

To reduce the loop currents flowing in cables due to magnetic fields (which can be extremely intense in the vicinity of some industrial processes, such as induction heating or magnetic stirring), cables must be run very close to the metal of their PEC throughout the length of their run.

PECs can also control higher frequencies. Figure 7.22 shows a variety of types of PECs, and ranks them by high-frequency performance. (Compare this to Figure 5.2 on page 106.)

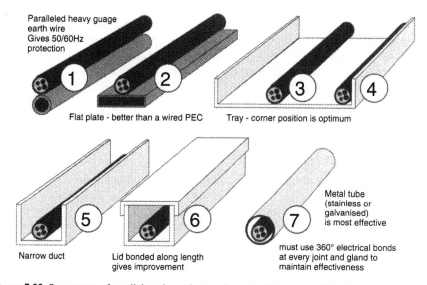

Figure 7.22 Some types of parallel earth conductors (in order of increasing HF effectiveness)

Cable trays are usually perforated with slots to make cable fixing easier, but these can detract from its high frequency performance (Figure 7.23). The problem is exactly the same as has been described in section 6.2.2 on shielding effectiveness: slots and gaps interrupt current flow and therefore increase the transfer impedance of the structure. Because of their open construction, ladder- and basket-type cable support systems are poor as PECs, except for the control of 50/60Hz disturbances. Even then they may not have a substantial enough metal cross-section, so a wired PEC may be required as well.

PECs must be electrically bonded to the local equipment earth at both of their ends, and to all their support structures and any other earthed metalwork at every available opportunity. This is so that they help to create a meshed earth structure, and it also helps the PEC to function effectively.

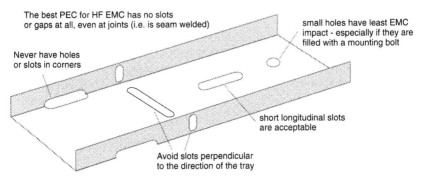

The best PEC for HF EMC has no slots
or gaps at all, even at joints (i.e. is seam welded)

small holes have least EMC
impact - especially if they are
filled with a mounting bolt

Never have holes
or slots in corners

short longitudinal slots
are acceptable

Avoid slots perpendicular
to the direction of the tray

Figure 7.23 Slots and gaps in structures used as PECs

Joints and end-terminations in PECs must be bonded using methods appropriate for the frequencies to be controlled (RF bonding techniques are described in section 5.2.2). Cable trays and rectangular conduits will need to make electrical bonds directly to the cabinet wall (or floor, top, or rear) using U-brackets or similar with multiple fixings. Round conduit can bond to the cabinet wall with circular glands, remembering to remove the paint first (to ensure a 360° bond) and apply corrosion protection. PEC conductors will need appropriately-sized and positioned earth terminals.

PECs may have to carry high continuous currents, and must be capable of handling the worst possible fault currents without overheating or other damage, so they must have an adequate metallic cross-section. Conductively coated plastic conduit or trunking will obviously not be adequate, for this reason, and if used will require a heavy-gauge copper wire PEC inside them to handle any heavy currents.

7.4.3.2   Using structural steelwork as a PEC

In a building fitted with an equipotential earth mesh any of the earth-meshed metalwork of the building may be used as a PEC (I-beam girders, building steel, etc.).

Because transfer impedance is affected by geometry, different positions relative to the structural member will have different high-frequency attributes, as shown by Figure 7.24. Running cables inside hollow metal structures is the best technique, but may be difficult to achieve and not easy to maintain. Running cables deep inside included angles on the outside of structural metalwork is a good practical alternative. Where there are too many cables for them all to be run in included angles, the noisiest should be run in one angle, and the most sensitive should be run in another. Less aggressive or less sensitive cable should use the less protected routes.

7.4.3.3   Providing electrically continuous PECs

Cables should run very close to their respective PECs at all points along their routes. Where a cable splits off from one cable route to take a different route, an electrically continuous PEC should be provided for it, and it should remain close to its PEC at all times. Figure 7.25 shows a variety of techniques for continuing a PEC where a cable enters or leaves a given cable tray. Similar considerations apply to other cable support types such as ducts or conduits.

7.4.3.4   Dealing with large gaps in trays and ducts

Breaks in cable trays and ducts often appear to be required (and sometimes for structural reasons are unavoidable) where cables pass through walls or floors. In such

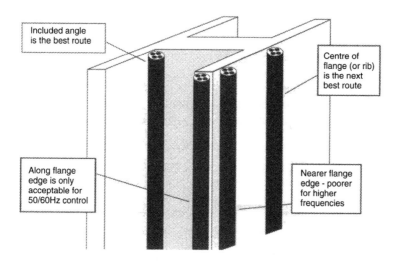

Figure 7.24 Using structural members as PECs (I-beam example)

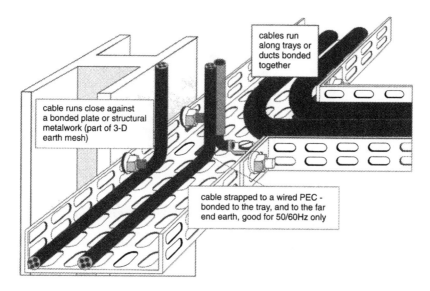

Figure 7.25 Branching PECs from the main structure

cases the PEC trays or ducts should run right up to both sides of the difficult region and the gap in the PEC bonded with at least one short wide braid strap or heavy-gauge earth wire (preferably more than one spread over the width of the tray). These bonding conductors (see section 5.2.2.4 on page 119) should be suitably rated to withstand the worst possible earth-fault or other current (which could be very high since equipotential mesh bonding is probably also restricted by the wall), and should follow exactly the route taken by the cables.

Such bonded-across gaps in PECs are weak points for high frequencies, so must be used with care. If high-frequency problems arise it may be possible to wrap metal "knitmesh" around the exposed lengths of the affected cables, bonding this additional overall screen to the cable tray (PEC) at both sides with a large P-clip, saddle-clamp, or similar. Different classes of cables (e.g. "noisy" and sensitive cables) should not be included in the same overall knitmesh. They should use different overall knitmesh shields, each bonded to the PEC at both sides of the gap.

Where it is difficult to install cable trays or conduits at all, a PEC may consist of heavy-gauge wire or a copper strip, strapped along the whole length of the cable. Such a PEC is only useful for controlling low frequencies, such as 50/60Hz and its first few harmonics, and the low frequency content of surges such as those from lightning.

7.4.3.5   *When a complete PEC is not possible*

It is sometimes difficult to use trays or conduit over the whole cable route. It is not good EMC practice for a cable to be run without a PEC, but it sometimes appears necessary where a new apparatus is being connected to a pre-existing installation. Figure 7.26 shows the preferred techniques for bonding cable screens as they leave or join a PEC. Unscreened cables should have any protective earth conductors bonded to the PEC at the point of joining. Where cables must leave the protection of a PEC, allowances should be made in the project plan for subsequent remedial EMC work such as the running of a wired PEC.

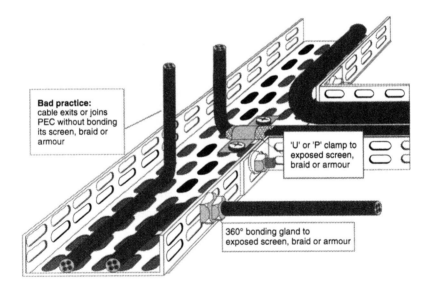

Bad practice:
cable exits or joins
PEC without bonding
its screen, braid or
armour

'U' or 'P' clamp to
exposed screen,
braid or armour

360° bonding gland to
exposed screen, braid or armour

Figure 7.26 Cables leaving or joining PECs

Chapter 8

Filtering

8.1 Attenuating noise at the interfaces

The purpose of a filter in the context of EMC is to prevent interference from travelling either into or out of equipment via its interfaces. This reduces conducted coupling directly, and also helps to reduce radiated coupling if the interference is radiated to or from the cables that connect to the interfaces.

8.1.1 The low pass filter

Typically, unwanted interference is at a higher frequency than the wanted signal that is connected via the interface. This is not universally true; in some cases the wanted signal may be a radio frequency and all disturbances both above and below this frequency should be rejected. Or, the wanted signal may be wideband, such as video or network data signals, and the interference to be rejected may occupy the same part of the spectrum. In general though, the interference frequency is above that of the wanted signal. This then calls for a low pass filter to provide the desired attenuation characteristic.

The conventional low pass filter is built from two elements: series inductance or resistance, and parallel capacitance. The minimum configuration (Figure 8.1(a)) is one or other of these, and combining them gives the L-filter (b), the T filter (c) or the pi filter (d). The choice of inductance or resistance for the series element is usually determined by the wanted signal current that has to be passed: power filters will typically be unable to stand more than a few ohms resistance, but signal filters might easily be able to cope

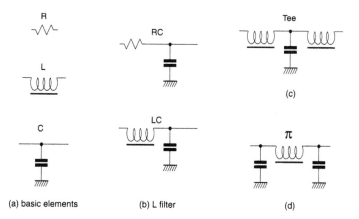

Figure 8.1 Variations of low pass filter

with kilohms. Resistance has the advantage that it absorbs interference energy and does not contribute to resonances, and is of course cheaper and smaller than inductance. Inductance on the other hand can provide high RF impedance with little DC or low frequency loss.

Choice of component types for a filter is determined by the frequency range that has to be attenuated. Non-ideal components have parasitic impedances that limit their effectiveness – capacitors have unwanted stray inductance, inductors have stray capacitance. High frequency performance demands that these parasitics are minimized, and this puts restrictions on the type of construction. In general, any single component can only achieve a maximum of 40–50dB of attenuation before its parasitics limit its performance, hence high performance filters must use multiple stages.

Capacitors create a low impedance to attenuate the interference frequencies, and are therefore most effective in a high impedance circuit. Inductors create a high impedance to the interference, and are most effective in a low impedance circuit.

8.1.2   Differential versus common mode in filters

At least as important as the configuration of the circuit, is the mode of interference that is being suppressed. Section 4.2.4 discussed the differences between differential and common mode. A filter must be designed to deal with the mode that is most relevant to the interference that is present; conversely, a filter that is attenuating the wrong mode will be ineffective, no matter how good it is. (This is actually quite a useful diagnostic device!)

Figure 8.2 shows the way that an L-configuration filter would be wired to attenuate (a) differential mode and (b) common mode interference on a signal pair. The differential mode interference is present between the conductors of the pair, and therefore the capacitor must be placed in parallel across the conductors and the inductor is in series with just one conductor. By contrast, the common mode interference appears on both conductors together, returning via the earth connection external to the desired circuit. Therefore, the differential mode filter configuration would not affect this mode at all – the capacitor is across two conductors which are carrying the same interference voltage and is therefore invisible to the interference, and the inductor is shorted by the lack of impedance in the opposite conductor.

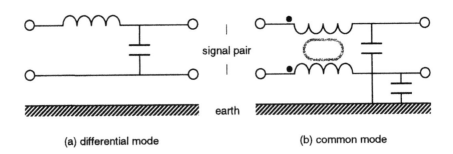

(a) differential mode   (b) common mode

Figure 8.2 Differential mode and common mode configurations

To make a common mode L filter, the series impedance must be present in *all* the signal conductors, and the capacitance must bridge between each signal conductor and the *earth reference*. Separate series inductors or resistors in each line can be used for common mode attenuation and will similarly be effective in differential mode. But it is also possible to implement all the inductors on a common core and wire them such that only the common mode current is affected. This has two crucial advantages:

- magnetic flux due to differential mode current cancels in the core, so a given core size can handle much larger signal or power currents and inductances without saturating, allowing higher attenuation;

- the windings are invisible to the signal current; therefore a filter can be built which attenuates in-band common mode interference without affecting the desired signal, which is in differential mode. This is of particular relevance to wideband signals.

The same convenient effect does not occur with the capacitors, which appear across each line and earth. This can have a serious effect on wideband circuits and often means that either such capacitors must have a very low value, or they cannot be used at all. For low frequency, low impedance signal and power circuits no such problem exists. Most of the commonly-used filtered D type connectors (see 8.3.1) use this principle: each pin has an in-line capacitor connected between it and the connector shell, which is earthed to the zone boundary.

The most important aspect of the common mode filter is that a connection to an appropriate earth point must be available for the capacitors. If it is not, then capacitors cannot be used and only series chokes or resistors are useable. Even if an earth connection is possible, the capacitors' effectiveness will be compromised if they are not connected via a low impedance to a good-quality RF earth. This normally means a direct, short bond to the chassis of the equipment being filtered.

A common mode filter using multi-winding chokes and low value capacitors can be designed to have negligible effect on differential mode currents, but you can instead design a filter to act in both differential mode and common mode as well, if this is desirable. Higher value parallel capacitors to earth will increasingly contribute to differential mode attenuation. The leakage inductance of the common mode choke will also appear in series with the differential mode circuit. Careful choke design and construction can optimize this leakage inductance value so that it provides the desired level of differential attenuation without contributing to choke saturation by, or attenuation of, the wanted differential currents.

8.1.3   Source and load impedances

The performance of any filter depends heavily on the impedance seen at its terminals. There are four relevant impedances for a simple single phase mains filter:

- differential mode (symmetrical) at the mains port;
- common mode (asymmetrical) at the mains port;
- differential mode (symmetrical) at the equipment port;
- common mode (asymmetrical) at the equipment port.

All these impedances will be complex and frequency dependent in real life, but most filters have their performance specified by tests done with 50Ω source and load impedances, which leads us straight to a very important point – filter specifications are optimistic when compared with their performance in reality.

Consider a typical supply filter, installed between the AC power supply and an AC–DC converter typical of the DC power supply of an electronic apparatus. The impedance of the AC supply can vary by as much as from 2 to 2,000Ω depending on the loads that are connected to it, the nature of the supply transformer and the wiring to the point of connection. The impedance is complex and is both time- and frequency-dependent. Research in the 1970s collected a wide variety of data on the domain within which the impedance might fall: an indicative diagram is shown in Figure 8.3, which gives the possible boundaries of the impedance domains for the residential public mains supply in the complex plane at three discrete frequencies, 10kHz, 100kHz and 30MHz [22].

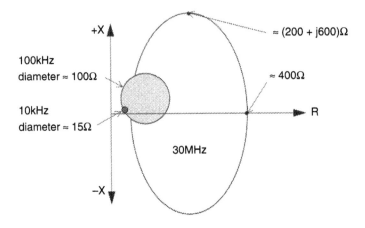

Figure 8.3 Impedance domains for the mains power supply

Looking from the filter into the equipment, the impedance of the AC–DC converter circuitry appears as a low impedance when the rectifiers are turned on near the peaks of the supply waveform, and as a high impedance at all other times. The situation is far from being the matched 50Ω / 50Ω set-up used to measure filter attenuation.

Filter specifications employ 50Ω source and load impedances because most RF test equipment uses 50Ω sources and loads and cables, and because the main specification standard (CISPR 17 [106]) requires this. For most practical uses of filters the specifications obtained by this method are at best optimistic, and at worst misleading. Filters made from inductors and capacitors are resonant circuits, and their performance and resonance can depend critically on their source and load impedances.

An expensive filter with excellent 50/50Ω performance may actually give *worse* results in practice than a cheaper one with a mediocre 50/50Ω specification.

8.1.3.1   The problem of resonant gain

The most sensitive to source and load impedances are supply filters with a single stage. They can easily provide gain, rather than attenuation, when operated with source and load impedances other than 50Ω. This gain usually appears in the 150kHz to 1MHz region and can be as much as 10 or 20dB, leading to the possibility that fitting an unsuitable mains filter can increase emissions and/or worsen susceptibility.

Filters with two or more stages are able to maintain an internal circuit node at an impedance which does not depend very much on the source and load impedances, so they are better able to provide a performance at least vaguely in line with their 50/50Ω specification. Of course, they are larger and cost more.

The easiest way to deal with the source/load impedance problem is only to use filters whose manufacturers specify differential mode (symmetrical) performance for both "matched" 50/50Ω and "mismatched" sources and loads. CISPR 17 requires that mismatched figures are taken with 0.1Ω source and 100Ω load, and vice versa. Draw an attenuation versus frequency curve consisting of the worst-case figures from each of these various curves, and use this as the filter's specification. An example of this graphical procedure is shown in Figure 8.4.

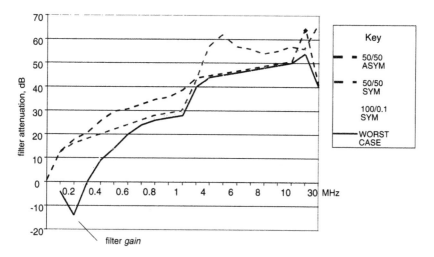

Figure 8.4 Deriving reliable filter attenuation figures from manufacturers' data

When filters are chosen using this technique to try to meet the predicted or actual requirements their performance can be at least as good as expected. When the 50/50Ω figures alone are used to predict filter performance the result is often disappointing.

The problem with this worst-case method using mismatched 0.1Ω/100Ω filter data is that the real impedances are somewhere between the worst case and the standard 50Ω termination, as Figure 8.3 shows. The effect is that the filter will almost certainly be over-specified, especially if performance below 1MHz is the main concern. The performance will also be different depending on the dominant interference path, common mode (asymmetric) or differential mode (symmetric), since the filter's attenuation is different for each mode.

If the data on impedance domains and coupling paths are actually known, then it is possible to analyse the effectiveness of any given filter using suitable software, to optimize its design [22]. This would be reasonable if the cost of the filter was a significant concern in the overall project budget. For system use though, this is rarely the case, and the measurements needed to determine the impedances are not likely to be cost effective. In this situation, using worst case manufacturer's figures for an off-the-shelf unit is a reasonable approach.

8.1.4   Layout and installation

Incorrect filter construction or mounting technique can easily compromise *radiated* emissions and immunity. Poor shielding can easily compromise *conducted* emissions and immunity. The correct way to view filtering and shielding is as a synergy, with each one complementing the other.

8.1.4.1   Locating the filter

A filter is rarely located anywhere else except at a zone boundary (see section 4.2.5 for a discussion of zoning). This is for two reasons:

- the filter is part of the protection offered by the zone barrier. Locating it at a distance from the barrier would allow cables between the filter and the barrier to breach this protection;

- a filter needs a high-integrity earth reference for good high-frequency operation. The zone barrier, which will usually be either a shielding wall in a cabinet or chamber or the entry to an earthing mesh, provides this directly. A large metal plate (at least 1m × 1m) bonded to the earth structure at the Single Point of Connection (SPC, see section 5.3.2.1) to a zone can also serve as such a reference.

An example of the application of these principles is given in the next section. In addition to this general rule, filters should be located as near as possible to the apparatus which is expected to be the source or victim of disturbances, to minimize the impedance of the connection. If, as a compromise, the filter is positioned outside the protected area or apparatus, the wiring between the filter and the protected area should be twisted and positioned close to the earthing structure. Figure 8.5 illustrates these concepts.

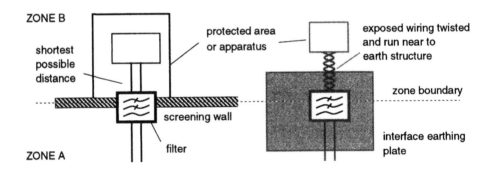

Figure 8.5 Filter location

8.1.4.2   Filter construction and mounting

The higher the frequency, the more a filter is compromised by RF leakage from its unfiltered side to its filtered side. Many engineers have been surprised by the ease with which RF will leak around a filter.

Where an external cable to be filtered enters a shielded enclosure or room, the filter should be fixed into the metal wall at the point of cable entry and RF bonded to the metalwork all around the aperture it fits into. Through-bulkhead filters are the best, since they maintain the integrity of the shield, but are often expensive to purchase and install. For mains supply filters, the IEC-320 inlet (Figure 8.6) is the most common commercial style of bulkhead filter for up to 10 amps, single phase, 230Vrms.

Because of commercial pressures, most mains filter manufacturers only specify their parts over the frequency range of the conducted emissions tests (up to 30MHz). The filter becomes progressively less effective above 30MHz and can compromise the shielding integrity of shielded enclosures, possibly causing problems with radiated electromagnetic disturbances. Good layout inside the enclosure will minimize high-

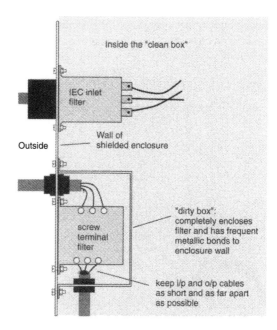

Figure 8.6 Mounting supply filters in
shielded enclosures

frequency coupling onto the internal (filtered) supply and hence control this possibility.

For high powers most commercially available filters use screw-terminal block or Faston connections, making bulkhead mounting impossible. Figure 8.6 also shows how to mount a screw-terminal filter using the "dirty-box" method. This encloses the filter in an individually shielded, segregated box within the main shielded enclosure. The general clean/dirty box arrangement is discussed in section 6.1.2.3 on page 136; here, the dirty box segregation is applied to just one filter. The filter input and output cables in the dirty box must be very short and far away from each other, but even so high frequencies may still leak across and ferrite sleeves (see 8.3.2, later) may be needed on either or both cables if performance above 30MHz appears to be compromised.

Filters sold as "room filters" generally deal with the problem of leakage due to cables by enclosing the filtered side in the metal filter box and passing them through their mounting base via a standard circular conduit fitting. Such filters are intended to be mounted directly on the external metal wall of a shielded enclosure (of any size) with only their filtered output appearing on the inside of the enclosure. This construction effectively shields the unfiltered from the filtered cables and so allows the filter to function effectively up to the highest frequencies. This is the method used by most of the supply filters intended for EMC test chamber applications.

8.1.4.3   The earth connection

All commercial filters are housed in metal bodies of some sort, with the body forming the filter's earth connection. An IEC inlet filter with a metal body installed in a shielded enclosure can only give a good account of itself at frequencies above a few tens of MHz *if* its body has a seamless construction *and* its body is RF bonded to the shielding metalwork. The same is true for any other metal-bodied filter, as shown in Figure 8.6.

The reason is that any inductance due to a wired earth connection will turn the filter into a *high-pass* configuration. The greater the inductance – that is, the longer the wire – the lower the frequency at which this becomes significant. Section 5.2.1 discusses the impedance of earthing wires in greater detail. In the context of filters, Figure 8.7 shows the measured effect – about 25dB difference at 15MHz – of two different lengths of earth wire to chassis on the attenuation of a simple single-stage mains filter.

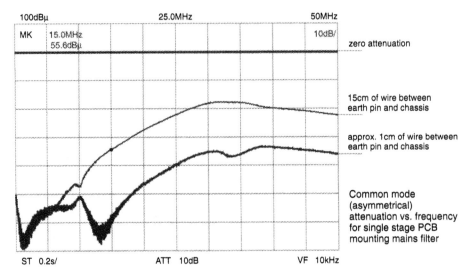

Figure 8.7 Comparison of filter earth bonding

Bonding the case directly to the chassis earth is the only sure way to realize a filter's published performance. The often-provided separate tag to the case is for safety purposes, not for use as an EMC connection.

8.1.4.4   *Wiring to filters*

Filtered and unfiltered cables must be kept rigorously segregated as they are always at least one Class of cable apart (see section 7.4 on cable classes and segregation). The rules on segregation generally advise between 150 and 300mm separation distance between adjacent classes, and 600mm or more between the most sensitive and most noisy classes. Of course, this may not be possible at the filter terminals themselves, since the filter body may be smaller. But in situations where there is no screen across the filter, input and output cables should be carefully dressed away from each other as they leave the filter terminals until the required separation distance is reached – screened cable, or conduit, may be needed close to the filter to help prevent leakage from the filtered to the unfiltered cable and vice-versa.

If a filter must be installed in-line in a cable tray or conduit acting as a PEC, then *all* the cables in that conduit must be filtered, otherwise coupling to and from the unfiltered cables will compromise the high frequency attenuation.

On no account should the installation technician be allowed or encouraged to loom together a filter's input and output wiring. This desire for neatness (Figure 8.8) is going too far and will render an expensive filter almost worthless!

8.2   Mains filters

Since the most common type of filter used by system designers is the mains filter, this section discusses these components in some depth.

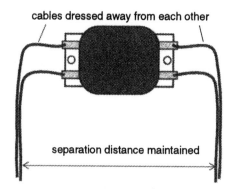
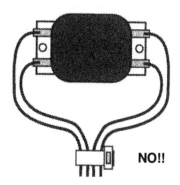

cables dressed away from each other

separation distance maintained

NO!!

Figure 8.8 Filter wiring

8.2.1 The operation of standard mains filters

A typical filter includes components to block both common mode and differential mode components. The common mode choke L (Figure 8.9) consists of two identical windings on a single high permeability, usually toroidal, core, configured so that differential (line-to-neutral) currents cancel each other. This allows high inductance values, typically 1–10mH, in a small volume without fear of choke saturation caused by the mains frequency supply current. The full inductance of each winding is available to attenuate common mode currents with respect to earth, but only the leakage inductance will attenuate differential mode interference.

Capacitors C_{X1} and C_{X2} attenuate differential mode only but can have fairly high values, 0.1 to 1µF being typical of smaller units. Either may be omitted depending on the detailed performance required, remembering that the differential source or load impedances may be too low for the capacitor to be useful. Capacitors C_{Y1} and C_{Y2} attenuate common mode interference and if C_{X2} is large, have no significant effect on differential mode.

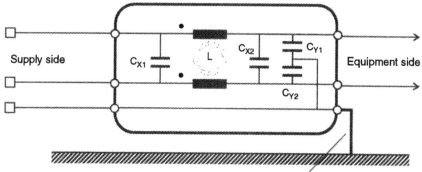

short, low-impedance connection to interface earth
reference – metal-to-metal at numerous points

Figure 8.9 The basic mains filter

Mains filters do not have a specific "input" side and an "output" side, although it is sometimes convenient to think of them in those terms. The filter is a passive device and will offer the same attenuation in both directions in a given circuit. Nevertheless, it is normal for filter terminals to be dedicated either to the mains port or to the equipment port. This is because of the likely impedances to be met at each side (see 8.1.3); the general rule is that capacitors should face a high impedance and inductors should face a low impedance.

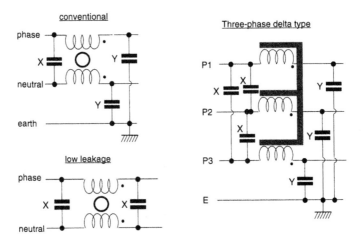

Figure 8.10 Variations on the theme of single-stage mains filters

8.2.2   Extending the performance of standard filters

8.2.2.1   Greater differential mode filtering

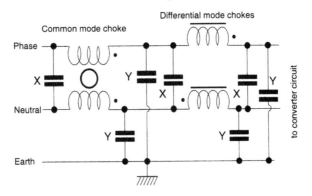

Figure 8.11 A typical 2-stage filter for a switched-mode power converter

Filters which have to cope with high levels of low-frequency differential mode noise (e.g. from switch-mode converters, phase angle power controllers, motor drives, and the like) often need more differential mode attenuation than can be achieved by X capacitors alone, and may need to use differential mode chokes as shown in Figure 8.11. These may be two separate chokes, or can be two windings on the same core with their senses reversed (in contrast to the similar-looking common mode choke). It is difficult to get much differential mode inductance into a small package, because of core saturation, so these filters tend to be larger and more expensive.

Adding further stages to the simple single-stage filter of Figure 8.10 is usually the most cost-effective way of improving a filter's performance. Parasitic reactances limit the attenuation capability of a single component. Also, as already mentioned a multi-stage filter is inherently less affected by extremes of source or load impedance.

8.2.2.2 Earth leakage current

The Y capacitors being connected to earth (Figure 8.9), result in a power frequency current flow through the earth terminal. This can be calculated (see also 4.4.1.1 on page 97) from

$$I_{lkg} \quad = \quad 2\pi F \cdot 1.1 \cdot V \cdot 1.2 \cdot C \text{ microamps} \qquad\qquad (8.1)$$

where V is the voltage across the Y capacitor(s) – e.g., 230V for the L-E connection
F is the power frequency, typically 50Hz
C is the total Y capacitance for a given connection, in microfarads
The factors 1.1 and 1.2 allow for maximum voltage (±10%) and capacitor (±20%) tolerances, respectively

Safety requirements may place limits on allowable earth leakage, particularly for portable or pluggable apparatus (see section 4.4.1.1). This in turn limits the total value of the Y capacitors. For medical apparatus, earth-leakage currents may be limited to such low levels that the use of any reasonable size of Y capacitors is impossible. Such filters need to use better inductors and/or more stages to improve common mode attenuation, and tend to be larger and more expensive for a given performance.

In large systems the earth leakages from many small Y capacitors can create significant earth currents. As well as the safety implications, these can cause earth voltage differences which impose hum and high levels of transients on cables between different equipments. Modern best-EMC-practices require equipotential three-dimensional meshed earth bonding (MESH-BN, see section 5.1.3.3), but many older installations do not achieve this so apparatus intended for use in such installations may benefit from the use of filters with small or even non-existent Y capacitors.

It is always best to use supply filters for which third-party safety approval certificates have been obtained and checked for authenticity, filter model and variant, temperature range, voltage and current ratings, and the application of the correct safety standard.

Filters sold for use on 50/60Hz may generally be used on power circuits ranging from DC to 400Hz with the same performance, but it is best to check with the manufacturers beforehand, and be aware of the implications for earth leakage and dissipation in the filter components.

8.2.2.3 Operating current

The current rating of a mains filter is principally determined by its inductor(s). These carry the peak current, an excess of which can cause three principal effects:

- overheating of the choke and the whole filter, with consequences for reliability and possibly safety

- saturation of the choke core, which causes an instantaneous loss of inductance and therefore of attenuation

- voltage drop across the filter, resulting in a lower supply voltage available to the connected equipment.

Clearly you need to choose a properly rated filter. This is not always simple: only a resistive load takes a purely sinusoidal current, whose peak and RMS values can be easily determined. Electronic equipment, low-energy lighting and motor drive inverters (for example) take a particularly non-sinusoidal waveform – see the discussion on mains harmonics in section 4.3. This can have a peak current very much higher (sometimes 5–10 times) than the rated RMS value. If there is any question of this peak current exceeding the filter's capability, then consult the filter's manufacturer to check the consequences. It may be necessary to measure or calculate the waveform of the current that will be drawn under worst case conditions, to gain enough information.

The filter manufacturer may publish a derating curve which applies to the RMS current and is relevant for overheating. An inductance saturation curve may also be given; this shows loss of inductance versus current and should be applied to the peak current, to check how effective the filter attenuation will remain. The same issues occur when using a filter on DC power, in which case the maximum DC current is applicable both for heating effect and saturation.

8.2.2.4   Overvoltages

Mains filter components are of course rated to withstand the nominal operating voltage, plus some safety factor. But the mains supply frequently carries transients which cause a brief overvoltage of several times the nominal. Chapter 9 looks at surge suppression in more detail. Before selecting a filter, you should consider the likely transient environment and decide whether the anticipated overvoltages will cause the filter too much stress. Filter capacitors are normally specified to IEC 384-14 or its European derivative EN 132400. This requires a pulsed overvoltage stress performance for X class capacitors depending on the sub-class (X1, X2 or X3) that is quoted, and a similar performance for Y capacitors. The construction of filter chokes is more variable; filter manufacturers will normally provide a maximum overvoltage figure for line-to-line and line-to-earth for a two second application.

A severe transient environment combined with a filter of low specification may prompt the need for external surge suppression (Figure 8.12). A metal-oxide varistor (MOV) network across the filter input terminals will usually give more than adequate protection, although the varistor(s) will need to be sized for the maximum anticipated surge energy.

The alternative placement of the varistor(s) on the equipment side of the filter protects the equipment and allows the varistor to be sized for a much lower surge level, since the filter attenuates the surge and provides a higher source impedance. On the other hand, the filter components remain unprotected. To make the best design decision, the expected transient environment, the filter capabilities and the cost and performance of various surge suppressors must all be considered.

Filters and varistors between phase and earth have implications for safety hi-pot testing. It is unrealistic to remove a filter for this sort of test and so it must be specified to survive the intended applied voltage; capacitors to EN 132400 are designed and rated with this in mind. Varistors will intentionally break down when this is applied, and must be temporarily removed from circuit to allow proper testing. Filters with large values of capacitance may be fitted with high-value "bleed" resistors to discharge these capacitors when power is removed to help protect personnel from shocks. These resistors can suffer damage or give false readings on high-pot tests. Some filter

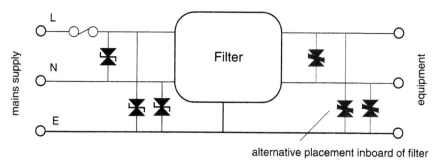

alternative placement inboard of filter

Figure 8.12 Application of varistors external to a filter

manufacturers provide identical filters with the bleed resistors removed – solely for use in safety testing. Also, remember that both varistors and filters have potentially hazardous failure modes, and should themselves be protected by a suitable fuse.

Varistors to earth are usually regarded with disfavour when it comes to safety assessment anyway, because of their effect of increased earth leakage as they wear out. It should normally only be necessary to use them in severe transient environments on permanently-wired equipment, in which case the earth leakage rules are relaxed (see 4.4.1.1).

8.3   Filtering other lines

Telecommunications, data and signal ports may all carry interference signals into and out of equipment. Reducing conducted or radiated emissions with differential mode filters is not practical if they are caused by the essential spectrum of the signal: screened cables and connectors will be required instead of or as well as filtering. *Common mode* filtering may however be applied to frequencies within the desired signal spectrum, since the wanted signals are differential rather than common mode. Common mode signals are always unwanted.

Power and 0V "bounce" noise in digital processing circuits is often overlooked when predicting the specification of signal filters. The wideband digital circuits used in many modern products (computers, PLCs, and increasingly in instrumentation) create a lot of internal noise. This tends to exit via I/O and other cables as common mode noise with a much higher frequency content and amplitude than would be expected from knowledge of the intended signals, and this should be considered when specifying a filter. Unfortunately, it is impossible to predict and can usually only be quantified by measurement, ideally by the manufacturer of the digital device, who should then take steps within his product to reduce these noise emissions to reasonable levels. System integrators and installers need to be aware, when considering filtering and shielding, that such spurious noises exist and are completely unrelated to the signals which are intended to be carried by the cables.

Filtering at interfaces is usually the responsibility of the equipment designer, but there are two areas where the systems engineer can make improvements if necessary: using filtered connector adaptors, and using ferrites, as illustrated in Figure 8.13.

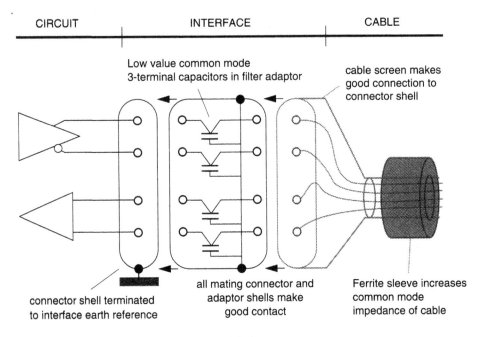

Figure 8.13 Using filter connector adaptors and ferrites

8.3.1   Filtered connector adaptors

Various makes of filter adaptors are available for common connector styles: the D sub-miniature series are the most common and popular. These adaptors are very simple, being nothing more than parallel capacitors between each line of the connector and the connector shell. As such, they will only work properly if used against a connector port on the equipment whose shell is terminated to a zone boundary, or interface earth reference. If there is no such connection – for instance, if the equipment is housed in a plastic case and the shell is unconnected – then the capacitors form a high-pass interconnection between all the pins on the connector. This is likely to affect the signal circuit, and won't provide any filtering.

The value of the capacitors is important: too high a value will also affect the desired signal, but too low a value will be ineffective for suppressing lower-frequency noise. Values between 47pF and 2200pF are typical. These capacitors appear in parallel with the capacitance of each conductor to the screen in the cable, which is a function of cable length and specification. To properly specify the use of such an adaptor, you need to know the maximum capacitance to earth that each circuit is able to drive. Multiply the cable length by the conductor capacitance to screen, and subtract this from the overall drive capacitance specification of the equipment to find the maximum allowable filter capacitance. For example:

> Belden 9305 cable, capacitance to screen 164pF/m, length 10m; maximum circuit drive capacitance 2000pF:

$$\text{Max. } C_{filter} = 2000 - (164 \times 10) = 360\text{pF} \qquad (8.2)$$

As with filtered connectors themselves, most adaptors can be supplied with a limited range of capacitor values. It may be necessary to check the voltage rating of the capacitors, especially if the signal line is intended to be isolated. Variations on the design of adaptor may include series chokes in one or both legs of each line, π or L-configurations, and varistor clamps in parallel with each capacitor.

8.3.2 Using ferrites

The most usual application of a ferrite component is as a sleeve over a cable to attenuate common mode currents on the cable. A particularly useful type of ferrite component is the split tube in a plastic "clip-on" housing. They are very easy to apply as a retro-fit to a cable (and to remove when found not to do much), and are available in a wide range of shapes and sizes, including flat types for ribbon cables.

8.3.2.1 The effect of magnetic material around a conductor

Current flowing through a conductor creates a magnetic field around it. Transfer of energy between the current and the magnetic field is effected through the "inductance" of the conductor – for a straight wire the self-inductance is typically 20nH per inch. In a non-permeable material, such as air or a plastic sheath, the magnetic field strength and magnetic flux density are equal. Placing a magnetically permeable material around the conductor increases the flux density for a given field strength and therefore increases the inductance.

Ferrite is such a material; its permeability is controlled by the exact composition of the different oxides that make it up (ferric, with typically nickel and zinc) and is heavily dependent on frequency. Also the permeability is complex and has both real and imaginary parts, which translate into both inductive and resistive components of the impedance "inserted" into the line passed through the ferrite (Figure 8.14). The ratio of these components varies with frequency – at the higher frequencies the resistive part dominates (the ferrite can be viewed as a frequency dependent resistor) and the assembly becomes lossy, so that RF energy is dissipated in the bulk of the material and resonances with stray capacitances are avoided or damped.

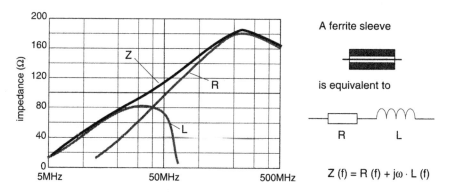

Figure 8.14 Typical ferrite impedance with frequency

8.3.2.2 The effect of cable currents

Elsewhere (section 7.1) the distinction between common and differential mode in cables is discussed. This distinction is particularly relevant to the application of ferrites.

Differential mode

The magnetic field produced by the intended "go" current in each circuit pair is substantially cancelled by the field produced by its equal and opposite "return" current, provided that the two conductors are adjacent. Therefore any magnetic material, such as a ferrite sleeve, placed around the whole cable will be invisible to these differential mode currents. This will be true however many pairs there are, as long as the total sum of differential mode currents in the cable harness is zero.

Placing a ferrite around a cable, then, has *no effect* on the differential mode signals carried within it.

Common mode

A cable will also carry currents in common mode, that is, all conductors have current flowing in the same direction. Normally, this is an unintended by-product of the cable connection, and the current amplitudes are often no more than a few microamps, but are usually the main cause of interference problems. The source of such currents for emissions is usually either

- earth-referred noise at the point of connection, which may have nothing to do with the signal(s) carried by the cable, or

- imbalance of the impedance to earth of the various signal and return circuits, so that part of the signal current returns through paths other than the cable harness.

A screened cable may also carry common mode currents if the screen is not properly terminated to a noise-free reference. Even though the currents may be small, they have a much greater interfering potential because their return path is essentially uncontrolled. Also, incoming RF or transient interference currents are invariably generated in common mode and convert to differential mode (and so affect circuit operation) due to differing impedances at the cable interfaces, or within the circuit.

Common mode currents on a cable *do* generate a net magnetic field around the cable. Therefore, a ferrite inserted around the cable will increase the cable's local impedance to these currents.

8.3.2.3   Circuit impedances

In the same way as was discussed in section 8.1.3 with respect to mains filters, when a ferrite is placed on a cable it functions between source (equipment) and load (cable) impedances. A quick glance at the equivalent circuit (Figure 8.15) shows that maximum attenuation due to the simple impedance divider will occur when these impedances are low. For cable interfaces, low source impedance means that the ferrite should be applied near to a capacitive filter to earth or to a good screen termination. For open or long cables, the RF common mode load impedance varies with frequency and cable length and termination: a quarter wavelength from an open circuit, the impedance is low, a few ohms or tens of ohms; a quarter wavelength from a short circuit, the impedance is high, a few hundred ohms. (This property of cables is well known to antenna designers but looks like magic to the rest of us.) Since you do not normally know the length and layout of any cable that will be attached to a particular interface, and since the impedance is frequency dependent anyway, it is usual to take an average value for the cable impedance, and 150Ω has become the norm. This is the *common mode* impedance: it has nothing to do with the cable's characteristic impedance or any differential circuit terminations.

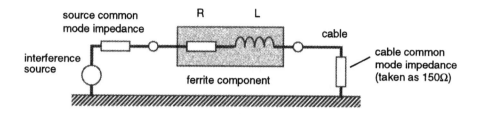

Figure 8.15 The equivalent circuit for ferrites

Ferrite impedances rarely exceed a few hundred ohms, and consequently the attenuation that can be expected from placing a ferrite over a cable is typically 6–10dB, with 20dB being achievable at certain frequencies where the cable shows a low impedance. Figure 8.16 shows the actual attenuation for two types of core at two different circuit impedances. Note the different vertical scales for the two plots.

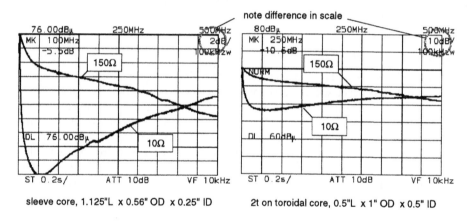

sleeve core, 1.125"L x 0.56" OD x 0.25" ID   2t on toroidal core, 0.5"L x 1" OD x 0.5" ID

Figure 8.16 Ferrite attenuation versus impedance

8.3.2.4 Choice and application

Size and shape

There are two rules of thumb in selecting a ferrite for highest impedance:

- where you have a choice of shape, longer is better than fatter;
- maximum impedance demands the maximum amount of material in a given volume.

The impedance for a given core material is proportional to the log of the ratio of outside to inside diameter but directly proportional to length. This means that for a certain volume (and weight) of ferrite, best performance will be obtained if the inside diameter fits the cable sheath snugly, and if the sleeve is made as long as possible. A string of sleeves is perfectly acceptable and will increase the impedance *pro rata*, though the law of diminishing returns sets in with respect to the attenuation.

Number of turns

Inductance can be increased by winding the cable more than one turn around a core; theoretically the inductance is increased proportional to the square of the number of turns, and at the low frequencies this does indeed increase the attenuation. But it is usual to want broadband performance from a ferrite suppressor and at higher frequencies other factors come into play. These are:

- the core geometry already referred to; the optimum shape is long and snugly-fitting on the cable, and this does not lend itself to multiple turns;

- more importantly, inter-turn capacitance, which appears as a parasitic component across the ferrite impedance and which reduces the self resonant frequency of the assembly.

The normal effect of multiple turns is to shift the frequency of maximum attenuation downwards. It will also increase the value of maximum attenuation achieved but not by as much as hoped. The source and load impedances are critical in determining the effect: the lower the impedances, the less the effect of parasitic capacitance. To gain improved attenuation at very high frequencies (say, above 300MHz) it is usually better to string more ferrite tubes or toroids along the set of conductors, with only one pass of the cables through each.

Capacitance to earth

Because a ferrite material is in fact a ceramic, it has a high permittivity as well as permeability, and hence will increase the capacitance to nearby objects of the cable on which it is placed. This property can be used to advantage especially within equipment. If the ferrite is placed next to an earthed metal surface, such as the chassis, an L-C filter is formed which uses the ferrite both as an inductor and as a distributed capacitor. This will improve the filtering properties compared to using the ferrite in free space. For best effect the cable should be against the ferrite inner surface and the ferrite itself should be flat against the chassis so that no air gaps exist; this can work well with ribbon cable assemblies.

Saturation

As with other types of ferrite, suppression cores can saturate if a high level of low-frequency current is passed through them. At saturation, the magnetic material no longer supports an increase in flux density and the effective permeability drops towards unity, so the attenuation effect of the core disappears. The great virtue of the common mode configuration is that low frequency currents cancel and the core is not subjected to the magnetic field they induce, but this only happens if the core is placed around a cable carrying both 'go' and 'return' currents. For three-phase star connected power all three phases and the neutral must pass through together. For heavy power cables, best performance from ferrites is achieved when all the relevant conductors are bunched tightly together and held in the centre of the toroid or tube, away from the walls. If the ferrite gets hot, it is not substantial enough or else the cables through it need better control.

If you must place a core around a single conductor (such as a power supply lead) or a cable carrying a net low frequency current, be sure that the current flowing does not exceed the core's capability; it is usually necessary to derive this from the generic material curves for a particular core geometry.

Chapter 9

Lightning and surge protection

9.1 The EMC problems of lightning

Safety is an overriding concern, so the appropriate standards for lightning protection for personnel safety and the protection of the structure must always be followed. However, the subject of this book is EMC, so only what is necessary to protect electronic equipment from the effects of lightning will be discussed in any detail. An adequate lightning protection system is assumed, and little detail will be given of lightning protection for reasons other than EMC.

Where safety is concerned it is necessary to take into account all reasonably foreseeable extremes, misuse, and fault conditions, and the design of basic lightning protection systems reflects this. Similar considerations may need to be applied where the failure or degraded operation of electronics due to the effects of lightning (or other EMC phenomena) could possibly have serious implications.

9.1.1 How lightning phenomena can affect electronic apparatus

The issues of lightning protection for electronic apparatus are addressed by IEC 61024-1 [99] and by Appendix C of BS 6651 [111]. Lightning can cause damage to electronic equipment in a number of ways:

- Resistively induced voltage. The resistance of the soil and of earthing networks, when subjected to intense lightning discharge currents (considered to be between 2kA and 200kA, with 1% of strokes exceeding 200kA) gives rise to potential differences between areas normally considered to be at the same "earth" potential, and this exposes electronics connected to these different areas to excessive voltages and/or currents. Long cables, and especially cables between structures, are particularly at risk of causing damage due to this effect, which is sometimes known as "ground potential rise" or "ground lift".

- Magnetically induced voltage. Excessive voltages may be induced into conductors and bonded earth structures due to the radiated magnetic fields from lightning discharge, for strikes at up to 100 metres distance, due to the rate of change of the discharge currents. A maximum rate of 200kA/μs is accepted for the arc channel itself, with lower values where the lightning discharge current is shared between a number of conductors. Even the pigtails traditionally used for bonding the screens of cables can present a serious risk to their equipment due to inductive voltage coupling (one reason why their use is no longer considered best practice and 360° screen bonding is preferred instead, refer to section 7.2.3.3 on page 170).

- Current injection from direct strike. Direct injection of the lightning main discharge current into exposed external equipment and cables such as antennas, security cameras, sensor heads, air-conditioning equipment, overhead lines, etc. The explosions and arcing flashovers associated with a direct strike to external equipment often results in damage to connected internal equipment, but often also causes damage to unrelated equipment by flashovers in shared cable routes or terminal cabinets. This problem is particularly of concern where potentially explosive atmospheres are involved.

- Electric field coupling. The whole area around a lightning strike that is about to happen can be exposed to electric fields of up to 500kV/m (the breakdown voltage of air) over an area of up to 100m from the eventual strike point, with fluctuating fields of 500kV/m.μs occurring during a strike. These fields will induce voltages and currents into conductors and devices, but are usually effectively dealt with by the measures taken to protect equipment from the other lightning threats.

- Lightning Electromagnetic Pulse (LEMP). This is a "far-field" phenomenon, and may be caused by cloud-to-cloud lightning as well as by distant cloud-to-ground lightning. It is usually only a problem for exposed external conductors [111], and is effectively dealt with by the measures taken to protect equipment from other lightning threats.

- Thermal and mechanical effects (e.g. shock waves in the air) of the intense energies associated with a lightning strike. These are more usually problems for the structure's fabric and the design of the lightning protection system itself.

- Multiplicity of the surges in a single "strike". A typical lightning event consists of many discharges (or "strokes"), of which the second one usually contains the most damaging energies. Multi-stroke flashes can exceed 10 strokes and last for over a second, which is of great importance in the design of software for error-correction and for the recovery of systems.

9.1.1.1   Assessing the criticality of the apparatus

Lightning damage to electronic equipment can cause safety problems to personnel and/or damage to the structure, usually through electrocution or fires, but sometimes because the equipment has a safety-related function.

Safety concerns such as fire and electrocution must be addressed as part of the normal health and safety at work procedures. For EMC we are concerned with the response of each item of electrical and electronic apparatus to the effects of lightning. Each item of apparatus should be assessed against the following criteria:

- (A)   catastrophic failure requiring replacement of the apparatus is acceptable

- (B)   the apparatus is required merely to survive the lightning event undamaged, with no concern about its functionality during the event

- (C)   the apparatus must continue to operate during a lightning event, although reduced performance is acceptable (the degree of reduction needs to be specified for each function)

- (D)   the apparatus must continue to operate without any reduction in performance during a lightning event (e.g. safety- or mission-critical equipment).

The same type of equipment (e.g. a DC power supply) may have a different functional requirement depending on:

- where it is used in a structure;
- how it is installed (its exposure);
- what it is used for (the criticality of its function).

Co-ordination is then required between these three:

a) the apparatus' functional criticality

b) the apparatus' ability to withstand lightning electromagnetic phenomena (especially voltage surges on earth, power, and signal cables)

c) the lightning electromagnetic phenomena that the installation exposes the apparatus to (especially voltage surges).

Where safety-critical apparatus is concerned, especially programmable equipment such as PLCs or computers, an assessment of its Safety Integrity Level according to IEC 61508-1 [101] (for example) is strongly recommended as a means of determining the ratio, or margin, between the withstand ability of the apparatus and its likely exposure to lightning phenomena.

Apparatus must therefore be designed and tested to achieve the required degree of protection and reliable functional performance depending upon its exposure to various lightning phenomena when installed as specified. This may require the use of surge protection devices (SPDs) to deal with both common-mode (e.g. earth potential differences) and differential-mode voltage surges.

9.1.2   Overview of design of a lightning protection system (LPS)

9.1.2.1   Planning and management of the design process

Any LPS design should always be done by an experienced and competent professional who involves all interested parties before, during, and after all stages of design including:

- architects
- utilities (gas, water, power, telephone, etc.)
- the owner's Fire and Safety Officers
- TV and radio installers
- the builders.

9.1.2.2   Basic design of an LPS

The design of a basic LPS for safety and protection of the structure typically requires:

- risk assessment based on lightning exposure and acceptability of consequential losses
- design of the air termination network and down conductors
- design of the earth termination network and earth electrodes
- bonding of the metalwork within a structure, and the metallic services entering a structure, to the LPS.

Special structures (e.g. tall buildings, buildings with flammable or high risk contents, fences and their gates, TV and radio masts, etc.) may require special provisions.

The possible utilization of metal parts of the structure (so-called "natural" components) as parts of the LPS should be foreseen in the design of the structure itself, but only used with the agreement of the owner and the structural engineer. All metal parts so used (metal sheets, metal parts of roof construction, gutters, ornamentation, railings, pipes, tanks, etc.) must meet specified minimum requirements. [55] points out that copper theft is a serious concern and puts an external LPS at risk; and that it is often difficult to persuade owners and their architects that an external LPS enhances the appearance of their building. For these reasons the use of natural components is preferred, although to use the technique successfully requires consideration right from the start of a building design.

9.1.2.3   Documenting and maintaining an LPS

Both [111] and [99] specify that records are required to be kept throughout the design and construction process. Certain procedures are also specified for the regular inspection, maintenance, and upkeep of the LPS, and records must be kept of these too. These records are generally required to meet the requirements of safety laws and insurers, but are also recommended for aspects of the LPS that concern the protection of electronic equipment.

9.2   Basic LPS design for safety and structural protection

9.2.1   Risk assessment

The unpredictability of occurrence of lightning is well-known, and the characteristics of a lightning event are also impossible to predict. A great deal of measurements on lightning have been made over many years, and lightning standards draw on these to make statistical assessments of the degree of lightning risk and the lightning strike characteristics.

Sadly, the old cliché "lightning never strikes twice in the same place" gives incorrect engineering guidance – if a structure is particularly exposed to lightning, for example through being taller than any surrounding structures, trees, or terrain, then it is more likely to be struck more often than its neighbouring buildings.

All structures should have had a risk assessment done according to the relevant lightning protection standard, and appropriate measures implemented according to the outcome of this assessment, to achieve the necessary or desired level of safety and protection of the structure. In the UK the relevant standard is BS 6651 [111], but other countries usually have their own standards which may use a different risk assessment philosophy.

When a risk assessment for safety purposes is carried out according to the various standards it may suggest that an LPS is not justifiable. Many domestic dwellings and smaller commercial premises may fall into this category, since they may not be at great risk of a strike, or contain critical electronic equipment or large numbers of people. The people living or working in those buildings may take a different view – for example they may wish to feel well-protected, or they may not wish their income or lifestyle to be harmed due to failure of their computers, telecommunications, central heating and air-conditioning, entertainment, security systems, etc.

Based on typical risks from other sources, BS 6651 sets the acceptable risk for lightning assessment at a 1 in 100,000 chance of a lightning strike per year (10^{-5}). But the judgement of whether to implement an LPS, and how far to go with its design, can never be totally based on a standard and must take into account the wishes of a fully-informed owner or user.

9.2.1.1 *Risk calculation*

A method of performing a lightning risk assessment is described in detail by BS 6651 [111] and [2]. The probability of a strike, P, is calculated from

$$P = A_E \cdot N_G \cdot 10^{-6} \tag{9.1}$$

where A_E is the effective collection area of the structure for lightning strikes, and N_G is the number of ground strikes/year for that geographical location. A lightning strike density map (sometimes called isokeraunic or isoceraunic maps) of the UK is provided in BS 6651. The factor of 10^{-6} in the equation occurs because the structure's collection area is given in m^2, whereas strikes/year from the map is given per km^2.

When P has been calculated it is multiplied by five weighting factors: A through E, to derive the final *risk factor* R.

Weighting factor A depends upon the use of the structure, B depends upon the type of construction, C depends upon the contents and consequential effects of any damage, D depends on the degree of isolation from other structures, and E depends upon the type of terrain. These weighting factors are simply read from the tables provided in the standard.

The final value of risk factor R is then compared with the "acceptable" risk factor of 10^{-5} to determine whether an LPS is needed. The guidance is as follows:

$R < 10^{-5}$	an LPS is not considered necessary
$10^{-5} < R < 10^{-4}$	an LPS is recommended
$R > 10^{-4}$	an LPS is considered necessary

Countries without their own lightning standards or codes of practice can use the IEC standard IEC 61024-1 [99], which has a broadly similar approach to establishing the degree of exposure and risk of a lightning strike, but instead of risk factors it calculates protection levels, numbered I to IV (with I being the toughest and IV being the least stringent).

9.2.2   Construction of an LPS

A basic LPS consists of an air termination network, a down conductor network, and an earth-termination network (Figure 9.1). It is possible to construct an isolated LPS that protects a structure whilst being separated from it, but the type of LPS described here is attached to the structure and bonded to its internal CBN.

The air termination network is intended to intercept the lightning strike and divert its currents via the down conductors to the earth termination network, thereby protecting the structure from lightning strikes.

An all-metal structure, made from substantial metalwork and adequately bonded to earth, can provide a perfect LPS without any additional LPS structure, and is also the best construction for protection of internal electronics from the effects of lightning. All-welded metal constructions are now common for explosives stores [55].

Many years ago the use of a single conductor, extending from an earth electrode to a point somewhat higher than the highest point on a structure was considered a good LPS. Partly this was based on the idea that the pointed end at the top would discharge the local field and prevent a strike, but it turns out that this does not happen and these

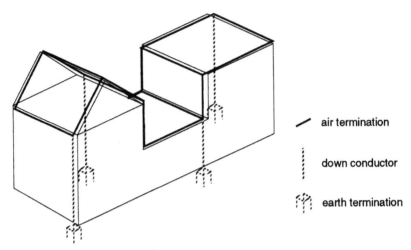

Figure 9.1 Components of building protection

LPSs just intercept the inevitable strikes before they strike the structure. Such single conductor systems are not recommended any more, partly because they tend to suffer from flashovers to the rest of the structure, partly because of the safety hazards caused by their strong voltage gradients at ground level, and partly because the intense fields surrounding them during a strike puts nearby electronic equipment at risk from induced voltages.

An LPS has to withstand extremes of weather, extremes of electrical, electromagnetic, and mechanical stress, and must last for many years. Consequently, corrosion is a major concern and only certain metals and combinations of metals are suitable. The electromagnetic force developed by the passage of a very high transient current is sufficient to tear or bend metal parts if they are inadequately sized or fastened. The specification of the components parts used to construct an LPS, and the construction methods and components used to install an LPS, are of great importance, and are covered in detail by [111] and [2].

9.2.2.1   The LPS air termination network

This can be a mesh arrangement of copper or aluminium conductors laid horizontally and vertically on the tops of roofs and the outsides of walls, with minimum spacings between conductors (specified by standards and/or codes of practice). "Natural" components such as gutters, railings, metal-clad roofs, etc. may be usefully pressed into service and can even take the place of a separately installed air termination network [111], [99], and [55].

Natural components used as part of an air termination network must meet various criteria so as not to be damaged by a strike, and are allowed to be protected by insulation or paint of no more than a given thickness. Metal-clad roofs and walls should always be bonded to the nearby LPS. Different types of air terminations create differently-shaped "zones of protection". Tall structures can suffer lightning strikes to their sides, in which case some or all of the down conductors may need to be designed as air-terminations too.

The air-termination network must have a closer mesh for more vulnerable structures, such as fuel or explosives stores, and we shall see later that the same is true where it is required to protect internal electronics better. An all-metal structure provides the optimum air-termination network.

Exposed apparatus such as antennae, radar and satellite dishes, security cameras, air-conditioning plant, water tanks, etc. can suffer from direct strikes giving rise to safety hazards and damage to the fabric of the structure (as well as its internal electronic apparatus). It is generally possible to protect these by an extension of the air-termination network, often by bonding them or their metallic support structure to the LPS, and sometimes (e.g. for antennae) by extending the air-termination into a vertical free-standing spike above the apparatus to create a suitable protective zone for it.

9.2.2.2   Down conductors

Down conductors provide a low-impedance path for lightning currents from the air termination network to the earth electrode system, and in general there should be several, equally spaced around the structure to share lightning current amongst themselves.

Many modern structures make extensive use of metal, and it is recommended to use their metal structures as "natural components" of the LPS, in addition to (or even in place of) external down-conductors. Metal structures such as radio masts or flagpoles may use their exposed metal structure as all or part of their air termination and down-conductor network simultaneously.

Metal-clad roofs and walls should be bonded to their nearby LPS. Concrete re-inforcement used as part of an LPS should make good electrical contact at the majority of their joints and crossovers, preferably by welding, or else by the use of tying wire.

Down conductors should be straight and vertical, fitted at least at the corners of a structure, equally spaced, and should provide the most direct route to the earth electrodes.

9.2.2.3   The earth termination network

This is the system of earth electrodes which dissipates the lightning currents into the mass of the soil and/or rock beneath the structure to be protected. All soils and rocks have finite conductivity, which compromises their performance as an earth mass, so care must be paid to the design, construction and maintenance of earth electrodes.

The earth termination network for a structure is generally required to provide an earth resistance of under 10Ω, although higher (or lower) resistances may be allowed (or needed) in special cases. The lightning standards and codes provide rules and formulae for designing different types of earth electrodes, including the special techniques which may be needed where soils or rocks are not very conductive.

It is interesting to apply Ohm's law to an LPS earth termination network: for a total resistance of 10Ω, and a peak lightning current of 200kA (only exceeded 1% of the time), it appears that the momentary rise in ground potential of the structure (relative to the mass of the earth at some distant point) could be as much as 2MV (2,000,000V). This is the "earth lift" for the entire structure. This simplistic calculation makes it obvious why it is important to have equipotential bonding within a structure, and why it is necessary to pay careful attention to people and cables near a structure which may be struck by lightning, especially those people actually entering or leaving the structure.

The finite conductivity of the ground means that there is a voltage gradient around a structure during a lightning strike. The design of the earth termination network needs to keep this to acceptable levels. A voltage difference between a person's feet of over a few thousand volts is considered hazardous to people, whereas just a few hundred volts can be dangerous to many animals. The difference is due to a typical agricultural animal's heart being between its front and back legs, and so more exposed to ground potential differences than a person's. Even so it may be good advice, when walking near a tree or structure during a thunderstorm, to take short steps!

Earth electrodes

Earth electrodes used to rely on buried metal plates, and/or on metal services such as water pipes. Plates have been found to suffer from corrosion, and service pipes are increasingly compromised by the Utilities' use of plastic pipes. A typical earth electrode these days consists of a copper alloy rod electrode deep-driven vertically into the soil (Figure 9.2(a)), sited at the foot of each individual down-conductor a metre or so from the boundary of the structure [1].

The reinforcement in concrete foundations (a little while after construction) can achieve a very low earth resistance, especially concrete pilings. This is known as a foundation earth electrode, and it requires the reinforcing bars to be welded, or at least reliably bound together with tying wire, at their crossing points.

Strip electrodes (Figure 9.2(b)) may also be used, especially to help reduce voltage gradients around a structure, when they are known as potential grading electrodes. A foundation strip electrode is a strip electrode laid in the trench cut for the foundations of a structure, before they are laid or poured.

(a) deep driven   (b) strip

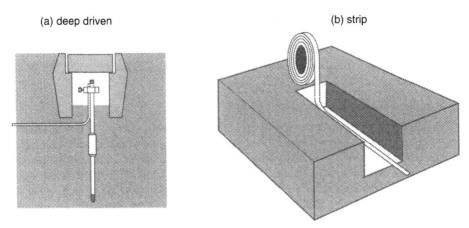

Figure 9.2 Earth electrodes

Another variant of the strip electrode is the ring earth electrode, which follows the complete perimeter of a structure at a given distance and is bonded to itself at its ends to form an unbroken ring, like the internal bonding ring conductors (BRCs) described in section 5.2.2.3 on page 117. Ring earth electrodes connect to all the other rod and strip electrodes. [111] requires the LPS for any structure over 20 metres high to use a ring earth electrode. Ring earth electrodes may be insulated and run above or below the soil surface just to interconnect other electrodes, or they may be un-insulated and buried at least 0.6 metres deep to act as an earth electrode as well. The ring earth electrodes of neighbouring structures should be interconnected to help level the voltage gradients in the ground between them.

All these electrodes need to be able to be tested (usually annually) so must have a single connection to the rest of the system, that may be broken, and often an earth reference electrode is also required to aid in testing. Rod electrodes are generally installed in "lightning pits" with removable covers to aid inspection.

9.2.2.4   Preventing side-flashes

During a lightning strike the high rates of change of current in down conductors (up to 100 kA/m.µs for a down conductor carrying 50% of the strike current) can experience very high voltages from top to bottom. These high voltages can "side-flash" to other metalwork, even through the fabric of the structure, usually causing cosmetic or structural damage and creating possible hazards to personnel. Prevention of side-flashing uses two techniques:

- isolation (through attention to clearance distances); or,
- bonding the down-conductors to the metalwork it might flash to.

For buildings of just a few storeys, say up to 15 metres high, with a properly designed LPS, it is usually enough to bond the lightning protection system (LPS) to the structure's internal common bonding network (CBN, see section 5.1.3.3) at ground level only. Structures higher than 20 metres should bond their non-LPS metalwork (e.g. structural metalwork, water pipes and other services, dry risers, lift shafts, ladders, staircases, railings, cable ducts, etc.) to their LPS at top and bottom, and at intervals of no more than 20 metres in-between [111]. [99] recommends that bonds between LPS and CBN take place where there is already a horizontal ring conductor which bonds the LPS down-conductors.

Bonding the LPS to the CBN does not result in huge lightning currents flowing through the internal earth mesh, because the natural tendency of lightning currents is to concentrate in the external LPS conductors, and the small unwanted currents that do flow are more than compensated for by the freedom from dangerous side-flashes. Improving a CBN into a MESH-BN actually discourages lightning currents from penetrating very far into a structure. The closer the mesh of the LPS (for instance, where its concrete reinforcement bars are used over the whole structure) the lower is the inductance of its down-conductor network, and the lower is the vertical voltage difference during a strike, reducing the risks of side-flashing.

Where bonding (the preferred method) is not used, appropriate isolation must be achieved. Clearances may require 2 metres of separation by air.

9.2.3   Bonding external cables and metallic services to earth

Metal entering or leaving a structure – whether cable sheathing, screening or armouring, or piping for electric power, gas, water, rain, steam, compressed air, or any other service – is a primary route for the injection of lightning arc channel currents, or "earth lift" voltages, into a structure. All such metal must be bonded as directly as possible to main earthing bars connected to the structure's bonding ring conductor (BRC). This must be done as near to the point at which the service enters or leaves the structure as possible (Figure 9.3). Dry risers should be similarly treated.

Where direct bonding is not possible, for whatever reason (e.g. when the corrosion would be too great, or when a type of LAN is used which only allows earth bonding of its screen at one end) surge protection devices (SPDs) should be fitted instead, at the same main earthing bars, dimensioned to withstand the anticipated surge currents (see section 9.3.6).

Fitting an SPD to earth is considered to be a form of earth bonding, although obviously the SPDs will allow part of the voltage surges through as a "let-through" voltage which subsequent apparatus should be able to withstand. Co-ordination is clearly required between SPD let-through voltages for a given current and the surge withstand (or "resistibility" as it is called by the ITU) of the apparatus they protect.

Power and signal inner conductors which obviously can't be bonded to earth should have SPDs from each of their inner conductors to their main earthing bars. A wide range of SPD devices is now available from a number of manufacturers for all sorts of data, signals, and power conductors.

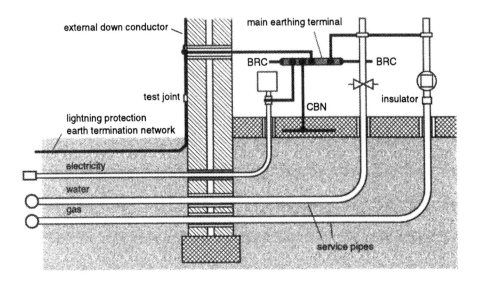

Figure 9.3 Bonding of services entering a building (after [111])

Metallic cables and services entering a structure should preferably have travelled underground for their entire length. Overhead cables and services are very exposed and there are special rules for dealing with these. Even telephone wires should also enter underground where possible, but most standards dodge the issue by stating that their safety is the responsibility of the appropriate telephone Utility.

9.3   Additional LPS measures to protect electronic apparatus

Having an LPS fully in accordance with the basic standards will not be sufficient to protect the electronic apparatus inside a structure from the direct or induced effects of a lightning strike, although it will be better than nothing at all.

Appendix C of BS 6651 [111] deals with the protection of electronic apparatus from the effects of lightning. IEC 61312-1 also covers protection of electronics (from a more German than British angle) and we also have IEC 60364-4-443 (European) and IEEE C62.41 and IEEE C62.64 (American), all with their own ways of dealing with this problem. Common threads emerge, however, and these will be described below.

This section describes the enhancements that should be made to a basic LPS to provide the desired level of protection for electronic apparatus. The existence of a basic LPS is assumed throughout.

9.3.1   Enhancing the LPS structure

A completely enclosed metal structure would create the ideal LPS – effectively an infinite number of horizontal and vertical conductors bonded together all around the structure. This would make a Faraday cage, or screened volume, with (ideally) no internal fields at all during a direct lightning strike. Where electronics absolutely must survive lightning or similar external surges a seam-welded all metal structure may turn out to be required.

In the absence of seam-welded metal structures, all authorities recommend increasing the number of vertical LPS down conductors to share the surge currents so that each down conductor carries less current and hence creates lower magnetic fields. This also reduces the magnetic fields inside the structure. (Lightning magnetic field is one of the main problems for electronics). "Natural components" such as metallic facades, conductive roof layers, concrete reinforcement bars, window and door frames, etc. may all be pressed into service as part of the enhanced LPS to further reduce its mesh size and hence the induced fields it emits when struck.

Where a structure is made all of reinforced concrete that uses welded or tied reinforcement bars, and when the reinforcement bars are also welded to the metal window and door frames that break into the concrete, using these reinforcement bars as part of an LPS can help create a very protected structure.

The use of buried ring earth electrodes or foundation strip electrodes is recommended (or implied) by many lightning references, where the protection of electronics is concerned. Ring electrodes are only suitable for use on their own where the soil conductivity is low enough. One of the main functions of a ring or electrode appears to be to improve the bonding of incoming cables and metallic services (rather than the reduction of earth resistance) and this is discussed later. [81], [74], and [77] recommend a foundation strip electrode instead of rod electrodes for new buildings.

Remote electronic sites are addressed by [132], which recommends using a ring electrode buried 0.75m under the surface and encircling the site at 0.65m distance. For small (e.g. roadside) cabinets [132] recommends burying the ring electrode 0.3 to 0.5 metres under the surface, underneath or just outside the foundation pad.

9.3.2   Protection of exposed equipment

Exposed equipment such as antennae, radar and satellite dishes, cameras, lights, etc. must be protected from direct strikes for their own sake. It is best to design nearby air terminations so they create zones of protection, intercepting the lightning strike before it reaches the exposed equipment. Metal support structures for exposed equipment could be used as part of this air termination network.

All exposed metal parts of equipment should also be bonded to the LPS, and where they can't be (e.g. whip antennae) they should be designed to withstand a direct strike or else be expendable. Exposed wiring should be run in metal conduit bonded to the LPS or else in a protected zone such as the inside angles of steel girders. Cables associated with metal masts should run inside them.

Where these measures still don't bring the surge voltage down to the immunity level of the equipment being protected, fit SPDs too. Bonds from SPDs to the LPS should be metal-to-metal where possible, or very short conductors indeed (a few inches at most).

9.3.3   Enhanced earthing and bonding

9.3.3.1   Improving the bonding between structures

Power or metallic signal cables passing between structures will inject severe surges, due to the huge differences in earth voltage that can exist between two structures when one of them is struck by lightning. They can also inject surge voltages caused by earth faults in one of the structures they are connected to.

To help protect the cables themselves, as well as the electronics in the structures they interconnect, the earthing systems of the structures should be interconnected by many parallel metallic paths, preferably forming a mesh. The cables should be laid within PECs (see section 7.4.3) – preferably totally enclosed metal conduits, ducts, trunking, etc. which are conducting from end to end, with their armour and/or screens bonded to their PECs at both ends. In turn, the PECs are bonded to the earth mesh between the structures and the equipotential earth bonding bars at the structures at both ends.

Metalwork and metal services (gas, water, etc.) passing between structures should be similarly bonded to the interconnecting earth mesh and the equipotential bonding bars at each end. Cables between structures should have any spare cores bonded to the earth at both ends, as long as those cores have sufficient cross-sectional area to carry the anticipated continuous and surge currents [3].

9.3.3.2   Improving the bonding of cables and services

One of the most important issues is that all external metallic cables (power, telephone, data, etc.) and services using metallic ducts or pipes (water, gas, steam, etc.) should enter/exit the structure to be protected in one small underground area and be bonded there to a single large equipotential bonding plate.

Recommendations for protecting telecommunication installations include the requirement that none of the bonds and SPDs for the cables and services should be more than 2 metres from the neutral of the main incoming supply disconnector, giving the entry area a maximum width of 4 metres [132]. This could be taken as good general advice for any electronic installation. This technique may be easiest to imagine as creating a "star point" between the structure to be protected and the rest of the world, and it helps prevent external surges from travelling through the structure.

Where something metallic cannot be bonded directly to earth, it should be bonded via SPDs also installed on the same single equipotential bonding plate. This plate should be inserted into the line of the ring earth electrode and bonded to it at both ends. It should also have additional multiple connections to the internal bonding ring conductor (BRC) of the structure's CBN, and to concrete reinforcing or foundation electrodes, as shown by Figure 9.4. Foundation strip electrodes or foundation earth electrodes may be used in place of the ring earth electrode.

The equipotential bonding plate to handle all the bonding and SPDs for all the incoming/outgoing cables and services will often need to be quite large, probably $> 1m^2$, similar to the internal "transient suppression plates" recommended by prEN 50174-2 [73]. For larger installations it may be acceptable to have a number of large plates side-by-side, each connected to the ring earth electrode at both ends.

Figure 9.5 shows the general principles of this technique.

A structure made of reinforced concrete could interconnect all its vertical and horizontal re-bars and use them as the "natural components" of a very effective LPS. The re-bars could be all that is needed for the down conductors and foundation earth

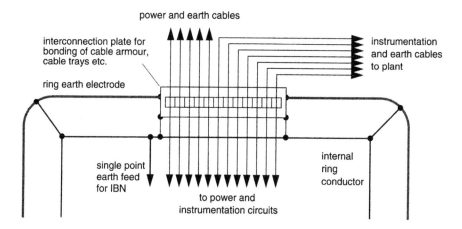

Figure 9.4 Bonding at ring earth

electrode, and give much better performance than an LPS made of copper conductors fixed to the outside and/or buried in the ground. The equipotential bonding plate for the "star point" of such a structure needs to be set into the vertical plane of the wall re-bars, and Figure 9.5 shows the principles involved.

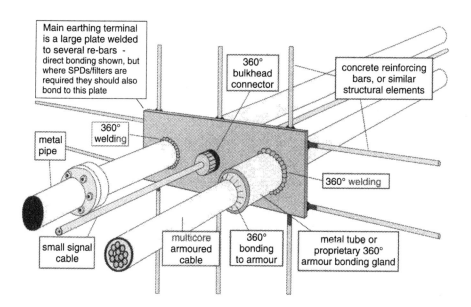

Figure 9.5 Bonding at entry/exit for maximum protection (after [132])

The mesh size of the LPS for this structure would be so small, and the number of down-conductors so large, that no re-bar would carry very large lightning currents and the current and field penetration inside the structure would be very low. The only structure significantly better than this for lightning protection would be one completely skinned in welded metal sheeting.

Where rod (or radial strip) electrodes are used, one rod (or a number of strips) should be positioned close to the equipotential bonding plate "star point" and bonded directly to it.

The two-ring approach

Bringing all the cables and services in to one small area may be difficult for existing structures, but it is considered the best design for new ones. When improving an older structure to provide added protection for electronics – and when it is not practical to route all the incoming/outgoing cables and metallic services via one small "star point" – a two-ring approach, as shown by Figure 9.6 is suggested.

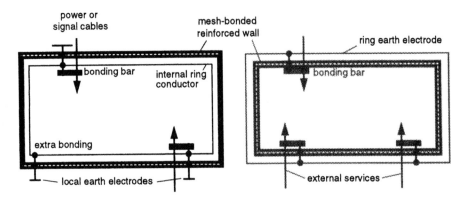

Figure 9.6 Two examples of two-ring bonding of disparate services [100]

For underground entry the two rings employed could be:

    a) a ring earth electrode plus a foundation earth electrode (reinforced concrete foundations)

    b) a ring earth electrode plus an internal bonding ring conductor (BRC)

    c) a foundation earth electrode (reinforced concrete) plus an internal BRC.

Cables and conductive parts entering/exiting a structure above ground level are not recommended as they are very exposed to direct strikes. [100] recommends that where this is unavoidable the two-ring bonding method is used again. Very sensitive structures such as fuel or explosives stores, or structures with large concentrations of electronics should not have any overhead entry of metallic cables or services at all.

For unavoidable above-ground entry the two rings employed could be:

    i) a horizontal ring conductor (bonding the down conductors together around the structure) plus mesh-bonded reinforcing bars in the wall (reinforced concrete walls)

    ii) a horizontal ring conductor plus an internal BRC

    iii) the mesh-bonded reinforcing bars (reinforced concrete walls) plus an internal BRC.

Where the re-bars in reinforced concrete are employed in the above schemes, they must be bonded into a mesh which completely surrounds the structure, to make a ring around it. Door and window frames weaken the effect, so should be welded to the cut ends of the vertical and horizontal re-bars prior to pouring the walls.

Where plastic pipes used for water, sewage, and other conductive services enter a protected structure, it might be worthwhile inserting (with the Utility's permission) a length of metal pipe (say a metre or two) where they cross the boundary of the structure, and bonding these metal pipes to the equipotential bonding plate.

9.3.3.3   Improving the bonding within a structure

Refer to section 5.3 for a thorough discussion of earth-bonding within a structure and a comparison of isolated mesh (MESH-IBN) versus fully meshed (MESH-BN) earth bonding systems.

As well as providing an equipotential earth, a MESH-BN inside a structure provides useful shielding against the magnetic fields (<1MHz) created by lightning currents. A mesh size with 3 to 4 metre spacing or less is recommended to help protect electronics from lightning effects [73]. Vertical meshes are said to give the greatest protection from typical lightning fields. Where there are no existing metallic components, large cross-section conductors or steel rods may be added. Where the 3-D earth mesh size has to be larger than 3 or 4 metres (e.g. shipping bays, display windows, etc.) sensitive electronic equipment should not be fitted near to that area unless it has been "hardened" to withstand higher levels of interference.

[41] argues that creating a MESH-BN with adequate performance at the frequencies required is expensive, and recommends MESH-IBNs instead. Star-earthed systems (IBNs and MESH-IBNs) are suitable for smaller IT systems, but to protect them from lightning magnetic fields their interconnecting cables must follow their star-earthing conductors at all times to avoid induction loops [100]. The MESH-BN system is preferred for larger or more open systems, or where services and cables enter the system at several points. Refer to section 5.3.2.1 for a discussion of the pros and cons of these two bonding methods, noting the fact that keeping an IBN isolated requires a high degree of control and inspection throughout the life of an installation, and that an accidentally compromised IBN (a single data cable brought in incorrectly) may suffer from very degraded reliability indeed.

9.3.4   Cable routing and screen bonding

Cables should avoid running near to lightning conductors, especially in vertical runs. It is always best-practice to route cables along members of the CBN, using them as a PEC (section 7.4.3) – but the outermost elements of the CBN (near the top and outsides of the structure) should not be used for cable routes because lightning currents tend to concentrate in them. The more highly meshed a CBN is, the closer to the perimeter the lightning currents will stay, freeing up more possible cable routes. However, this restriction depends upon the quality of the PEC provided by the CBN – cables run in completely enclosed metal ducts providing a high-quality very low-impedance PEC (e.g. round conduit) can usually be routed anywhere with no problems [129].

[37] reports tests with simulated lightning strikes on real installations and measuring the surge voltages induced onto the inner conductors of instrumentation signal cables. Fitting a PEC reduced the surge voltage dramatically. Lower impedance PECs (e.g. going from a single parallel conductor, through woven armouring, to a fully enclosing conduit) lowered the surge voltages in the conductors even though increasing the surge currents in the PEC. The overall improvement achieved was from 2.3kV for

the cable with no PEC to under 1V for an armoured cable in a fully-enclosed conduit – both bonded to their local earth at both ends.

Bonding cable screens to earth at one end only (see the discussion in section 7.2.4.1 on page 171) can result in flashover, with consequent damage and safety hazards during a lightning strike [1]. Cable screens should be bonded at both ends. Where direct bonding at both ends is not permitted (for a good engineering reason, not merely a mistaken desire to avoid "ground loops"), SPDs should be fitted instead to the local earth bonding plate.

Of course, bonding cable screens at both ends runs the risk of overheating the screen with earth-loop or surge currents. Most earth-loop currents are at power frequency and the solution is to run the cable over a low resistance PEC, as discussed in section 7.4.3. Annex D of IEC 61312-1 [100] gives a useful formula for calculating the minimum cross-sectional area (csa) for cable screens so they will withstand lightning surge currents.

When cables of any sort interconnecting two items of equipment follow different routes between them, magnetic fields due to lightning and other interfering phenomena (especially power-frequency fields) will drive circulating currents around the loops they create. This problem is easily avoided by running all cables along the same route, while maintaining adequate separation by distance or screening according to cable class (see section 7.4.2.3).

9.3.5   Use of isolation techniques and fibre-optics

Galvanic isolation used on its own for lightning protection of interconnections between different structures must be able to withstand more than 100kV, easily achieved by using:

- metal-free fibre-optics
- wireless, microwave, laser, or infra-red communications.

The electronic devices at the ends of the fibre or other communications links are sensitive and must be well protected from lightning events and other types of interference.

When fibre-optic cables employ metal draw wires, metal waterproofing, or metal armour, these need to be stripped well back at point of entry (possibly by as much as several metres). Alternatively, bond the fibre-optic's metalwork (directly or via SPDs) to the equipotential bonding plate at the "star point" on entry, and nowhere else.

Ordinary PCB-mounted opto-isolators can't withstand 100kV, but a combination of earth bonding and SPDs can reduce the surge voltages to as low as is required (usually 500V) to allow standard opto-couplers to be used without fear of damage.

9.3.6   Zoning and surge protection

9.3.6.1   The purpose of zoning

Generally speaking: in a properly earth-meshed and lightning-protected structure the central volume is the one least exposed to the effects of lightning. So this is the best place to install the most sensitive apparatus. Installation of electronics should be avoided on roofs, on the top floors of buildings (especially tall ones), near to outside walls, near to outside corners, near to down conductors, or near to tall structures such as masts, chimneys, etc. However, in all-metal structures resembling screened rooms location of equipment is not generally of any concern.

The concept of "zoning" in general has already been described in section 4.2.5.1 on page 90. The general idea is to identify or create "zones" within structures where there is less exposure to some or all of the effects of lightning, and to co-ordinate these with the immunity characteristics of the equipment installed in them. As is clear from the foregoing, some areas of a structure are better protected from lightning than others, and it is possible to extend the more protected zones by careful design of the LPS, earth bonding, and cabling techniques.

The idea is that equipment should only be installed in the zone that is most appropriate for its immunity to lightning effects. Although an obviously desirable approach, it may be difficult to achieve in practice since few of the immunity standards harmonized under the EMC Directive require any lightning related tests to be carried out. The telecommunications industry takes a more organized and professional view of this issue [133]. This requires the calculation of failure probabilities based upon the "resistibility" of the equipment concerned and the protection measures implemented for the zone it is used in. Telecommunication resistibility is specified by ITU Recommendations K.20, 21, and 22.

Appendix C of BS 6651 [111] has a zoning approach, which begins with a modified form of lightning risk assessment (see 9.2.1). A new probability of strike (P) is calculated, using different rules for the calculation of the effective collection area (now called A_e) which now takes full account of all connections to external metallic services, cables, etc. Once a new value of P has been calculated from $P = A_e \cdot N_g \cdot 10^{-6}$, it is multiplied by a new set of weighting factors to calculate the risk:

$$R = P \times F \times G \times H \qquad (9.2)$$

where F depends on the type of structure; G depends on the degree of isolation from other structures, trees, etc.; and H depends on the type of terrain. The standard includes tabular values for these factors. The final step in this new risk assessment is to correlate the calculated new value for R with the consequential loss rating, using the table provided (Table 9.1), to classify the exposure level. If no risk assessment has been done, a value for R of 0.6 should be used.

Table 9.1 Classification of exposure level (extracted from BS6651:1992 Appendix C)

Consequential loss rating	Exposure level			
	R<0.005	R = 0.005 to 0.0499	R = 0.05 to 0.499	R>0.5
1	Negligible	Negligible	Low	Medium
2	Negligible	Low	Medium	High
3	Low	Medium	High	High
4	Medium	High	High	High

Lightning transients progressing through the power network of a structure become attenuated. Appendix C of BS 6651 and IEEE C62.41-1991 [140] both identify three distinct zones with differing surge exposure categories: C (most severe), B, and A (least severe).

Category C: Power conductors outside a structure; supply side of main incoming LV
                  distribution   board/switchgear;   load   side   of   distribution

board/switchgear for outgoing mains cables such as to another structure or external apparatus such as transformers, pumps, external lights, etc.

Category B: Power conductors inside a structure: between load side of incoming distribution board and supply side of socket outlet or fused spur, or within apparatus not fed from a wall socket, or sub-distribution boards located within 20 metres cable run of Cat C, or plug-in equipment or fused spur within 20m cable run of Cat C.

Category A: Power conductors to plug-in equipment or fused spur located more than 20 metres cable run from Cat C and/or 10 metres from Cat B (Category A may not exist in smaller structures).

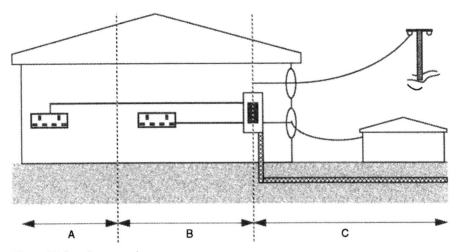

Figure 9.7 Location categories

All data/signal cables (e.g. telephone wires) entering a structure from outside are Category C over their whole length, except internal to the structure after suppression by SPDs, since surges in these lines are time-stretched over distance, rather than attenuated as happens with power cables.

9.3.6.2   Selecting SPD ratings according to zone

For each of the above exposure categories, Appendix C of BS 6651 provides a table listing the surge protection device test specifications required, for each of the risk assessment levels: high, medium, or low. An example is given for mains cables in the zone covered by Category C.

Table 9.2 Location Category C (mains) (extracted from BS6651:1992 Appendix C)

System exposure	Peak voltage (kV)	Peak current (kA)
Low	6	3
Medium	10	5
High	20	10

The surge current test levels in this table may be applied to the rating of direct earth bonds as well as to SPDs, although well-made earth bonds are either metal-to-metal or very short lengths of thick wire so should not have problems with currents of this size. The surge voltage and current test levels in the table can be used for testing electronic apparatus for its ability to withstand lightning surges on its mains inputs in the appropriate zones. The shape of the test voltage and current waveforms is also specified by BS 6651, and corresponds to the "combination wave generator" which has been used for many years and is now on the list of IEC EMC immunity test standards as IEC 61000-4-5 [93] (see section 10.2.4).

The exposure levels suggested by Appendix C of BS 6651 are based on lightning risk assessment only. If transients of other origin are present, consider upgrading any SPDs and/or specifying apparatus with higher surge immunity. For example, if in an industrial area the risk assessment suggests that an SPD suitable for a medium exposure level is appropriate, the presence of inductive switching transients may make a high exposure level SPD more appropriate. In these circumstances, specialist or manufacturer's advice should be sought.

9.3.6.3   Co-ordinating SPDs with the equipment they are to protect

SPDs function better and are most cost-effective when applied to a carefully designed and constructed protective structure, such as one which follows the above guidelines. They need to be chosen on the basis of the relevant standards or codes of practice (as described above) for the current waveforms and peak values of current and energy they must handle. To be able to co-ordinate them with the equipment they are intended to protect also requires data on their "let-through" voltages.

SPDs can be connected between a power (or signal) conductor and earth to suppress line-to-earth surges, or between two power (or signal) conductors to suppress line-to-line surges (Figure 9.8). Apparatus will have different immunity ratings for line-to-earth than for line-to-line surges, so the SPDs used for the relevant modes may not be the same.

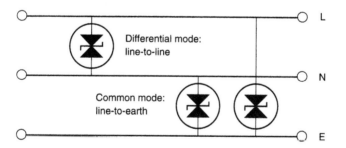

Figure 9.8 Applications of SPDs: common mode and differential mode

9.3.6.4   Types of SPD

Surge arrestors are variable resistance devices, whose resistance is a function of the applied voltage. They are designed so that they provide a clamping effect when the voltage across them exceeds a certain level, rather like the action of a zener diode.

There are four basic types of SPD:

- Gas discharge tube (GDT), essentially just a spark gap, slow but very high power and negligible leakage
- Metal-oxide varistor (MOV), a bulk semiconductor, fast and less rugged than a GDT
- Avalanche devices, a semiconductor device with a zener type of action, very fast indeed but not very high power
- Thyristor devices, another type of semiconductor device, slow but will handle high currents.

The GDT and thyristor devices have to be self-triggered before clamping, and during the initial period they let through surge voltages which may be potentially damaging. They also have a foldback characteristic, which means that once they "fire off" and are carrying current, the voltage across them drops to well below the level they were previously able to block. Careful design is needed to make sure that, if connected to a source of DC current, they do not remain in continuous conduction, once triggered by the surge. This characteristic is not shared by the MOV and Avalanche devices, but their disadvantage is a higher clamping voltage, and hence dissipation, for a given normal system operating voltage.

Figure 9.9 indicates the voltage versus time waveforms of these four types of SPD when exposed to the leading edge of a typical surge test waveform.

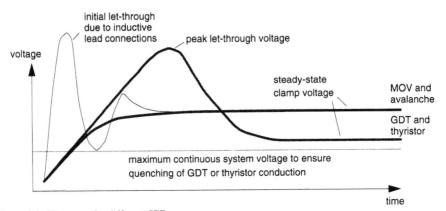

Figure 9.9 V/t curves for different SPDs

Modern SPDs used in installations often use a combination of different types to achieve the overall performance desired [20] [25]. MOV types used to be considered rather short-lived, but in recent years they have been developed so that they now appear to be very robust as long as they are rated correctly for their anticipated worst-case surge currents.

Some of the test standards for SPDs are becoming tougher too, for instance IEC 61643-1 stresses the SPD under test with 20 surges at its maximum rated current, and then checks that its electrical parameters have not changed by more than 10%. If the lightning risk analysis has been done correctly each SPD will only have to carry its full rated current very occasionally, and 20 full-current surges should take quite a while to occur.

· All SPDs are prone to failure, however, so whenever SPDs are fitted they should be inspected and checked regularly and replaced if found to be degraded or failed.

Consequently a number of SPD manufacturers are now providing SPD units fitted with indicators which warn of degraded performance and impending failure, and which also indicate a failed device.

9.3.6.5   Installing SPDs

As has been described above, SPDs are installed first at the equipotential bonding plate (or main earthing terminal in less sensitive structures) on all incoming or outgoing conductors that are not directly bonded to the earth at the same place. Other SPDs may be fitted at lightning protection zone boundaries in a similar way within the structure, and some may be fitted at the equipment itself. What they all have in common is a need to keep their lead lengths short, and to ensure that their earthing is relative to what is being protected [3]. (Connecting to a separate "SPD earth" is counterproductive and defeats the protection.) The issue is exactly the same as was described in section 8.1.4 on filter installation.

The total "let-through" voltage of an SPD is the combination of its own voltage clamping action plus the transient voltage drop in its connecting leads due to their inductance (usually reckoned to be 1µH/metre). For the rates of change of current associated with lightning strikes, if an SPD has an earth lead longer than 1 metre (or is connected to an imperfect CBN) very much greater voltage surges can be "let through" to the equipment supposedly being protected. For example, tests reported in [3] showed that on the 6kV 3kA test an SPD gave a let-through of 630 volts when it was installed correctly (leads < 250mm long and bound together), but 2 metre long bound leads let through a total of 1,200V, and unbound 2 metre leads let through 2,300V.

In general, SPDs for lightning suppression should have lead lengths of 500mm or less. Although large diameter wires are slightly better than thinner ones, by far the most important concern is to achieve the shortest lead length. Binding the leads to/from an SPD together also helps reduce lead inductance (and hence let-through voltage) considerably. Always make sure that SPD protected cables are segregated from any cables which have not been protected by SPDs by at least 150mm on a parallel run. MESH-BNs (see section 5.1.3.3) provide much lower surge impedance than ordinary CBNs, and this helps SPDs to achieve let-through voltages which are closer to their specifications.

When SPDs are correctly fitted to signal and data lines at an item of equipment, their earth bonds will be very short and connect to the chassis (local earth) of the equipment itself. Unfortunately, the neutral conductor in the mains supply is earthed many metres away (for other good EMC reasons) so when a surge is clamped by the signal's SPD it will inject a current into the chassis, raising its voltage considerably with respect to the remote neutral earthing point. This could create a surge voltage problem for the mains supply. A similar problem can occur when the SPDs are fitted locally to the equipment's mains terminals, with the resulting overvoltage appearing on its communication cables. So when fitting SPDs to any cables at an item of equipment, it generally means that all inputs and outputs and power cables need to be fitted with SPDs too [18]. Clearly this could be very expensive for equipment with large numbers of cables.

However, when equipment is correctly installed on an SRPP or bonding mat (see section 5.3.1) which is part of a MESH-BN installation, the impedance between the neutral earth connection and the local chassis of the equipment could be made so low that these problems would not arise and no additional SPDs would be needed, an example of how valuable a good earth bonding structure can be.

9.3.6.6   SPDs for special purposes

Connecting the types of wired-in SPDs usually used for mains suppression to signals and data would destroy the signal, through too much leakage, or too much capacitance, or impedance mismatch, or a variety of other problems. However, special SPDs are available fitted with connectors intended to plug directly into a wide variety of digital and analogue signal cables and be mounted to earth bonding plates. Radio antenna and high-speed data lines use matched transmission-line type SPDs with RF connectors.

Special SPD products are often sold via distribution for fitting to items of equipment, such as radio receivers, CCTV cables, telephone cables, LAN cables or computers. These may be of limited power rating and intended only for the "last-ditch" protection of equipment after the bulk of the surge has been suppressed by sturdier SPDs fitted at the point where the external cables first entered the structure.

Note that data is lost during data/signal/telephone SPD activity, so communications protocols that detect and re-send lost data will usually be needed.

SPDs should not be fitted in rooms where there is risk of fire or explosion unless special precautions are taken, such as a suitable enclosure [73].

9.3.6.7   Fusing SPDs

All SPDs fail eventually, and since the majority use metal-oxide-varistor types (whose failure mode is to leak increasingly and finally to go short-circuit), those that are placed across power conductors need to be fused.

If the fuse is dedicated to the SPD alone (see (a) in Figure 9.10), when it opens during the surge event that kills the SPD, the protected equipment may be exposed to the remaining parts of the surge and damaged. Afterwards, even if the protected equipment is undamaged, it has lost its surge protection and so is very exposed to the next surge that comes along.

If the fuse is in the common line that also goes to the protected equipment (see (b) in the figure below), then the opening of the fuse due to SPD failure will remove the power from the equipment, which may not be acceptable in some applications.

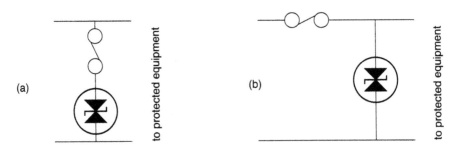

Figure 9.10 Two different fusing arrangements for SPDs

The arrangement of Figure 9.10(b) is preferred for domestic premises, where no inspections and preventative maintenance of SPDs can be expected, since the continued opening of fuses will inform the user that there is something wrong that he should get fixed. Where a system of inspections and preventative maintenance is in place, if the recent models of SPDs with "health" indicators are fitted either fusing arrangement may be chosen.

9.3.7   Protecting from non-lightning surges

Surges can originate from other sources than lightning, and most of the problems they cause are overvoltages. Considerable research has been undertaken to determine the amplitudes, rate of occurrence and other characteristics of transients and surges in various environments and much data is now available [29], [30], [140].

Externally-generated surges are especially common on incoming HV or MV power supplies, caused by the switching of large reactive loads, or load shedding by HV or MV switchgear or in the wider distribution network.

External non-lightning surge sources also include telephone and data lines outside structures, usually due to shorting to mains cables when a vehicle knocks down a utility pole, or when a mechanical digger cuts through an underground cable conduit.

Internally generated surge transients also occur. Large on/off controlled DC or AC motors can emit very significant amounts of stored energy when switched off; and surges can also be created by large power-factor correction capacitors, fuse-opening (typically double the peak value of the nominal mains voltage), and earth faults. At the more extreme end, a superconducting magnet in an MRI scanner can emit 1MJ of stored energy when its field collapses (which can happen at unpredictable times).

An early assessment of where external surges are likely to originate from and their parameters allows their suppression by some of the lightning protection techniques (especially earth-bonding and fitting of SPDs at the external ring earth electrode or MET).

Internally-generated surges are best controlled by segregating high power and sensitive equipment and their cables and power supplies (section 7.4.2), and providing a good low-impedance MESH-BN or a number of MESH-IBNs (section 5.1.3.3). But where surges originate within an area (or "lightning protection zone") it may be difficult to stop the other apparatus in that area/zone from being exposed to them, and SPD techniques may be needed.

Remember that the surge exposure levels suggested by lightning protection standards are only based on lightning risk assessments. If other surge transients are present it may be necessary to upgrade one or more of the lightning protection measures, or one or more of the other best-practices described in other parts of this book. If (for a given area) the lightning risk assessment suggests that SPDs or equipment suitable for a medium surge exposure level is appropriate, the presence of additional surges may make it more appropriate to specify SPDs or equipment rated for a high level of exposure. In these circumstances it may be best to seek specialist or manufacturer's advice.

Chapter 10

In situ testing

It is clear that in-situ testing can only be performed by experienced organisations that have a wide knowledge of both testing, use of the radio spectrum and judgement of test results in the light of equipment use at several locations. If in-situ testing is allowed in general it may create situations that can be considered dangerous.

– Association of Competent Bodies, May 1997

There are a number of differences between testing equipment for EMC at a test laboratory, and testing systems or installations in their place of use, i.e. in situ[†]. A fair amount of information (including the applicable international standards) is available to describe the former case (see for instance [13]), but much less has been written to cover the special requirements of in-situ tests. This chapter will briefly cover the standard test methods and will concentrate on those aspects and problems that distinguish their in-situ variants. The important issue of the test plan has already been discussed in section 3.2.

There is some debate as to whether testing in situ can be used to represent the compliance status of systems which are not tested on a test site. CISPR 11[‡] [66] states explicitly that

Measurement results obtained for an equipment measured in its place of use and not on a test site shall relate to that installation only, and shall not be considered representative of any other installation and so shall not be used for the purpose of statistical assessment.

In contrast, CISPR 16-2 [104] suggests that where a given system has been tested at three or more representative locations, the results may be considered representative of all sites with similar systems for the purposes of determining compliance. The US FCC Rules [138] have a similar condition. But in any case, for compliance with the EMC Directive via the Article 10.2 route, a manufacturer may write his TCF around whatever degree of in-situ testing he wishes, if his chosen Competent Body agrees.

10.1 Emissions

10.1.1 CISPR instrumentation and transducers

10.1.1.1 The measuring receiver

The fundamental instrument for measuring radiated and conducted RF emissions is the measuring receiver. The requirements for this instrument are defined in CISPR 16-1,

---

† "in situ" may mean either the manufacturer's or the user's premises, or any other location not being a controlled test environment.
‡ Throughout this chapter the CISPR documents are referred to for consistency; CISPR 11 and 22 are equivalent to the European standards EN 55011 and EN 55022.

"Specification for radio disturbance and immunity measuring apparatus and methods: Part 1: Radio disturbance and immunity measuring apparatus" [103]. The CISPR specification includes:

- bandwidth and frequency range
- detector characteristic for pulsed signals
- sinusoidal input accuracy
- input impedance and VSWR.

These parameters are summarized in Table 10.1, which shows the requirements versus frequency. Four frequency bands are defined. Bands A and B (9kHz to 30MHz) are usually applied to conducted emissions, and bands C and D (30MHz to 1000MHz) are usually for radiated emissions; but this is not universally fixed, and there are circumstances where radiated emissions are measured below 30MHz and conducted above this. At the time of writing, no CISPR requirements have been published for emissions measurements above 1GHz. This will change in the future, as the subject is being actively considered and various draft requirements have been circulated for comment. Other standards, particularly for military and aerospace applications, have been requiring tests above 1GHz for some time.

Table 10.1 CISPR measuring receiver requirements

	Band A	Band B	Band C	Band D
Frequency range	9kHz–150kHz	150kHz–30MHz	30MHz–300MHz	300MHz–1000MHz
6dB bandwidth	220Hz	9kHz	120kHz	
Sinusoidal input accuracy	±2dB			
Input impedance and VSWR	50Ω, < 2:1 with no attenuation, < 1.2:1 with > 10dB			

Two characteristics which are important for any receiver, sensitivity and frequency accuracy, are not defined in the CISPR specification except indirectly. The sensitivity requirement is stated as a maximum error (1dB) that may be introduced by background noise or spurious responses. No specification is given for frequency accuracy, although in practice good quality receivers will have more than adequate accuracy for most testing purposes.

The high cost of measuring receivers is at least partly due to the unusually severe specification that is placed on their linearity and amplitude accuracy. An EMC test is inherently a measurement of a quantity with unknown parameters; particularly, the bandwidth and nature of modulation of the signal(s) being measured are unknown. They may be narrowband, stable, single frequency signals such as microprocessor clock harmonics, or they may be broadband impulsive noise that spreads over a very wide frequency spectrum such as from DC motors. In between is quasi-impulsive noise with periodic frequency components such as from switch mode power supplies, digital data signals or variable speed inverter drives, or mains-related broadband pulses such as from rectifiers and thyristor switches. A real EUT is likely to have a mixture of several of these, as also is the environment in which it is measured (if not in a screened test chamber). The receiver must be able to measure all of these components, simultaneously if necessary, without degradation in its accuracy or linearity. This demand is not placed on any other type of radio receiver.

Two particular aspects of its specification are peculiar to a CISPR emissions measuring receiver: its bandwidth, and its detectors.

Bandwidth

Table 10.1 shows the preferred CISPR bandwidths for each of frequency bands A–D. The bandwidth specification is important because, for an interference signal whose spectrum is wider than the measurement bandwidth, the indicated signal is proportional to the measurement bandwidth. Since the interference spectrum is generally unknown, control of the measurement bandwidth is needed. Military EMC tests historically have dealt with this issue by requiring tests at different bandwidths and consequent reporting of results as "narrowband" and "broadband" emissions. CISPR takes the opposite approach of defining a single bandwidth for all measurements in a particular frequency range[†]. The bandwidth is actually defined as the difference in frequency between upper and lower –6dB points in the receiver's frequency response (see Figure 10.1).

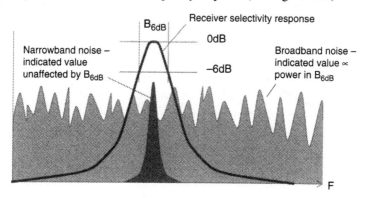

Figure 10.1 Measurement bandwidth

Detectors

CISPR 16-1 defines three principal detector types, peak, quasi-peak and average. For the present, only the latter two are used for final compliance tests, although the peak detector is often used in practice for speed of response. A continuous, unmodulated signal will give the same indicated value on all three. However, many interference sources are pulsed or modulated. CISPR has found it necessary to weight the measurement of such interference according to how severe its disturbing potential is, and this has resulted in the specification of the quasi-peak and average detectors.

The time constants of the quasi-peak detector are given in Table 10.2 and the characteristic response that these produce is shown in Figure 10.2. From this you can see that impulsive noise will give a lower indicated output as its repetition rate is reduced. Maximum output is reached as the impulse repetition rate approaches the measuring bandwidth. The average detector simply gives the average value of the pulse which, for a very narrow pulse, is directly proportional to repetition rate, whereas the peak detector gives an output which is independent of the pulse or modulation characteristics of the signal – provided that the signal appears for long enough in the measuring bandwidth for its peak value to be captured. CISPR emissions standards are structured so that the limit with the average detector is 10–13dB below that with the

† So does the latest revision of DEF STAN 59-41 [117].

quasi peak. This means that the average limit is more significant for continuous or high repetition rate pulsed interference [35], while the quasi peak limit effectively applies to low repetition rate pulsed interference. This applies only to conducted emissions below 30MHz – there is no average limit for radiated tests in either CISPR 11 or 22.

Table 10.2 CISPR QP time constants

	Band A	**Band B**	**Band C/D**
Charge τ_1, ms	45	1	1
Discharge τ_2, ms	500	160	550
Overload factor, dB	24	30	43.5

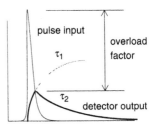

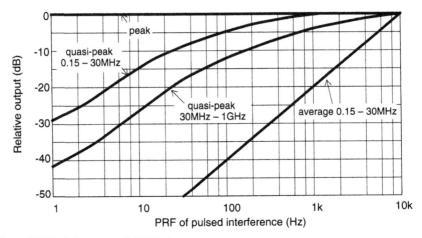

Figure 10.2 Relative output of CISPR detectors

The justification for weighting of pulsed interference is that such noise, when applied to broadcast reception, is subjectively more disturbing the higher its repetition rate, and therefore low repetition rate noise can be treated more leniently. The assumption of a particular weighting curve does not hold for digital broadcast or data transmission, and a new weighting function needs to be developed to cover this [53].

10.1.1.2 The spectrum analyser

A frequently-used alternative to the measuring receiver is the spectrum analyser. Several suppliers produce an EMC-specific version of their standard spectrum analysers, that is the instrument includes the required CISPR bandwidths and the quasi peak detector (average detection can be implemented, with care, in any spectrum analyser simply by reducing the video bandwidth [39]).

Measuring receivers normally use on-board or external software control to make best use of their facilities. A spectrum analyser can be used stand-alone, or an external software package can control it. The major advantage of the analyser is that it can show a large part of the spectrum in a single sweep, in near real time, whereas a conventional receiver requires software control and a sweep of several seconds to achieve this (top-of-the-range receivers may combine both functions). This can be especially useful in diagnostic and pre-compliance work although its value in full compliance testing is less marked.

A spectrum analyser has a number of disadvantages compared to a measuring receiver which have a bearing on in-situ testing. These revolve around the extreme sensitivity of its input, and its lack of discrimination of high-level wideband input signals.

Input destruction

In order to achieve the capability of near-instantaneous sweep across GHz, the analyser has no tuned filtering at its input. If no attenuation is switched in, the input connector is directly connected to an extremely sensitive semiconductor mixer diode. This device can be destroyed very easily by quite low-level transient impulses, certainly within the scope of what might appear on a direct supply connection. The analyser front-end is therefore not robust, and should never be connected directly to the output of a conducted emissions transducer such as a LISN or voltage probe. A transient limiter is an essential accessory which must be interposed between the two. This is particularly important when making in-situ conducted measurements, since direct connection to a mains supply via a voltage probe is often needed, and no isolation from mains-borne transients is provided by this connection.

The measuring receiver is more robust in this sense, since it has front-end selectivity which reduces the energy of transients, as well as protection devices of its own. Considering the cost of receivers though, many test engineers will use an external limiter even with a receiver.

Input overload

Allied to the issue of lack of selectivity, the spectrum analyser can also be overloaded by wideband pulsed noise, or by other high-level signals on its input. This results in an artificially low indicated value, and can also generate spurious signals through intermodulation in the analyser's signal processing chain. If the test operator does not realize that the analyser is being overloaded, the measurements may be logged as if they were real. Since the high-level signals that are responsible for the overload can appear outside the frequency range under observation – pulsed noise below 150kHz is a common phenomenon – their presence may often remain unknown.

The use of tuned input circuits to provide selectivity in the measuring receiver overcomes this problem, and to some extent this can also be applied to the analyser, through a device known as a "preselector", which provides a degree of external selectivity. The problem is more acute for conducted emissions rather than radiated emissions testing, since wideband pulsed noise is more common on mains ports of all kinds of apparatus, and so measuring receivers are normally to be preferred for these tests. It is probably also true that with a good quality spectrum analyser an unusually high level of pulsed noise would be needed to cause overload, and most AMN/LISNs (see next section) include high-pass filters which remove the worst offending part of the spectrum, below 9kHz.

10.1.1.3 The AMN/LISN

Conducted emissions measurements are most commonly made on the mains supply port of the equipment under test (EUT), principally because the mains supply offers the most direct route for conducted disturbances to reach other mains-powered apparatus, and because mains supply wiring is inherently a relatively efficient radiating antenna at the frequencies of the conducted tests. The third edition of EN 55022 (CISPR 22) [68], published at the beginning of 1998, includes a requirement for a conducted emissions test on telecommunication ports [64] but these have yet to be widely applied. A set of widely harmonized mains port limits have been published (see Figure 10.3); the Class A and B levels (the thicker lines) are common to most CISPR-based standards.

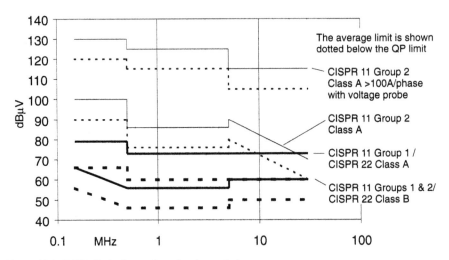

Figure 10.3 CISPR limits for conducted mains emissions

These limits assume a defined RF impedance presented to the EUT's mains connection by the test set-up. In general, the mains impedance can vary over a wide range depending on the circumstances of the individual test location. Since we are measuring a voltage across this impedance, and since the EUT has a source impedance which is unknown but may be quite high, in order to make a test which is repeatable between different laboratories the mains impedance must be stabilized. This is the function of the Artificial Mains Network (AMN, to give it its CISPR term) or Line Impedance Stabilizing Network (LISN, which was originally a US term applying to several different types of network).

The requirements of the AMN/LISN are defined in CISPR 16-1. Several variants are presented but one in particular has become the norm for standard tests: this is known as the 50Ω/50μH + 5Ω AMN/LISN. The principal requirement is that it should present an impedance between each line and the earth reference point which is equivalent to 50Ω in parallel with 50μH, for the frequency range from 150kHz to 30MHz. For tests below 150kHz, a 5Ω resistance appears effectively in series with the 50μH inductance. Figure 10.4 shows the basic circuit of such a network and Figure 10.5 shows the impedance–frequency response given in CISPR 16-1 to which it is calibrated. An identical network referenced to earth is needed for each line, hence two networks are needed for a single phase supply, three or four for a three-phase supply. The high-pass filter shown in the circuit is not mandatory, but most commercial AMN/LISNs implement it; it is helpful in reducing low-frequency (50Hz and harmonics) feedthrough to the measuring receiver and consequent overload.

The AMN/LISN must also couple the signal to be measured, with a known and preferably low insertion loss, to the input of the measuring receiver; and it should attenuate signals that are coupled into the test set-up from the mains supply. These functions are carried out by the other components in the network.

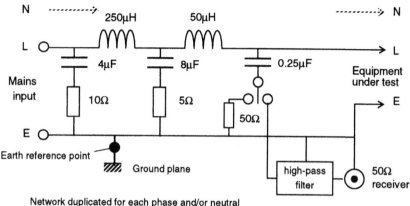

Figure 10.4 Circuit of the CISPR AMN/LISN

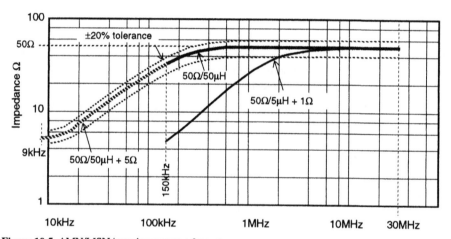

Figure 10.5 AMN/LISN impedance versus frequency

CISPR 16-1 also defines a 50Ω/5μH + 1Ω network, whose impedance curve is also shown in Figure 10.5. This is said to be suitable for currents up to 500A and so might be relevant for in-situ testing. However it is not specifically referenced in any of the usual CISPR standards, it is not widely available, and consequently it is rarely if ever used in commercial tests.

The most important aspect of using the AMN/LISN is that its earth reference point must be solidly bonded to the ground plane which forms part of the test set-up. This is covered in greater detail in section 10.1.2.1. You will also notice that, according to the standard circuit, each network presents about 12μF capacitance between the phase and earth. In the case of the live line, the full 230V AC at 50Hz appears across this capacitance, and this causes a current of nearly 0.9A to flow in the earth conductor. This current can easily be fatal if it is not properly returned to mains safety earth, and therefore *secure and reliable safety earth bonding is vital when using an AMN/LISN*. A regular testing regime for the security of the earth bond is recommended (with records kept), the equipment should be marked by the manufacturer with the mandatory safety warning about high earth leakage currents, and the earth should never be disconnected without first isolating the AMN/LISN from all poles of its mains supply, including neutral. Use of an AMN/LISN should be restricted to appropriately safety-trained and competent staff, and all others should be prevented from touching or using it.

A secondary result of this earth current is that the mains supply to the AMN/LISN cannot be protected by an earth leakage or residual current circuit breaker (RCD). If an RCD-protected supply has to be used, or personnel safety cannot be ensured by adequate bonding, then the mains supply should be passed through an isolation transformer. This will not affect the RF properties of the test, though it will of course limit the current that can be supplied, which may be a problem for power convertors (such as the DC power supplies found in most modern apparatus) as they may saturate the isolating transformer with the very large peak currents they draw each half-cycle. This can distort the mains waveform and may lead to erroneous EMC test results. Consequently, where an isolating transformer is to be used to supply an EMC test, it may need to be rated for a very much higher VA than would at first seem to be necessary. (Power-factor-corrected supplies that meet EN 61000-3-2 draw substantially sinusoidal supply current and so will not cause this problem).

10.1.1.4 The voltage probe

Of particular interest for in-situ tests is the alternative option provided in CISPR 16-1 to use a voltage probe. The AMN/LISN discussed above is designed to pass the full mains supply current; it must be inserted in series with the mains supply. This can be inconvenient and sometimes impossible, if

- the supply current drawn by the EUT is greater than the rating of the AMN/LISN, or
- the mains supply cannot easily be interrupted, either electrically or physically, to insert an AMN/LISN for the test.

In either of these cases it is necessary to use the voltage probe. This is a very straightforward and simple device (Figure 10.6), consisting of a series resistor, a 50Hz AC blocking capacitor and an inductor to provide a low impedance to 50Hz at the receiver connection. The resistor gives a typically 30dB insertion loss which must be corrected for to arrive at the proper measured value. The probe is connected between the measurement point (phase or neutral connection) and a local reference earth point, discussed further in 10.1.2.1.

The voltage probe is inserted across the mains connection rather than in series with it and therefore is unaffected by mains current. Its disadvantage is that it cannot stabilize the RF impedance; its own impedance is high enough not to affect the measured circuit, but it does not substitute for the actual (unknown) RF impedance of the supply at the point of measurement. Therefore, strictly speaking, its only use should be for in-situ measurements where the measurement results are relevant only to that particular installation, and should not be taken to represent the same EUT used at different locations (see the beginning of this chapter).

10.1.1.5 Antennas

Radiated field tests require a measuring antenna to convert the incident field strength to a voltage which can be measured by the receiver or spectrum analyser. The limits for radiated emissions are quoted in terms of dBμV/m (electric field strength) for the common frequency range of 30–1000MHz (Figure 10.7); those few standards which require magnetic field tests below 30MHz quote their limits in dBμA/m. EMC antennas must be as broadband as possible in order to cover a wide frequency range in a single

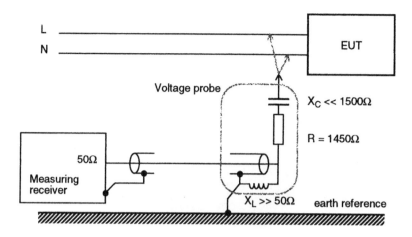

Figure 10.6 The voltage probe

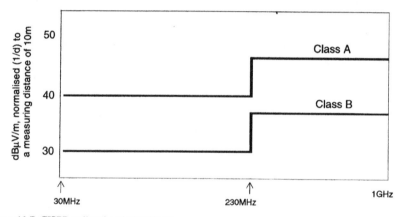

Figure 10.7 CISPR radiated emission limits

sweep. The antennas which are now commercially available and widely used (Figure 10.8) fall into three types:

- loops, for the range 9kHz–30MHz (Bands A and B)
- biconicals, log periodics or the hybrid BiLog, for the range 30MHz–1GHz (Bands C and D)
- horns, for above 1GHz (some log periodics or BiLogs can be stretched to 2GHz).

The antenna factor is the most important property for EMC measurements because it determines the voltage at the antenna terminals for a given incident field. Each calibrated broadband antenna will be supplied with a table of its antenna factor versus frequency. This will apply for an incident field in the same plane of polarization as the antenna, and with the antenna terminals loaded with its specified impedance (usually 50Ω).

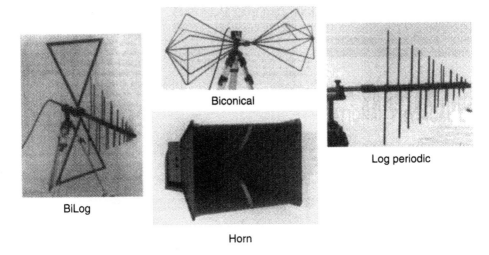

Figure 10.8 Antenna types for radiated tests

Note that in virtually all circumstances a cable is used to connect the antenna to the measuring instrument. The attenuation of this cable (which is also frequency-dependent) must be allowed for in calculating the field strength from the actual measured voltage. Modern software-controlled spectrum analysers and receivers allow for the antenna factor plus cable loss to correct the measured reading before it is displayed, so that the display can be calibrated directly in terms of field strength. The antenna factor AF and cable attenuation A are added to the indicated voltage V to give the true field strength value:

$$E \ (dB\mu V/m) \quad = \quad V \ (dB\mu V) + AF \ (dB/m) + A \ (dB) \quad\quad\quad (10.1)$$

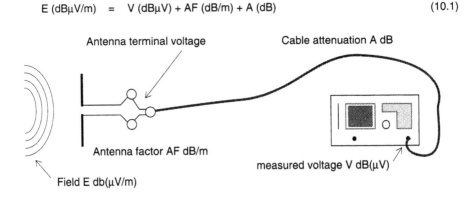

Figure 10.9 Using the antenna factor

The loop

The loop antenna is a coil which is sensitive to magnetic fields and is shielded against electric fields. Measurements below 30MHz are in the near field for most of the frequency range, and magnetic field measurements give greater repeatability than electric field, since the magnetic field is largely undisturbed by nearby objects (including the antenna). The magnetic field component perpendicular to the plane of the loop induces a voltage across the coil which is proportional to frequency according to Faraday's law. The voltage is also proportional to the area of the loop – the larger it is, the more sensitive is the antenna.

Few standards require a loop measurement for apparatus. It can though be very useful for in-situ measurements to determine where particular sources of interference are located and their area of influence. The loop is quite directional (it has been used for direction-finding since the early days of radio); a sharp null in response occurs in a direction orthogonal to the axis of the loop.

By contrast to the loop, both CISPR 11 and CISPR 22 require radiated emissions testing from 30MHz to 1GHz. These tests can use either a combination of biconical and log periodic, or a single BiLog, to cover the whole range. These antennas are a sophisticated variation on the electric dipole, modified to extend their frequency range.

The biconical

The standard antenna for the 30–300MHz frequency range is the biconical. This is a development of the simple half-wave dipole in which the elements have been formed into a conical shape, which dramatically increases the operating bandwidth. The usual form is a skeleton construction which maintains the antenna's electrical properties while minimizing weight and wind resistance. The directivity and plane of polarization remain the same as the dipole.

Different versions of the biconical, with minor variations in construction, cover either 30–200MHz or 30–300MHz. The sensitivity is greatest at the half-wave resonant frequency, around 75MHz, at which the AF is 5–6dB/m. Either side of this the AF increases rapidly to 15–20dB/m, making the measurement less sensitive. Because the antenna is balanced, it must be interfaced to the unbalanced 50Ω coax cable via a balun (balance-to-unbalance transformer) mounted at the centre of its elements.

Adequacy of the balun is one of the most important characteristics of an accurate measurement antenna. Any imbalance results in considerable sensitivity of the antenna to nearby conducting objects, especially the ground plane and the antenna feed cable.

The log periodic

The log periodic structure (so called because the element dimensions and spacing are logarithmically periodic with frequency) is commonly used for the higher frequencies up to and beyond 1GHz. With this structure a small amount of directivity is available, which gives some gain and hence a lower AF than would be the case for a simple dipole. The structure is essentially an array of tuned dipoles which is fed via a transmission line formed by the twin booms, which removes the need for a separate balun.

The directivity means that the antenna factor for signals impinging from off axis, such as from ground plane reflections, is slightly different from that for direct signals (CISPR 16-1 allows no more than 1dB difference). Also, the operational part of the antenna (known as the "phase centre") shifts with frequency as different elements become active, so that the effective distance to the EUT also changes with frequency. For the sake of uniformity, distance measurements are always taken to a reference point in the middle of the boom. The effect of a shifting phase centre is most marked at close separation distances, e.g. 3m.

Most commercially-used log periodics cover the frequency range 200–1000MHz or 300–1000MHz. The type can be designed to cover a very wide range, e.g. 80–1000MHz or 200MHz–5GHz, but it becomes unwieldy if extended to lower frequencies.

The BiLog

The BiLog was designed in 1993 by the University of York. It is essentially a combination of a biconical and log periodic with the element dimensions being substantially the same, the biconical elements having been collapsed into a flat "bow-tie" shape. This approach means that the elements must be driven via a balun at low frequencies and via the boom transmission line at high frequencies.

The BiLog's benefit is that a single antenna covers the whole frequency range from 30 to 1000MHz. This is extremely valuable to test houses as the full compliance measurement can be performed in one sweep, saving time spent changing antennas, with the associated introduction of unreliability due to changed set-ups, damaged connectors and so on. It is larger and more cumbersome than either biconical and log periodic, for which reason it is less popular for in-situ measurements. However a more recent development is the X-wing BiLog, whose biconical section is extended to give an improved low frequency response. This has allowed a more compact version to be produced which is as portable as the other types.

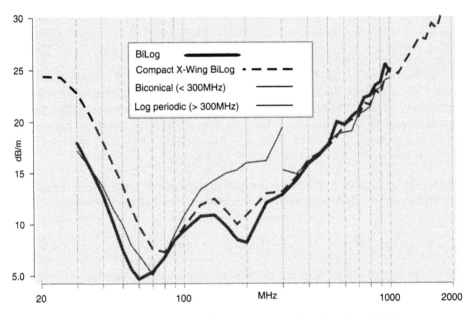

Figure 10.10 Typical antenna factors versus frequency (courtesy Schaffner-Chase EMC)

10.1.2   Conducted test methods

To understand the issues surrounding conducted emissions tests, it helps to be familiar with the equivalent circuit. Figure 10.11 shows a generalized circuit which is applicable to any type of equipment under test, whether it is a component, apparatus or system. There are four significant parts to this circuit: the AMN/LISN, the EUT, the mains cable between the two, and the earth reference. In the case of a test in a test laboratory, the earth reference is a conductive plane whose minimum dimension must be substantially larger than the EUT. For in-situ testing, the earth reference arrangement may be more flexible.

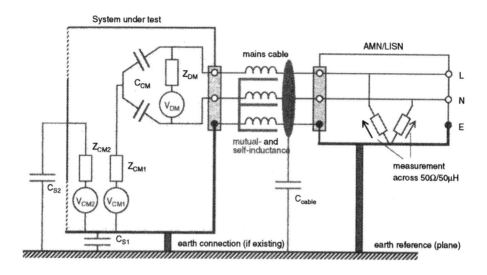

Figure 10.11 The equivalent circuit for conducted tests

The EUT can be represented as having three distinct equivalent noise sources. These are:

- the differential mode source V_{DM} across live and neutral: caused for example by harmonics of the power supply switching current, due to rectifiers or semiconductor switches;

- the common mode source V_{CM1} between the structure of the equipment and the mains connection; this will normally be capacitively coupled through C_{CM}, since the mains is isolated from the chassis, and represents the noise sources within the equipment that can be contained by shielding and mains filtering;

- the common mode source V_{CM2} which represents noise sources that are capacitively coupled to the environment by leakage through the shielding or through lack of shielding, shown as C_{S2}.

The EUT may have a direct earth connection, in which case V_{CM1} and V_{CM2} are isolated from each other and only V_{CM1} is measured. If it does not, or if the earth connection has a high impedance, then the metalwork is coupled to the earth reference via C_{S1} and the green-and-yellow safety earth in the mains cable (assuming it is safety Class I apparatus). In this case, both common mode sources are relevant; the amplitude of the voltage due to V_{CM2} is affected in a complex manner through C_{S1}, C_{S2} and the mutual inductances within the mains cable.

The mains cable itself can also affect the measurement of common mode voltages. If it is run close to the earth reference then C_{cable} can be large enough to affect the load impedance presented by the AMN/LISN at higher frequencies, or it may introduce its own resonances when combined with the cable's inductance or other inductances in the EUT.

The differential mode voltage V_{DM} is largely unaffected by the stray capacitances or by variations in the mains cable. It can be measured quite successfully without serious uncertainties being introduced by layout issues. Unfortunately, there is no way of isolating it from the other sources with a conventional AMN/LISN, and experience suggests that there are very few types of EUT in which it is the dominant source across the whole frequency range.

10.1.2.1 Important issues in the conducted test

Armed with this understanding of the equivalent circuit, we can now explore some of the aspects which are important to achieve a successful and repeatable test. The features which most affect the coupling of the interference are the earthing of the equipment under test, the stray capacitance between the EUT and the earth reference, and the layout of the mains cable. These aspects are all closely defined in the test set-up for laboratory tests, particularly in CISPR 22, but there is much less definition for an in-situ test. It is also vital that the AMN/LISN should make a good low-inductance connection to the earth reference.

Clearly the earth reference is a crucial factor in all of these. In test laboratories it is defined by the ground plane as part of the test set-up, but in-situ earthing has to reflect the existing situation. CISPR 16-2 recommends as follows:

> The existing ground at the place of installation should be used as reference ground. This should be selected by taking high frequency (RF) criteria into consideration. Generally, this is accomplished by connecting the EUT via wide straps, with a length-to-width ratio not exceeding 3, to structural conductive parts of buildings that are connected to earth ground. These include metallic water pipes, central heating pipes, lightning wires to earth ground, concrete reinforcing steel and steel beams.

> In general, the safety and neutral conductors of the power installation are not suitable as reference ground as these may carry extraneous disturbance voltages and can have undefined RF impedances.

> If no suitable reference ground is available in the surroundings of the test object or at the place of measurement, sufficiently large conductive structures such as metal foils, metal sheets or wire meshes set up in the proximity can be used as reference ground for measurement.

There is no hint as to what size of structure is "sufficiently large", and this is usually the difficulty when it comes to implementing a test ground plane on site. Consideration of the equivalent circuit suggests that it needs to be large enough to stabilize C_{S1}, C_{S2} and C_{cable}, in other words that it should underlie those parts of the EUT which are relevant for these stray capacitances. Where these are not obvious the largest practicable structure is used by default. However, there is also a prohibition in some documents against using a test ground plane at all, if it is not a permanent part of the installation.

A proper choice of earth reference is necessary both for the AMN/LISN test and the voltage probe test; in either case the reference connection of the transducer must be bonded via a short, wide strap to the chosen reference point. Lengths of green-and-yellow wire are not adequate! Choice of the point of measurement will be dictated by the circumstances of the test, but the normal choice will be at the boundary of the system where the power supply connects to it; typically this would be the terminals of a power outlet or a supply transformer that is dedicated for the system.

If the route of the mains cable leading to the point of measurement is well defined, for instance run in a fixed conduit, there is no problem. Sometimes this is not the case and the mains cable is loosely bundled or coiled on the floor when the test is undertaken. As should be clear, this is not the optimum situation and some steps should be taken to fix and record the layout of the cable, at least for the duration of the test, in what is hopefully going to be representative of its operational state. Because the stray capacitance C_{S2} is often dependent on other (signal) cables, the layout of these should be fixed and recorded similarly.

10.1.2.2 Test procedures

Practically none of the test standards provide adequate guidance for in-situ testing, so site-specific test plans have to be developed. Many decisions have to be taken by test engineers on the spot, taking into account the local situation and specific aspects of the EUT. Nevertheless, some basic practices which apply to conducted tests in the laboratory will also apply here:

- wherever possible, an ambient scan is taken at the chosen test point with the EUT switched off or quiescent. Situations where it is impossible to eliminate the EUT from the measurement (usually because it cannot be switched off) are severely disadvantaged;

- a peak detector sweep is made with a reasonably fast scan speed, taking into account the EUT cycle time, to identify parts of the spectrum in which EUT emissions are most significant, and a list of these frequencies is created;

- these frequencies are re-tested individually with quasi peak and average detectors to make the comparison with the limits, modifying the EUT's operation to maximize the emissions if this is relevant;

- this procedure is repeated for all phases at each location to be measured.

10.1.3 Radiated test methods

Conventional radiated emissions compliance tests as described in CISPR 11 and 22 for performance at a test laboratory have the following features:

- the test is performed on an Open Area Test Site (OATS) with a defined site attenuation characteristic

- the measurement antenna is placed at a fixed distance from the EUT

- the test site includes a ground plane between the EUT and the antenna

- the antenna must be scanned in height to find the height at which the measured emissions are maximum

- the EUT must be rotated in azimuth to find the direction of maximum emission

- the EUT set-up (particularly cable layout) and operating mode must be varied to find the configuration that gives maximum emission

- all test must be performed for both horizontal and vertical polarization of the antenna.

Each of these features is now discussed in turn, relating to the methods for in-situ tests. A further standard with some relevance is ETS 300 127, "Radiated emission testing of physically large telecommunication systems" [76]. Although this does not refer directly to in-situ testing, some of its requirements may be applied. A draft amendment to CISPR 22 relating to user installation testing [107] was circulated in late 1998, and some of the comments in this chapter have regard to that document. Another standard explicitly for in-situ testing is under consideration (prEN 50217), but is not sufficiently mature to be reviewed here.

10.1.3.1   Test site and measurement distance

Site attenuation is the overall loss between two antennas on a measuring site, spaced at the measurement distance. It can be obtained theoretically, and theoretical figures versus frequency are given in CISPR 16-1 Annex G for the two polarizations, three common measurement distances and two different types of antenna. Normalized site attenuation (NSA) is the figure obtained when antenna factors are corrected for. Figure 10.12 gives the theoretical curves for 10m and 3m distances. The criterion for acceptability of a measurement site is that the measured NSA should not depart from the theoretical value by more than ±4dB across the useable frequency range.

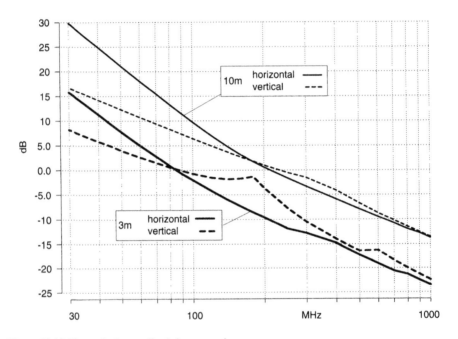

Figure 10.12 Theoretical normalized site attenuation

Imperfections which cause these departures are related to irregularities in the ground plane, inadequate size of the ground plane, and reflecting or absorbing objects within close proximity of the site. While a laboratory test site can control these factors, an in-situ test cannot; it is very rare to find a site where the absence of reflecting or absorbing objects is ensured. So, NSA is not a measure which has any real relevance to in-situ tests.

Measurement distance

Radiated emissions limits are set for a specific measurement distance, normally 3m, 10m or 30m, between the reference point on the antenna and the boundary of the EUT. 30m tests are rarely used in practice because of the size of the site needed and because of ambient noise considerations. Limits for 30m could be corrected to 10m by increasing them by a factor of 9.5dB, assuming a linear 1/d relationship for the radiated signal. Similarly, 10m limits could be corrected to 3m in the same way. This practice is allowed by CISPR 22 *but not by CISPR 11*, which allows closer measurement distances but does not permit correction of the limit values – although a recent amendment to CISPR 11 (A1:1999 to the third edition) has changed this.

As may be imagined, this restriction in CISPR 11 is the source of some controversy. It has its roots in the observation that the linear 1/d relationship does not hold in many circumstances, the closer in to the EUT the measurement is made. This is due to near-field effects (at 30MHz the near field/far field transition distance is 1.6m, and near field coupling may extend out some distance beyond this) or, with large EUTs, the problem that a close-in antenna is "seeing" only a portion of the whole EUT.

But most in-situ measurements are severely restricted in the measurement distance they can employ, because of local features. Although CISPR 11 envisages measurements at the boundary of a premises at any distance between 30m and 100m for Group 2 Class A equipment (CISPR 22 allows 10m), in most circumstances in-situ measurements have to be made *inside a building*, at much closer ranges to the EUT. A practical way around this problem would be to make cross-checking measurements of some particular emissions at different distances, to see whether the 1/d relationship holds in the situation of interest. If it does, then the rest of the measurements might have corrected limits applied. Though not strictly in accordance with CISPR 11, and of questionable technical merit since the same distance law does not necessarily hold for all emissions, it may be acceptable as part of a self-certification exercise if the manufacturer's EMC expert agrees, and it may also be acceptable to a competent body assessing a TCF. It is always recommended to discuss such issues with the appropriate experts or competent bodies before carrying out the testing.

10.1.3.2   Ground plane and height scan

The OATS features a ground plane as an integral part of its layout. The EUT emits radiation in all directions: some reaches the measuring antenna directly and some is reflected from the ground. The purpose of the ground plane is to stabilize the reflected wave. If the ground between the two were left untreated, variations in reflectivity between test sites and on a daily basis at the same test site, due for instance to rainfall, would cause variability in the measurements of the same EUT under otherwise identical conditions. The alternative, of testing an EUT in free space, has not (until recently[†]) been considered practical.

Placing a metal plane on the ground between the antenna and EUT ensures that the amplitude of the reflected wave is known – assuming no losses in the plane, the reflected wave is slightly less than the direct wave, because of the path difference (Figure 10.13). Provided that otherwise the open area is truly open, i.e. no other reflecting objects exist, only the direct and ground-reflected waves are received by the antenna and the measurement set-up is stable and repeatable.

Note the significant difference in function between the *ground plane* for radiated tests and the *earth reference plane* for conducted tests. The ground plane acts only as a reflector for incident waves. Its electrical connection is irrelevant; except for subsidiary requirements of electrical safety, it could float in space and still fulfil its purpose. The earth reference plane is an integral part of the conducted test circuit and must be connected directly to the AMN/LISN reference terminal in order to work.

The height scan

The disadvantage introduced by the ground plane is that the relative phases of the direct and reflected waves vary with frequency. At some frequencies, they will be in antiphase and the received signal will suffer a null. To get around this, the antenna is scanned in

† There is now a proposal within CISPR to standardize an alternative method to the OATS using a fully anechoic room (FAR), which approximates to free space. At the time of writing this proposal has not matured into a published standard.

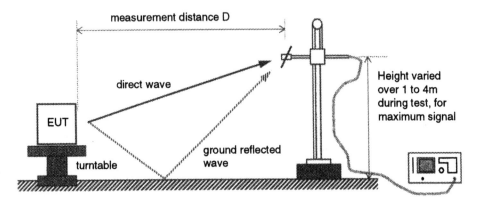

Figure 10.13 Reflection from the ground plane

height, from 1 to 4m (for 3 and 10m measurement distances) and the maximum indicated value is taken for the measurement. The effect of the height scan is to vary the differential path length over more than $\lambda/2$ for most frequencies. This will then ensure that nulls are eliminated.

For in-situ tests, a ground plane is rarely practical, and because of the existence of other reflecting objects, implementing one would not significantly improve measurement repeatability. It can also be argued that an in-situ test should measure the actual emissions rather than an artificially modified situation. If, then, a ground plane is not used, then neither is a height scan needed for the purpose of removing nulls. However, since the EUT radiation pattern will vary with height, it is still preferable to make a height scan at each measurement position to find the maximum. Any restrictions on the achievable height scan should be noted in the test report.

The combined effect of ground plane and height scan is to result in a measured value which is 5–6dB higher than the direct wave alone. It could be said that if only the direct wave is being measured, without a ground plane, then the limits that are applied should be reduced by 5dB to take this into account. This suggestion has not been seriously pursued and is not found in any CISPR documents that directly specify in-situ tests.

10.1.3.3 EUT azimuth

When equipment is tested on an OATS, it must be rotated in azimuth (that is, in the horizontal plane) in order to find the maximum direction of emission at each frequency. The same principle applies with in-situ testing, except that the EUT cannot normally be rotated. Therefore the measurement antenna has to be moved around the boundary of the EUT instead.

CISPR 11 requires that the number of measurements made in azimuth are "as great as reasonably practical", with at least four measurements in orthogonal directions, and in the direction of any existing radio systems which may be adversely affected. The four orthogonal directions should be viewed as the absolute minimum, with more being taken where the local topography allows it. ETS 300 127 [76], which refers to laboratory testing of large telecom systems where a turntable is impractical, requires eight equally spaced radials; but it also allows the measurement to be performed in one direction only, if the EUT meets the specified limits minus 10dB, and provided the chosen direction is that which is likely to produce the worst case emissions. Justifying this latter condition would be difficult for an in-situ test.

In practice, questions of accessibility will determine what directions are actually possible. It is not necessary to maintain exactly the same measurement distance in each direction.

10.1.3.4 EUT configuration

On a test site, the EUT has to be configured in a representative build state and operated in its typical operating mode. The configuration should then be varied, keeping it representative, in order to find that which gives the maximum emissions.

For in-situ measurements, this requirement is relaxed, because the measurement is of an actual operating system that is installed in a particular configuration in its place of use. Any variation of configuration that may be necessary therefore depends on how flexible the equipment is. This has to be a matter for the experience of the test engineer and the requirements of the client. However, it is usually necessary to explore the effect of operating mode (for instance, the operating profile of motors supplied by variable speed drives) in order to be sure that the most disturbing state has been found.

10.1.3.5 Antenna polarization

In contrast to all the variables discussed above, the simple instruction to repeat the measurement for both antenna polarizations is quite straightforward. Usually this will be done in the initial stages at each chosen measurement point, and if there is a marked difference between the results then the polarization with the higher result at each frequency is investigated fully.

10.1.4   Practical aspects of in-situ emissions tests

The basic feature of an emissions test is that the EUT is operated in its normal operating mode and its interference emissions are measured across a frequency spectrum, and using measuring instrumentation and methods, defined by the standard that is being applied. The differences that must be addressed when an in-situ test is being performed are:

- can the standard measurement method be applied successfully and correctly; if not, what are the implications of necessary departures from the standard method?

- can the emissions that are due to the EUT be successfully distinguished from those that are caused by the background environment?

These issues can be dealt with relatively easily when a single product is tested at a test laboratory, but they assume much greater significance when a system or installation is being tested in the field.

10.1.4.1 Departures from the standard method

The previous few pages have catalogued the major areas in which emissions tests in situ might differ from emissions tests on a test site. In many cases these differences are authorized by the standards themselves. If they are not, it is up to the test engineer to justify the course of action taken in a particular case, and record it in the test report. The greater the departure from the standard method, the more detailed and thorough should be its justification, and the more understanding of basic principles of EMC coupling and testing is demanded of the person doing the justification.

For example, a common issue is the choice of position of the measurement antenna around the EUT for radiated tests. The determining factor is usually accessibility, as has been noted above. But the test engineer must appreciate the impact of variations in the standard measuring distance, proximity to metallic objects both of the antenna and the EUT, the effect of the flooring, and the issues surrounding antenna height and polarization, in order to be able to assess the impact of a particular position on the test result. All these factors, along with their influence on the final decision, must be noted in the report. Similar issues apply to the choice of transducer and measurement point for conducted emissions.

10.1.4.2 The problem of ambients

Ambient signals are those from other transmitters or unintentional emitters such as industrial machinery, which mask the signals emitted by the EUT. On any open site they cannot be avoided, and for in-situ measurements they can be especially problematic (Figure 10.14). Continuous narrowband signals can be identified and tagged as such in the test report; transient signals may require multiple measurement sweeps to eliminate them. Continuous broadband noise above the limit is the most difficult to deal with, since you can neither work around it nor wait for a break in it.

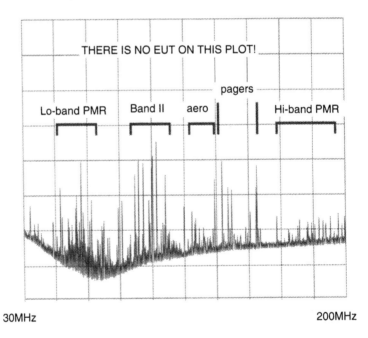

Figure 10.14 Ambient signals in a radiated measurement

Emissions standards recognize the problem of ambient signals and in general require, rather optimistically, that the test site ambients should be below the limits by at least 6dB. When they exceed this, CISPR 22 recommends testing at a closer distance such that the limit level can be increased – as discussed earlier, CISPR 11 does not allow this. This is usually only practical in areas of low signal strength where the ambients are only a few dB above the limits. CISPR 22 also suggests interpolating between adjacent readings, but this is a dangerous practice if the emissions are narrowband. CISPR 11 includes a formula in its annex C which assumes that the ambient signal can be subtracted from the EUT emission, but this is clearly stated only to be valid where the unwanted signal is from a stable AM or FM transmitter with a total amplitude up to twice the EUT emission, and if the EUT emission is itself stable in amplitude and frequency – a rare combination.

A proposed amendment to CISPR 16-2 [105] attempts to address the problem of ambients from another angle (Figure 10.15). This distinguishes between broadband and narrowband EUT emissions in the presence of broadband or narrowband ambient noise. If both the ambient noise and the EUT emissions are narrowband, a suitably narrow measurement bandwidth is recommended, with use of the peak detector. The measurement bandwidth should not be so low as to suppress the modulation spectra of the EUT emission. If the EUT noise is broadband, the measurement cannot be made directly underneath a narrowband ambient but can be taken either side, and the expected actual level interpolated.

When the ambient disturbance is broadband, bandwidth discrimination is not possible, but a narrowband EUT emission may be extracted by using the average detector with a narrower measuring bandwidth that maximizes the EUT disturbance-to-ambient ratio. The average detector should reduce the broadband level without affecting the desired EUT narrowband signal, as long as the EUT signal is not severely amplitude or pulse modulated; if it is, some error will result.

Broadband EUT disturbances in the presence of broadband ambients cannot be directly measured, although if their levels are similar (say, within 10dB) it is possible to estimate the EUT emission through superposition, using the peak detector.

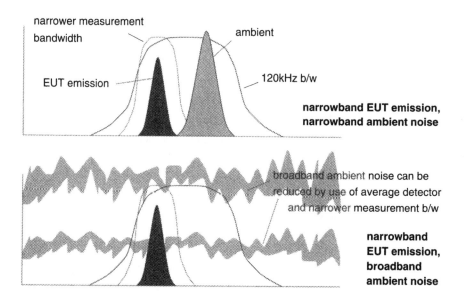

Figure 10.15 Ambient discrimination through use of detectors and bandwidths

10.1.4.3 The in-situ test engineer's kit

The practical demands of site testing – and particularly the expensive consequences of inadequate preparation – mean that an engineer who specializes in this activity soon develops a "travel kit" which goes on all site visits. Hofmann [32] has described such a kit and the following list is derived in large part from his paper. Apart from the basic measurement instruments such as spectrum analyser or measuring receiver and laptop computer control system, the following "accessories" can be considered vital:

- camera and film to document measurement locations – a digital camera is a useful variant for ease of production of test reports
- 15 or 30m tape measure
- masking tape to mark antenna positions
- multi-outlet power extension leads, spare mains leads for the instruments
- clipboard, pen and paper for notes (some test engineers develop a pro-forma notepad to record the salient details of each measurement)
- toolkit containing at least flathead and crosshead screwdrivers, wire strippers, pliers and cutters, insulating tape
- roll of flexible conductive fabric or foil, for use as a portable earth reference plane
- several calibrated and marked antenna cables
- an assortment of other RF cables for connecting probes, etc.
- BNC-N and vice-versa adaptors, 10dB and 20dB attenuator pads
- assorted field probes
- coaxial SPDT switch for rapid comparison of signals from two sources, e.g. antenna and probe
- collapsible tripod and mast extenders
- collapsible biconical and log-periodic antennas, loop and horn antennas, depending on frequency range for investigation
- RF preamplifier
- voltage probe with crocodile clip adapters
- portable AMN/LISN and transient limiter
- battery-operated comb generator reference source with radiated and conducted outputs, to double-check the analyser/receiver operation
- portable AM–FM radio and/or scanner
- magnetic field survey meter, if the site investigation is to include LF EMC
- clip-on ferrite suppression cores, with various sizes to suit a range of round and flat cables
- a selection of filters (signal and power), shielding materials (e.g. 0.2mm tinned steel sheet and the means to cut it and solder it), an assortment of EMC gaskets, connector backshells, and glands
- spare batteries.

10.2 Immunity

10.2.1   Practical aspects of immunity tests

In contrast to emissions tests, the basic feature of an immunity test is that the EUT is subjected to external interference phenomena generated by the tester, and its performance is monitored for unusual or unacceptable operation. But, as in the case of emissions, not all the standard methods can be properly applied in-situ. So the issues that have to be addressed for in-situ immunity testing are:

- is it practical and legal to generate the necessary disturbance levels, and do special precautions need to be taken?
- will the disturbances generated by the tests interfere with or upset other equipment than the EUT?
- can the standard measurement method be applied successfully and correctly; if not, what are the implications of necessary departures from the standard method?

The latter question is identical to that already raised by emissions tests, and has been commented on in 10.1.4.1. The question of precautions against interfering with other apparatus, either within the installation or belonging to an innocent third party, is in some respects the reciprocal of the problem of ambients. Since it presents different problems for different phenomena, it is treated separately for each, below.

10.2.2   Electrostatic discharge

10.2.2.1   The ESD phenomenon

Electrostatic discharge is a source of transient upset which most typically occurs when a person or other body that has been charged to a high potential by movement across an insulating surface then touches an earthed piece of equipment, thereby discharging through the equipment. Currents of tens of amps can flow for a short period with a very fast (sub-nanosecond) risetime. Even though it may have low energy and be conducted to ground through the equipment case, such a current pulse couples very easily into the internal circuitry and can disrupt its operation.

When two objects with a high electrostatic potential difference approach, the air gap between them breaks down and the charge difference is equalized by the current flow across the resultant ionized path. The current route is completed by stray capacitance between the objects and their surroundings, and the inductance and resistance of the path (Figure 10.16).

The actual current waveform is complex and depends on many variables. These include speed and angle of approach, and environmental conditions, as well as the effect of the distributed circuit reactances. At lower voltages a "precursor" spike due to the local area of the source (a finger or a metal tool) is produced which has a very fast risetime, of the order of a few hundred picoseconds. This spike, although low in energy, can be more damaging to the operation of fast digital equipment than the bulk discharge that follows it, which may have a risetime of 5–15 nanoseconds.

The discharge current will take the route of least inductance. If the case is well bonded to ground then this will be the natural sink point. If it is not, or if it is non-conductive, then other routes may be via the connecting cables. Because the discharge edge has an extremely fast risetime (sub-nanosecond), stray capacitive coupling is essentially transparent to it, whilst even short ground connectors of a few nH will

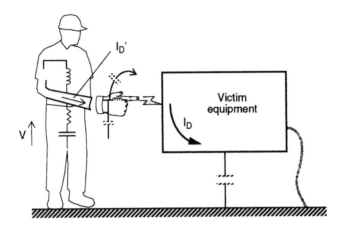

Figure 10.16 The ESD event

present a high impedance. Apertures in a conductive enclosure will also result in intense local magnetic fields which will couple to the internal circuits.

10.2.2.2 The test method

The standard test for ESD immunity is defined in IEC 61000-4-2 [89]. This requires a well-defined test set-up with a ground reference plane and horizontal and vertical coupling planes; the EUT is placed a specified distance from the reference plane and discharges are applied to it from a calibrated generator, at specified test levels. The generator voltage is developed with respect to the reference plane.

Some product or generic standards call up different versions of the basic ESD test standard, and this book describes the issue (1995) which was current at the time of writing. It is generally agreed that this issue is more realistic, comprehensive, and tougher than previous issues of this basic standard. A successful test to the 1995 issue thus gives confidence that a successful result would have been achieved if an earlier issue had been employed, but this cannot be guaranteed.

The standard defines two methods of application, known as contact discharge and air discharge, as shown in Figure 10.17. In the air discharge method, the point of the discharge generator is brought close to the EUT until an arc occurs between the two. The human body is modelled as a 150pF capacitor charged to the test voltage, and discharged into the EUT through a 150Ω resistor and a discharge tip which simulates the dimensions of an outstretched finger. The return path is completed by a strap from the test capacitor to the ground plane. For maximum repeatability the tip should approach the EUT as fast as possible and with the probe at right angles to the surface.

To improve repeatability the contact discharge method requires a test probe in direct contact with the EUT or with a nearby coupling plane, and the test discharge current is initiated using a high-voltage vacuum relay. This avoids the unpredictability associated with variable approach speed and environmental conditions. The tip itself is sharply pointed to allow a clean contact with the EUT surface by penetrating any surface corrosion or coatings which are not intended to have insulating properties. Coatings which are declared to be insulating are subjected to the air discharge only.

Note that the air and contact discharge methods are not alternatives but

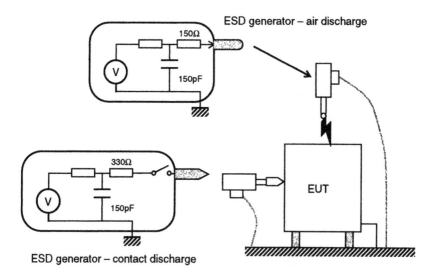

Figure 10.17 ESD test methods

complementary. A correlation between the results of the two methods has not been possible, and it cannot be assumed that their test severities are equivalent, although this is the implication of their widespread use in product standards.

10.2.2.3   In-situ tests

IEC 61000-4-2 gives some limited guidance for in-situ ESD tests, which it calls "post-installation" tests. As with the conducted emissions test, the most important aspect is the provision of an adequate ground reference connection. The standard requires a specific plane placed on the floor of the installation about 10cm from the EUT. It should be of copper or aluminium not less than 0.25mm thick, with dimensions 0.3m wide and 2m length where possible. It is connected to the protective earthing system, or to the earth terminal of the EUT, if this exists. The discharge return cable of the ESD generator is then connected to this reference plane at a point close to the EUT (Figure 10.18 shows a typical set-up).

Choice of application points

Contact discharges should be applied to metallic points accessible to the operator or user during normal use. This excludes servicing operations (such as changing pcbs) that would be carried out by trained maintenance personnel who can be expected to take proper anti-static precautions. A draft amendment has been proposed [90] which offers a definition of accessible parts. In brief, excluded from testing would be:

- points that are only accessible under maintenance
- points that are accessible under customer's service
- points and surfaces that are no longer accessible after installation
- contact pins of connectors that have a metallic connector shell – contacts within a plastic connector should be tested by air discharge only
- contact pins which are ESD sensitive because of functional reasons and that have an ESD warning label

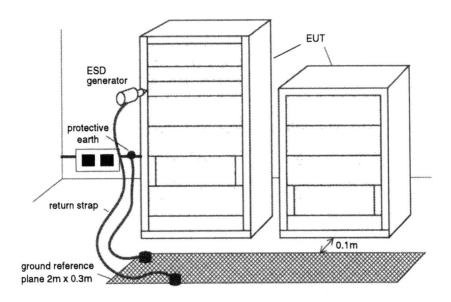

Figure 10.18 Typical in-situ ESD test set-up (after IEC 61000-4-2)

- accessible parts that are only accessible when the equipment is not fully installed.

Air discharges are attempted at slots, seams, insulated switch and keypad apertures and other areas where a discharge would be expected to occur but which cannot be adequately tested by the contact method. The discharge voltage is maintained on the round tip as it is approached at right angles to the EUT as fast as possible without causing damage. The actual points of application must be chosen experimentally by investigating all around the boundary of the EUT with both direct and indirect discharges. If the EUT is satisfactorily immune at the scheduled test levels, these should be increased to attempt to find the worst case application points.

Testing at lower and higher levels

It is well known that on this test, upsets can occur at some stress levels but not occur at higher stress levels. Testing solely at the maximum stress level specified by the product or generic standard may thus provide a misleading result. A fraction of the maximum test level should be applied at first, rising in increasing fractions until the maximum is achieved.

It may also be desired to apply an even higher test level to determine the "margin" that has been achieved, but this should only be undertaken where the possibility of damage to the equipment under test and consequent financial or time losses have been fully evaluated and found acceptable.

Quick testing method to identify worst-case points

Most ESD test "guns" have a rapid-fire option for air discharge, in which their output relay is disabled and their internal high-voltage generator connected directly to their probe tip. They will then discharge as soon as their tip charges sufficiently to break down the air gap to the equipment under test, resulting in a rapid succession of discharges. Varying the distance of the probe from the tested point varies the breakdown voltage, allowing a whole series of tests over the full range of voltages to be carried out in a few seconds. This method may be used to quickly identify the most susceptible test points and voltages, so that they may be more properly tested according to the standard, and is especially useful where there are a large number of such points.

The maximum distance of the tip from the tested point should not exceed that which is seen to cause air breakdown at the maximum test voltage, unless some overtesting is required. Because the test method is not carefully controlled, it is likely that some discharges will exceed the maximum level, so care should be taken (which might include not using this time-saving method) where damage to the tested unit is possible and must be avoided.

Effect on other equipment

The ESD pulse has low energy but very wide bandwidth, and it couples via the loop formed by the ESD generator, the EUT, the ground reference plane and the generator's return strap. Therefore it radiates very effectively. Equipment in the vicinity, especially digital processing equipment with poor immunity, may be upset by the pulse. Effects are possible to a distance of 10m but unlikely beyond that, unless a very sensitive system is involved. Clicks will also be heard on radio reception over a wide radius, but as they are single pulses with a low repetition rate they are unlikely to be intrusive.

If possible, the tests should be done in an area removed from other operating equipment, or if this is not possible, nearby equipment should be powered down. In the worst case where neither of these are possible, operators of adjacent equipment should be warned to expect unreliable operation for the duration of the test.

10.2.3   Electrical fast transient bursts

10.2.3.1 The EFT phenomenon

When a current carrying circuit is interrupted by an unsuppressed air gap switch (such as a contactor or relay) the resultant voltage rise across the gap causes a repetitive ignition and extinction of an arc discharge – the so-called "showering arc" effect. The nature of the arc is determined by the magnitude of interrupted current, the circuit and load inductance and stray capacitance, and the rate of separation of the contacts [33]. It generally results in a burst of fast, low energy current transients which couple along the supply circuit or are radiated from the conductors on either side of the switch. The burst repetition rate varies from 10kHz to 1MHz.

The duration and rise time of these transients is short compared to the travel time in building wiring systems. Typically the rise time of each strike is less than 10ns and the duration is tens to hundreds of ns; transit time is roughly 5ns/m. Thus the supply wiring appears as a transmission line, so that the line characteristic impedance determines the transient source impedance, and the transients themselves are easily attenuated by distance. As a result, these fast switching transients have low energy and are only a threat to local susceptible equipment. But since there are many sources of such transients in the typical electrical installation, and because fast-rise events have considerable upset potential for digital circuits (though little or no potential for damage), equipment reliability will be enhanced by taking them into account.

While switching operations are the most widespread source, certain other similar sources – particularly arc welding and unsuppressed DC motors – can be particularly aggressive to equipment in their neighbourhood.

10.2.3.2   Test method

The immunity test against EFT bursts is specified in IEC 61000-4-4 [92]. This calls for
a generator with a particular pulse waveform, known as the 5/50ns waveshape because
of its rise and fall times. The pulse is repeated in a burst at a repetition rate of 5kHz for
a duration of 15ms, and this burst is repeated every 300ms for the test duration of one
minute. Various lines have this burst signal applied, the combinations, polarities and
levels being dictated by the applicable product or generic standard.

The test is intended to demonstrate the immunity of equipment when subjected to
types of transient interference as described above. It is *not* intended to directly simulate
such interference; this would be impossible because of the wide variety of interference
frequencies, waveforms and coupling routes. Rather, the rationale for applying this test
is that equipment which survives it has been found to be more robust in the presence of
real sources of interference.

Coupling methods

Because the transients have a very high frequency spectral content, the test set-up and
coupling methods must use RF techniques as with the electrostatic discharge set-up. A
ground reference plane is mandatory and the test waveform generator and coupling
network must be referred to it, as should the protective earth. It should project beyond
the EUT by at least 10cm on all sides. As with the ESD test, the EUT must only be
earthed according to its normal installation practice; free standing items should be
spaced from the ground plane by an insulating support 10cm thick, while table-top
equipment should be 0.8m above it. The basic principles of the laboratory EFT test are
shown in Figure 10.19.

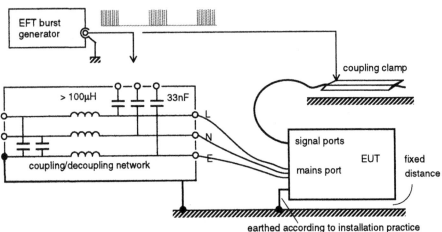

Figure 10.19 EFT test set-up in laboratory

Power leads (less than 1m long) are connected to the test pulse generator via a
coupling/decoupling network which capacitively couples the transient burst onto each
individual line. The coupling point is isolated from the supply by an L-C decoupling
network in each line. This includes the protective earth conductor, which has transients
applied to it as well. Thus unless the EUT's enclosure is normally separately bonded to
installation earth, it too will see transient voltages.

There is some confusion amongst test laboratories as to the power conductors and their combinations to be tested, due to the different forms of words used in the generic/product standards and in IEC 61000-4-4. In some situations it may be that one power conductor or combination of conductors is less or more susceptible to the test than others, and this can lead to complaints of over-testing or under-testing. To be inclusive, and possibly overtest, all the phase conductors should be tested individually (e.g. L, N, E), and then all the phase conductors should be tested together (L+N), and then all the phase conductors and the unit's earth should be tested (L+N+E).

I/O cables are subjected to transient coupling via a capacitive coupling clamp spaced 10cm above the ground plane, which sandwiches the cable in question between two flat metal plates, both of which carry the test pulse burst referred to the ground plane. The distributed capacitance of this arrangement is 50–200pF. Note that because the voltage is applied through a capacitive divider (coupling clamp capacitance or coupling network capacitance in series with the EUT's capacitance to ground) the actual voltage applied to the EUT will be less than the peak open circuit voltage, which is specified as the test level.

10.2.3.3 In-situ testing

"Post-installation" tests are covered in reasonable detail in IEC 61000-4-4. The most significant difference compared to the laboratory test is that the burst generator is coupled directly into the power supply terminals, and the protective earth terminal, at the EUT. No coupling/decoupling network is used, though a 33nF blocking capacitor may be needed. A reference ground plane of similar construction to that used for the ESD test, but this time of dimensions 1m × 1m, is laid "near" the EUT and connected to safety earth at the power supply mains outlet; the burst generator is located on, and bonded to, this plane, and connected to the EUT terminals by a lead of less than 1m length.

I/O lines are tested with the capacitive clamp as for the standard method. If there is insufficient room to use the clamp, it can be replaced by a length of conductive tape or foil wrapped around the cable loom under test, or by discrete 100pF capacitors connected to the terminals. IEC 61000-4-4 accepts that the results from each of these methods might be different, and suggests that the test levels might be amended "in order to take significant installation characteristics into consideration". As well as the coupling method, further variability is likely to be introduced by the detail of the connection of the burst generator to earth reference in the vicinity of the coupling point, since this earth reference is undefined except by the installation layout.

Although the fast transient burst test is a useful and widely referenced method of determining immunity to an important phenomenon, the variations and site-specific issues inherent in an in-situ test make it unreliable as a means of declaring compliance. Any testing that is done should be meticulously reported, including a detailed and careful justification for the layout and choice of coupling method employed.

Interference with other equipment

The lack of any form of decoupling of the injected pulses, either onto the power terminals or onto I/O cables, makes this test quite aggressive with respect to other apparatus that is not being tested. Any equipment which shares the same power supply will see the same pulses, attenuated only by their distance from the point of injection. The same is true of ancillary apparatus that is connected to any I/O cables under test. Any other cables that run near to the cables under test must be considered as themselves being subject to the test, since crosstalk coupling will be effective for these wideband interference pulses.

Computers and programmable electronic systems such as PLCs and "intelligent" controllers, which have poor immunity can suffer upsets from EFT tests, with possible consequences as outlined above for ESD testing. Thus, as with the ESD test, nearby equipment and that which is connected to the tested circuit should be powered down. If this is impossible, operators of adjacent equipment should be warned to expect unreliable operation for the duration of the test.

10.2.4   Surges

High energy transients are generally a result of lightning coupling to the supply network, or are due to major power system disturbances such as fault clearing or capacitor bank switching. Lightning can produce very powerful surges by the following mechanisms (see also Chapter 9):

- direct strike to primary or secondary circuits: the latter can be expected to destroy protective devices and connected equipment; the former will pass through the service transformers either by capacitive or transformer coupling or insulation breakdown;

- indirect cloud-to-earth or cloud-to-cloud strikes create fields which induce voltages in all conductors;

- earth current flow from nearby cloud-to-earth discharges couples into the supply earthing network via common impedance paths;

- primary surge arrestor operation or flashover in the internal building wiring causes dV/dt transients.

Fault clearance upstream in the distribution network produces transients with energy proportional to $0.5 \cdot L \cdot I^2$ trapped in the power system inductance. The energy will depend on the let-through current of the clearance device (fuse or circuit breaker) which can be hundreds of amps in residential or commercial circuits, and higher for some industrial supplies. Power factor correction capacitor switching operations generate damped oscillations at very low frequency (typically kHz) lasting for several hundred microseconds.

10.2.4.1 The surge test

The test defined in IEC 61000-4-5 [93] is intended to demonstrate the immunity of equipment when subjected to high energy surge voltages from the above effects. The transients are coupled into the power, I/O and telecommunication lines. The surge generator called up in the test has a combination of current and voltage waveforms specified, since protective devices in the EUT (or if they are absent, flashover or component breakdown) will inherently switch from high to low impedance as they operate. The values of the generator's circuit elements are defined so that the generator delivers a 1.2/50μs voltage surge across a high-resistance load (more than 100Ω) and an 8/20μs current surge into a short circuit. These waveforms must be maintained with a coupling/decoupling network in place, but are not specified with the EUT itself connected.

Three different source impedances are also recommended, depending on the application of the test voltage and the expected operating conditions of the EUT. The effective output impedance of the generator itself, defined as the ratio of peak open circuit output voltage to peak short circuit output current, is 2Ω. Additional resistors of 10 or 40Ω are added in series to increase the effective source impedance as specified in the standard.

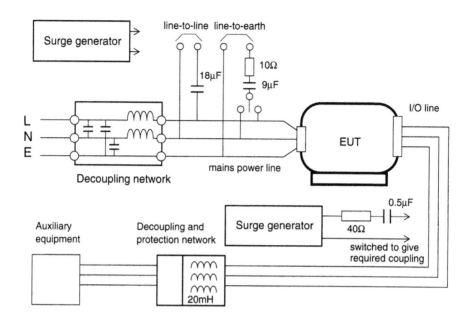

Figure 10.20 The IEC 61000-4-5 surge test

10.2.4.2 Surge coupling and application

High energy surges are applied to the power port between phases and from phase to earth (Figure 10.20). For input/output lines, again both line-to-line and line-to-earth surges are applied, but from a higher impedance. 2Ω represents the differential source impedance of the power supply network, 12Ω represents the line-to-earth power network impedance while 42Ω represents the source impedance both line-to-line and line-to-earth of all other lines.

Power line surges are applied via a coupling/decoupling network incorporating a back filter, which avoids adverse effects on other equipment powered from the same supply, and provides sufficient impedance to allow the surge voltage to be fully developed. For line-to-line coupling the generator output must float, though for line-to-earth coupling it can be earthed. A 10Ω resistor is included in series with the output for line-to-earth coupling.

I/O line surges are applied in series with a 40Ω resistor either via capacitive coupling with a decoupling filter facing any necessary auxiliary equipment, or by spark-gap coupling where the bandwidth of the I/O line is wide enough that capacitive coupling would affect its operation. Note that testing with differential mode surges is only intended for unbalanced circuits in critical applications, i.e. ports intended for process measurement and control. Long-distance telecommunication lines require special treatment, involving the CCITT $10 \times 700\mu s$ waveform applied at up to 4kV.

The purpose of the surge immunity test at equipment level is to ensure that the equipment can withstand a specified level of transient interference without failure or upset. In this case it is assumed that the equipment is fitted with surge protection devices (varistors, zeners etc.). Typically such devices have low average power ratings, even though they can dissipate or handle high instantaneous currents or energies. So the maximum repetition rate of applied surges will normally be limited by the capabilities of the devices in use, and a maximum of 10 surges (5 positive and 5 negative) is recommended for any one test procedure. Over-enthusiastic testing may lead to premature and unnecessary damage to the equipment, with possible consequential damage also occurring. Because of this latter risk, some test houses take the precaution of physically isolating the EUT during the test. In any case, the EUT should be disconnected from other equipment where possible and the whole set-up should be well insulated to prevent flashover.

Testing at higher and lower levels

Upsets can occur at some stress levels but not occur at higher stress levels. Testing solely at the maximum stress level specified by the product or generic standard may thus provide a misleading result. A fraction of the maximum test level should be applied at first, increasing until the maximum is achieved. Typically, surge testing starts at 500V and increases in 500V steps. It may also be desired to apply an even higher test level to determine the "margin" that has been achieved, but this should only be undertaken where the possibility of damage to the equipment under test and consequent financial or time losses have been fully evaluated and found acceptable.

Where surge protective devices are fitted in an apparatus, it is advisable that tests are carried out at stress levels just below, and just above, the levels at which the protective devices are intended to operate. This is because actual surges with levels less than the maximum are much more likely, and it is sometimes easier to design a surge protection device that is more effective the higher the surge level. So it has been found in practice that apparatus which meets the maximum surge test levels can suffer field failure with less powerful surges. For example, a gas discharge tube (GDT) device will only trigger when a voltage threshold has been passed. When the maximum surge test level is applied, the GDT only lets-through an overvoltage for a few microseconds before it triggers and protects the circuit. But a lower level of surge voltage may be insufficient to trigger the GDT, allowing the surge to be applied to the circuit for its full duration and possibly causing damage.

10.2.4.3   Surging in situ

IEC 61000-4-5 does not give any guidance for in-situ tests, in contrast to the previous two standards. This may indicate that such testing is not expected to be commonplace.

The surge waveform has a much lower bandwidth than either the ESD or EFT test pulses, and so radiative coupling is rarely a problem. The major issue is that the surge is a high energy destructive transient, and therefore it is essential to implement decoupling (or back filter) networks to ensure that the energy is only applied to the equipment under test. Trying to apply the surge without these networks onto a power supply that was shared by several systems, would be inviting disaster. (It is also debatable whether a client who has already expended considerable resources to install an expensive system, would want that system deliberately exposed to the potential damage that a surge test can cause.)

If the connections to the potential EUT can be broken to insert suitable coupling/decoupling networks, and any ancillary or adjacent apparatus can be properly isolated from the surge, then surge testing in-situ is feasible, with all the above caveats. Otherwise it is not.

10.2.5   Radiated and conducted RF

The question of radio frequency immunity testing has probably caused more controversy and discussion in the EMC field than any other phenomenon. It is undoubtedly necessary, since there is a proliferation of portable and fixed radio transmitters which can cause field strengths high enough to upset electronic equipment. It is unfortunately also expensive to do, at least according to the standard methods. There is thus considerable pressure on test engineers, particularly in manufacturing companies, to find shortcuts in the RF immunity test. Some of these are already half-established and it is worth discussing them in some detail to analyse their advantages and shortcomings.

To do this it is necessary to understand the basic mechanism for RF susceptibility. Radiated RF fields are developed by all intentional transmitters as a consequence of their function; close in to the transmitters the fields are intense but they decay proportionally to distance. An introduction to the properties of RF fields can be found in section 4.2.3 on page 84. The fields couple with susceptible equipment by two mechanisms:

- direct interaction of the field with the circuit wiring and printed circuit traces inside the equipment;

- induction of common mode currents onto cables, which are then coupled into the equipment by conduction through the cable port interfaces.

Other chapters of this book have described these mechanisms, and particularly how to defeat them, in more detail. They are to a large extent the reciprocal of the coupling mechanisms that are tested in the RF emissions tests outlined in section 10.1. The differences in approach that exist between RF emissions and immunity tests have to do partly with concerns of spectrum management and partly with the historical bias of the relevant standards committees. Reciprocity of coupling can be assumed, because the coupling paths are linear; but the sources and victims of the disturbances are not, and in general they cannot be assumed to be the same, which adds a degree of complication to any effort to harmonize test methods between the two.

Test methods have developed independently for different sectors. The most established standard methods are related to the military/aerospace sector (DEF STAN 59-41, MIL STD 461, RTCA DO160); the automotive sector (ISO/IEC 11452, SAE J1113); and the commercial sector (IEC 61000-4-3 and -4-6, which developed from the IEC 801 series, originally written for industrial process control and instrumentation). Although there are a few points of contact between these, they are mostly separate, with little commonality in the methods or equipment. They all have in common an approach which applies conducted testing at the lower frequencies and radiated at the higher, although in some cases there is considerable overlap of the ranges. The rationale for this is that cables, being electrically long, are the dominant route for coupling at low frequencies; once the cable becomes much longer than a wavelength (above, say, 200MHz) its efficiency as a coupling path declines, and coupling with the equipment enclosure becomes dominant.

Another factor that these standards have in common is that none of them allow for in-situ testing.

The methods of applying the conducted signal or generating the radiated field are mostly different, but it is generally true to say that developing a uniform, calibrated radiated field is harder at low frequencies than injecting current or voltage onto a cable. It is the goal of each standard to devise a test method which both simulates the real environment for its particular application sector adequately, and yet allows minimum uncertainties so that the test is repeatable between test houses and at different times, on the same EUT. It is this latter aspect which is hardest to achieve for RF immunity, and which accounts for much of the complexity and expense of the various methods.

This section will look at the standard methods, mostly in the IEC documents, for conducted and radiated tests, and then discuss how these may be applied, where possible, in situ, and what alternatives might exist.

10.2.5.1 Conducted methods

Disturbances can be induced directly onto cables by two principal methods – voltage coupling, which requires a connection to be made to each conductor, or current coupling, in which a wideband current probe is clamped around the cable, and which requires no direct connection.

Direct voltage injection

This is the preferred method in IEC 61000-4-6 [94], and it employs a specific coupling-decoupling network (CDN). Different CDNs are required for each type of cable, since the coupling is invasive, and this has restricted the use of the technique to commonly-employed cables, particularly power cables. It is particularly *un*suitable for unscreened cables carrying wideband data. The CDN is designed to establish a 150Ω common mode impedance at the port which is connected to the EUT, and to isolate this port from the end of the cable away from the EUT, i.e. the power supply or ancillary equipment. Figure 10.21 shows the set-up for a three-core mains cable.

The CDN is similar but not identical to the CDN used for transient coupling, and the AMN/LISN used for conducted emissions tests. The main difference is in the 150Ω impedance, with coupling to the lines in common mode only. The requirement for proper bonding to the ground reference plane is as important as in the other tests.

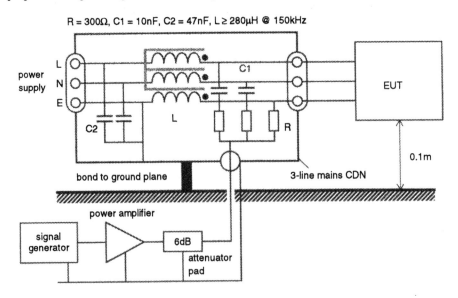

Figure 10.21 Direct injection set-up according to IEC 61000-4-6

The standard requires a test level to be set up into a 150Ω calibration load, rather than into the EUT itself. The power that is needed to maintain the specified level is then re-played across the frequency range with the EUT in place. The major advantage of the CDN method is the low power required for standard test levels – a test of 10V emf can be achieved by a power amplifier of 7W rating. CDNs are specified, and can be built, to give a common mode impedance within $\pm20\Omega$ of 150Ω from 150kHz to 26MHz, and over a wider tolerance up to 80MHz. The test can under certain circumstances be extended up to 230MHz, if properly characterized CDNs are available – most commercial ones do not extend this high.

Bulk current injection

As an alternative to voltage injection, IEC 61000-4-6 allows the use of a current injection probe, over the same frequency range. This method has become known as Bulk Current Injection (BCI). Very similar, but not identical, methods are specified in the military standard DEF STAN 59-41, test DCS02 [118], and the automotive components standard ISO 11452-4 [108]. ISO 11451-4, the companion standard for whole vehicle tests, also describes the BCI method. The automotive standards give a frequency range from 1MHz to 400MHz, while the military specification extends from 50kHz to 400MHz.

The BCI probe is simply a current transformer whose secondary is the cable under test, and into whose primary is fed the disturbance signal. It is halved and hinged, so that it can be clamped over the cable harness without having to break into it.

There are two ways of applying the BCI test. Because the EUT's common mode impedance is unknown and frequency-dependent, simply attempting to apply a fixed level of current would result in overtesting at some frequencies and under-testing at others. One method, known as the "substitution method", calibrates the power required to generate the required disturbance current into a fixed impedance calibration jig, and then requires this power profile to be repeated with the EUT in place. This is the only method allowed in IEC 61000-4-6 and DEF STAN 59-41 DCS02. The induced current can be monitored by a separate probe placed between the injection probe and the EUT (see Figure 10.22), so that it can be limited if it exceeds some higher level than the expected test level, if the EUT impedance drops to a low level at some frequency.

The second method, allowed by ISO 11452-4, is known as the "closed loop" method. In this case the monitor probe records the actual injected current, which is increased at each frequency until either the maximum test current level is reached, or the maximum power to the injection probe is reached.

In all cases the test set-up must be carefully controlled with both the cable under test and the EUT mounted a set height above the ground reference plane. Especially at the higher frequencies, the cable layout and the distance between the probes and the EUT are critical; large variations occur for small changes in probe position. The impedance at the far end of the tested cable has to be stabilized in some way; for the military and automotive tests, it is usual to connect the actual ancillary items mounted on the ground plane as they would be in the real vehicle, on the assumption that this provides a reasonably realistic situation. For tests on commercial products, ancillary equipment is often ill-defined and instead some method of impedance stabilisation to 150Ω is required.

Although it requires more power for a defined injection level than direct voltage injection (because of the power lost within the probe), the great merit of the BCI method is that it can be used on any cable that can pass through the aperture of the chosen probe. This makes it highly suitable for in-situ tests.

EM-Clamp injection

A similar transducer to the BCI probe is the EM-Clamp, which is the third alternative

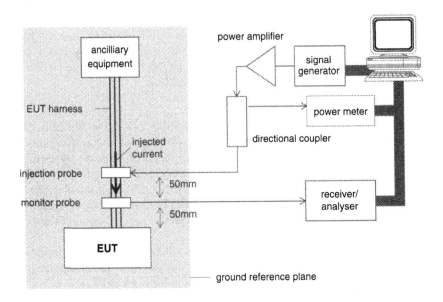

Figure 10.22 Bulk current injection set-up

specified in IEC 61000-4-6. This operates along similar lines but couples both inductively and capacitively at the same time; through the use of ferrite absorber material it allows a certain degree of isolation from the ancillary equipment end of the cable. It is considerably bulkier and much longer than the BCI probe and although it is frequently used in laboratory tests, tends to be inconvenient for in-situ purposes, where the cables to be tested are usually less accessible.

10.2.5.2 Radiated methods

The basic approach to radiated RF testing is to develop a constant, uniform field over a specific volume in which the EUT is immersed. Provided that the field level is uniform and the EUT's coupling to it is invariant, the assumption is made that the test will repeatably and representatively exercise all the EUT's susceptibilities that need to be tested. Actually achieving an adequately controlled field and coupling is the goal towards which all standard radiated immunity test methods strive.

The most thoroughly specified method of developing the field is from an antenna in a test chamber, and this is described next. Alternatives do exist for equipment tests – the TEM cell, the GTEM cell, or the stripline – but these have no relevance to large system testing and are not pursued further here. The reverberating chamber method has many fans (*sic*) and is particularly suited to developing high field levels, but it also has several functional and operational disadvantages and is yet to be specified in any commercial standards, although a draft document is in existence.

Antenna and screened anechoic chamber

IEC 61000-4-3 [91] describes the requirements and methods for a radiated field test from 80MHz to 1GHz using an antenna. After many years in which all standards assumed a closed-loop control method, in which the field at the EUT was monitored and controlled during the test, this approach has finally been scrapped in favour of the substitution method. Exactly as with the conducted immunity substitution method, the field strength is first calibrated over a uniform area without the EUT present; and then the power level which gives this field strength is re-played with the EUT in place and aligned with the uniform area, in the various orientations that have to be investigated (Figure 10.23).

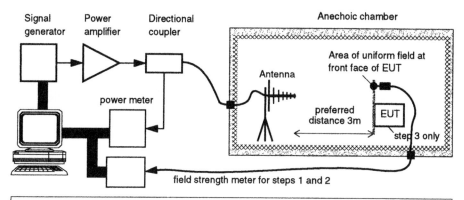

Procedure:
1   Calibrate field uniformity without EUT
2   Calibrate required field level, **without** EUT
3   Re-play power vs. frequency for all orientations and polarizations, **with** EUT

Figure 10.23 Generating an RF field in an anechoic chamber

The rationale for this method is that the presence of an EUT inevitably distorts the field, especially at frequencies where the EUT is large compared with a wavelength, and so it is impossible to specify a precise applied field strength with the closed loop method. If the monitoring probe is placed in one position next to the EUT, the field at any other position is unknown but is likely to be different. This leads to the possibility of severe over- or under-testing if the probe is positioned at a point which exhibits a field null or peak at any given frequency.

The benefit of the substitution method is that the field strength at any given point around the EUT is irrelevant for the test outcome. What matters is the field strength for which the set-up is calibrated in the EUT's absence; distortions introduced by the EUT are therefore circumvented.

In practice, there are a number of major difficulties with this method, the most important being that it relies on an area of uniform field strength being generated within which the EUT will be placed. The larger the EUT, the larger this area must be. The standard specifies the limits of uniformity to be −0, +6dB over three quarters of the points at which it is measured. It is specified in this asymmetrical way so that under-testing will not occur at any point, but by implication, over-testing of up to 6dB (i.e. 2 times, or 20V/m instead of 10V/m) can still occur. Even with this apparently wide margin, a very expensive test facility is needed. The test must be performed inside a screened chamber to avoid interfering with other radio services. The screened room walls introduce reflections which, if untreated, would create deep nulls in the field pattern over distances of a half wavelength, i.e. 1.875m at 80MHz down to 15cm at 1GHz, which would completely demolish the field uniformity. Anechoic material (carbon foam or ferrite tiles) which is effective over the whole frequency range must line all six surfaces of the chamber, and this greatly increases the facility cost.

But there is more to come. The size of the uniform area also depends on the directivity of the antenna that is used. To maximize the area, the antenna needs to be as far away as possible from the plane in which the field uniformity is measured. This also has the beneficial effect that interaction between the EUT and antenna, which would modify the applied field from that which was achieved in calibration, is minimized. But the further away the antenna is, the more power is needed for a given field strength. The power is actually proportional to the square of the separation distance, so that the power needed for a 3m distance (preferred in the standard) will be nine times that needed at 1m. 10V/m at 3m will require typically 100 watts from 80MHz to 1GHz with a conventional BiLog antenna (Figure 10.24 – note how the power requirement climbs sharply at lower frequencies). A broadband power amplifier covering the same frequency range at this power level is a further major expense.

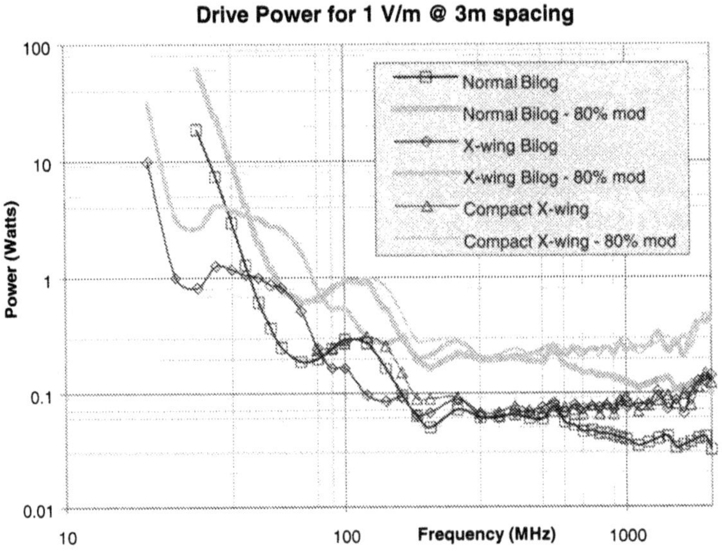

Drive Power for 1 V/m @ 3m spacing

Legend:
- Normal Bilog
- Normal Bilog - 80% mod
- X-wing Bilog
- X-wing Bilog - 80% mod
- Compact X-wing
- Compact X-wing - 80% mod

Figure 10.24 Required power versus frequency for different antennas (source: Schaffner-Chase EMC)

Subsidiary requirements of the test method are:

- modulation must be applied to the test signal, since real disturbances are likely to be modulated in one way or another, and this affects the EUT's response; the default is 1kHz sinusoidal at 80% modulation depth. This increases the actual peak of the applied level by 1.8 times (18V/m rather than 10V/m) and demands an extra 5.1dB from the power amplifier over the unmodulated level

- the test signal must be scanned or stepped over the frequency range of 80–1000MHz at a rate slow enough for the EUT to respond. For swept measurements the standard requires a sweep rate no faster than 1.5×10^{-3} decades per second. For stepped testing, which is the norm for computer-controlled tests, the step size should be no more than 1% of the previous frequency (i.e. logarithmically increasing with frequency), but there is no explicit requirement for dwell time at each frequency; applying the above sweep rate works out to nearly three seconds per step and an overall minimum time of around 15 minutes per full sweep

- the EUT must be illuminated with both horizontal and vertical polarization on all four sides, or all six sides if it can be used in any orientation, so that a total of eight or twelve sweeps are required.

The first two points apply also to the conducted test.

10.2.5.3 In-situ tests

Reading through all the above requirements and methods, you might be forgiven for wondering how they can be applied in situ. The answer, of course, is that they can't, at least for the radiated tests.

The main issue with radiated testing in situ is that it is illegal (at least in the UK, though it may not be in other countries, even in Europe). Since the installation in situ is by definition not inside a screened chamber, any radiated test would be generating unacceptably high levels of disturbance signal which could affect other radio equipment over quite a wide area, certainly exceeding the boundaries of the installation. It might be possible to apply for a special test licence to do this on a site-specific basis, but the grant of such a licence would be subject to inevitable bureaucracy (*pace* the Radiocommunications Agency) and might contain conditions which would make it unworkable in practice.

High levels of radiated field can also have implications for human health. Guidelines on acceptable levels of field to which personnel can be exposed are contained in [135] and other documents, which should be consulted if this question arises in an RF immunity testing context.

But some form of RF immunity testing is a necessity if the installation is to be shown to be compliant, either with the immunity requirements of the EMC Directive, or with operational safety and reliability requirements in the case of a contractual EMC obligation. Therefore, some way must be found to get around the legal restriction. There are three main possibilities, which have been explored by competent bodies who have to approve Technical Construction Files, and they tend to be used in combination.

Immunity tests of modules, combined with a coupling analysis

RF immunity testing is a pre-requisite of apparatus that is to be CE marked, and if the results of such tests are available they can be fed into an EMC analysis for the apparatus when it is incorporated into a system or installation. If the apparatus (such as a chart recorder or temperature controller) is to be built into a screened enclosure, the known shielding effectiveness may be used (with caution, i.e. a safety margin of 10–20dB) to justify a reduction in the immunity level for that apparatus if necessary. The usual difficulty with this approach at present is that sufficient information is not available on which to base a sound analysis, or the equipment is supplied without an adequate form of EMC compliance statement – see section 3.4.2 on procurement specifications for a further discussion.

Limited-frequency RF testing using pre-existing licensed transmitters

In many cases the main portable transmitter frequencies that will be present in the EM environment on the installation site, and which therefore pose statistically the greatest threat to the system, are known. Examples are:

- security and plant engineers' communication transmitters
- cellular mobile phones

- CB and radio amateur transmitters.

These transmitters are available to the site operator and are licensed for use on the site, and can therefore be used for testing easily and cheaply without restrictions. Keying such a portable device while holding it near to potentially susceptible parts of the system under test may not be particularly "scientific", but it can be effective at exposing weak areas of the installation, and pointing out the need for remedial measures. The disadvantage is that testing with these devices does not characterize the EUT's response to other frequencies that may be introduced throughout the lifetime of the installation, for example by the emergency services, and therefore such tests should not be relied upon on their own.

Some effort should be made when applying this form of ad-hoc testing to characterize the field strength generated by the device and applied to the EUT. This can be done quite easily with a broadband field strength meter positioned at various distances and orientations from the transmitter, but beware of errors introduced by near field effects such as interactions with personnel and nearby objects. Also, beware of the intelligence of digital cellphones, which adapt their transmitted output according to the path loss, which can change very rapidly as the unit is moved around the EUT; checks of the actual emitted field strength, and/or "engineering fixes" to ensure maximum output, are essential with these devices.

Substitution of conducted immunity tests

Because the conducted immunity test methods, particularly BCI, are relatively easy to perform, they can be applied to many cables on the system under test over as wide a frequency range as possible. Typically, the BCI method can be applied down to 50kHz and up to 400MHz as in the military tests if it is felt necessary. Since the cables for the system are already terminated in their actual impedances, the artificial test set-up of the laboratory BCI test need not apply; all that is necessary is for a pre-calibrated probe system to be used on each cable in its actual position that might act as a coupling route for local radiated fields. It is the test engineer's responsibility to select appropriate cables based on knowledge of the signals they carry, the equipment they connect to and their routes between the equipment. The levels of injected current do not correlate directly with radiated field levels; different authorities can quote anywhere between 1 and 10mA as giving about the same stress level as a 1V/m field.

Strictly speaking, though, conducted immunity tests could be as serious a source of radiated disturbances as is a radiated test. The cable on which the currents are induced will act as an antenna, albeit not as efficiently as an intentional one. The legality of performing conducted tests outside a screened room is questionable, and a test engineer who takes the responsibility of doing so should be aware of the potential for affecting legitimate radio users.

One possible way to reduce the magnitude of this problem is to fit a number of clip-on ferrite suppressors to the cable on the "other" side of the BCI probe. The probe should be fitted close to the unit to be tested in any case, and the ferrites should help prevent the rest of the cable from radiating too much. The reduction achieved is unpredictable, due to the influence of other cables and metalwork; the method is also modifying the system configuration, which could be unacceptable for system-wide testing. A broadband omnidirectional field probe could be used to measure the actual fields "leaked" by the test method and the consequent reduction by applying ferrites to the untested cable lengths.

Systems EMC procedures checklist

A.1 Company Procedures

Must:

Contain EMC compliance management and control procedures as described in this book, added to and modified by company circumstances and experience, such as EMC test results and experiences of actual interference events.

Describe the EMC testing regime to be followed which will range from simple D-I-Y tests to partial or full tests on the factory floor, at the final installation, or in an independent testing laboratory, or any mixture of these methods.

Include detailed work instructions for:

- Assessing the operational electromagnetic environment, and deciding on the functional degradations allowed during EM disturbances and the amount of emissions (usually part of the tender submission and contract approval process)
- Specifying the EM performance of bought-in electrical/electronic/ programmable units, and those designed in-house
- Obtaining electrical/electronic unit supplier's EMC instructions, and following them in every detail
- Specifying earth and screen termination methods
- Specifying cable classes (and standard cable types, e.g. by manufacturers' part numbers)
- Backplate/chassis layout rules and minimum cable spacings (by cable class)
- Segregation of items of apparatus and other EMC protective measures
- Providing all this information on the assembly drawings and installation manuals
- Storing of drawings, design rationales and test results, to be able to show "due diligence" in EMC compliance up to ten years later.

A.2 Designers

Must:

Follow the company's procedures and work instruction for the EMC practices concerned.

Take care only to purchase units with adequate EMC performance for the intended environment and use. Obtain and check suppliers' EMC documentation. Obtain all manufacturers' instructions for EMC and follow them. If in doubt get more information and expert advice.

Communicate all the detailed assembly requirements clearly on all assembly drawings, for example

- Identify the earthing and screening methods
- Identify the cable classes (and the types and grades of cables to be used)
- Draw backplate/chassis layout positions of major items and those critical for EMC
- Specify segregation distances from other apparatus and any other protective measures required by the installation (e.g. filtering, surge protection, earthing)
- Show on the assembly and installation drawings the routes to be used for the different cable classes.

Remember some electrical/electronic units may have special EMC-related needs, such as: cable screens only connected at one end; or cable screens connected to other than earthed chassis (e.g. connected to an insulated terminal); or special cable types (e.g. double-screened). In every such instance the assembly drawings must be clearly marked with exactly what is required using clear sketches or descriptions in words.

A.3 Assemblers

Must:

Follow the company's procedures and work instructions for the EMC practices concerned.

Reject all drawings which are incomplete, for instance any with an earth connection without an associated bonding method, or any with a cable class (or its route) undefined.

Follow the designers' drawings exactly, including:

- Following earthing and screen-terminating methods
- Using the correct cable types for the different classes
- Laying out the electronic and electrical units on the backplate or chassis according to the assembly drawings
- Routing the segregated cable classes exactly as shown on the assembly drawings
- Following any other special instructions on the assembly drawings.

Any necessary changes must be agreed with designers and shown by the final issue of the drawings. If there are any doubts about anything on the drawings, ask.

A.4 User and Installation Manuals

These should encourage good EMC practices to be continued in the user's system or installation, including:

- Identifying the cable classes for all user connections
- Specifying the cables and connectors to be used (e.g. twisted pair, braid or foil, etc., manufacturers' part numbers preferred).

These should also recommend following IEC 61000-5-2:1998 or other appropriate installation EMC standard, including:

- Segregating the cable classes from each other
- Cable routing
- Installation earth bonding and meshing
- Cable screen 360° bonding at both ends
- The use of parallel earth conductors (PECs).

These manuals should also describe any special installation instructions for achieving EMC, such as those derived from an electronic unit manufacturer's special installation instructions for EMC.

Manuals may also need to specify who is allowed to install (e.g. what skills, experience, or qualifications they must have). In some cases these have to be your own installation personnel.

Finally, user manuals must state that EMC compliance cannot be assured if their instructions are not followed, or if installation, commissioning, servicing or modifications are carried out by non-approved or incompetent persons. Offer assistance with queries or difficulties.

It may be useful to duplicate all the company's own drawings of earthing and screening methods, and the rules for choosing and routing cable classes, so that the user is quite clear about what to do and is therefore unable to claim you didn't provide him with the necessary detailed EMC installation information.

A.5 Installation, commissioning, service

Follow the company's procedures and work instructions for the EMC practices concerned.

Reject all drawings which are incomplete, e.g. any with an earth connection without an associated method, or any with a cable class (or its route) undefined.

Follow the designers' drawings and installation instructions exactly, including:

- Following earthing and screen-terminating methods
- Using the correct cable types for the different classes
- Maintaining the intended segregation between items of apparatus and their cables
- Routing the segregated cable classes exactly as required
- Using parallel earth conductors (PECs) of the correct type, where required
- Following any special instructions on the assembly drawings and installation manuals
- Not altering or modifying anything which is important for achieving EM performance.

Installation, commissioning and service staff may need to be trained in identifying and dealing with interference problems, and may need a basic kit of equipment and tools to allow them to apply remedial EMC measures during their work.

Any necessary changes must be agreed with designers and shown by the final issue of the drawings. If there are any doubts about anything on the drawings or in the installation and commissioning instructions, ask.

Appendix B

Determining performance criteria

(This appendix is based on reference [102])

The effects and consequences of the different electromagnetic phenomena on electronic systems and installations are directly related to the different functions carried out by the various electronic instrumentation and control systems, and to the processes involved. In the evaluation of these effects and consequences, it is useful to identify the 'main functions' of the apparatus. The following functions are considered of particular relevance to electronic equipment and systems:

- Protection and teleprotection
- On-line processing and regulation
- Counting
- Command and control
- Supervision
- Man-machine interface
- Alarms
- Data transmission and telecommunication (analogue and digital)
- Data acquisition and storage
- Metering
- Off-line processing
- Monitoring
- Self-diagnosis

Combinations of different functions are generally present in most items of apparatus.

Depending on the types of electromagnetic phenomena (conducted or radiated, low or high frequency) present in the intended operational environment, the resulting interference may be limited to one function or to an unforeseeable number of functions.

Table B.1 summarizes the typical performance criteria for industrial control systems, versus the threats which were discussed in Chapter 4. A short discussion follows of the different functions in the context of this type of system, together with their possible degradations due to electromagnetic phenomena. For each function, considerations on the relevance of such degradation, in terms of consequences to the process, are given, taking into account the electromagnetic phenomena typically encountered in heavy industrial apparatus.

Although this example covers heavy industrial apparatus, similar tables and discussions may readily be created for other types of systems and installations using the following example as a guide. This procedure is recommended as a way of helping to determine the performance criteria, always listing what the user reasonably expects to be achieved given the use to which the equipment is put and the quality of personnel who will use it. It is recommended that the performance criteria are contractually agreed with the user (along with the parameters of the user's electromagnetic environment), although this is probably more likely to occur for larger, high-exposure, or otherwise critical projects.

Table B.1 Summary of functional performance versus threats (in descending order of criticality) – heavy industrial application

Functions (*)	Functional requirements versus the nature of the electromagnetic threat		
	Continuous phenomena	Transient phenomena with high probabilities	Transient phenomena with low probabilities
Protection, teleprotection, and safety (**)	**Normal performance** within the limits set by the apparatus's functional and functional safety specifications		
On-line control of processes			
Counting and metrology			
Command and control			Short delay (1)
Supervision			Temporary loss, self-recovered (2)
Man–machine interface			Stop and reset, manual restoration is acceptable
Alarm		Short delay (3), temporary wrong indication	
Data transmission and telecommunication (***)		No loss, possible bit error rate degradation (4)	Temporary loss (4)
Data acquisition and storage		Temporary degradation (2) (5)	
Metering		Temporary degradation, self-recovered without affecting metering accuracy	
Off-line processing		Temporary degradation (5)	Temporary loss and reset (5)
Monitoring		Temporary degradation	Temporary loss
Self-diagnosis		Temporary loss, self-recovered within the system's diagnostic cycle	

Notes:

*	For the application of the assessment criteria to apparatus with multiple functions, as well as for concurrent functions (e.g. supervision and monitoring), the performance related to the most critical function applies.
**	For teleprotection using power line carrier the "normal performance" during the switching of HV isolators may need an appropriate validation procedure.
***	Only where used in automation and control systems as an auxiliary function to other ones, e.g. to implement co-ordination.
1	A delay of a duration which is insignificant compared to the time constant of the controlled process is acceptable.
2	Temporary loss of data acquisition, and deviation in event scheduling time is acceptable, but the correct sequence of events shall be maintained.
3	With respect to the degree of urgency (not to the process).
4	Temporary bit error rate degradation can affect the communication efficiency; automatic restoration of any stoppage of the communication is mandatory.
5	No effect on stored data or processing accuracy.

Protection and teleprotection

Protection is of particular relevance to heavy industrial apparatus, and this function is considered to include safety and security.

Protection involves the detection of abnormal conditions and the taking of appropriate actions. The precision and rapidity of electronic protection equipment shall not be subjected to degradation of performance as a consequence of electromagnetic phenomena, such as:

- lack of protection function, with the consequences of critical conditions (including damage of power system components)

- delay in the protection operation, with consequent overstress of power system components

- spurious operation, with unavailability of the process or discontinuous working conditions, (depending on the type of apparatus).

Any degradation of protection function or safety is unacceptable; consequently, the requirement of total immunity against electromagnetic phenomena, with proper margins, is mandatory in achieving compliance with both the EMC and Machinery Directives, as well as a host of other laws and the bylaws of all Professional Institutions.

Where apparatus is intended to perform some safety function it is recommended that IEC 61508 [101] be applied in full. This covers all aspects of apparatus which are involved with functional safety, (including EMC) and requires that the integrity requirements for a safety function be detailed in the manufacturer's specification.

On-line processing and regulation

On-line processing and regulation systems give the process working condition as defined by control/telecontrol systems or by operators. The optimum running of the process is achieved by these functions.

Degradation of on-line processing and regulation could occur due to lack of immunity of the equipment and related input/output interfaces and process instrumentation involved. A possible consequence is unnecessary stress or damage of the process and degradation of performance.

The immunity of on-line processing and regulation systems to electromagnetic phenomena, including transient phenomena with low probability of occurrence, is of particular importance.

Counting and metrology

The function of counting a product being manufactured, or its energy usage, or of measuring its parameters such as weight or dimension, can be of particular significance due to the contractual aspects which may be involved.

This applies for example to traditional watt-hour meters for electrical energy measurement, and to similar equipment based on advanced technology, having a capability for setting operating conditions and storing data.

This function is required to be highly reliable, and therefore immunity against all continuous and transient phenomena is mandatory. Separate legal requirements and Directives apply to such apparatus in the EU.

Command and Control

Command and control functions are important to all operating conditions of industrial apparatus, including partial operation of the apparatus or temporary out of service conditions.

Insufficient reliability of the command and control functions due to lack of immunity could result in:

- improper operation of electrical equipment, involving safety aspects
- wrong operation sequence or procedure, with possible damage/overstress of the controlled equipment
- unavailability of the process equipment and thus of part or all of the process.

Command and control units are required to operate properly in actual environmental conditions, e.g. for continuous phenomena or high occurrence transient phenomena.

Electromagnetic phenomena with a low probability of occurrence, having only minor influence, may be observed on a control system and accepted. For example: a delay in the execution of a command may be insignificant compared to the time constant of the controlled process, so that the main function is not affected. But spurious operation of controlled switchgear is never acceptable.

Supervision

Supervisory systems collect data from the process and related equipment for diagnostic purposes, for maintenance programme purposes and for evaluation of the process. They generally do not interact with the process itself.

The performance degradation or temporary unavailability of the supervision system causes loss of information on the process and deviation in the event scheduling time. These effects can occasionally be accepted, e.g. in the case of transient phenomena with low probability of occurrence affecting acquisition of cycling measurements; but the sequence of acquisition of event data should not be affected.

Man-machine interface

The man-machine interface function allows the operators to interact directly with the process from operator desks or to manage the information from the apparatus. Control and regulation systems interface with the process, and the manual command to process equipment has higher priority.

This function may be activated by the operator by using this interface; high priority commands to the running process are generally available, and are given by the operator through the use of dedicated devices.

The absolute immunity of this function to transient phenomena having a low probability of occurrence is not generally essential, since the operator presence allows manual restoration. Obviously, this begs the question of what is understood by "low probability": manual restoration once a week (or even once a day) may be acceptable, but once an hour would probably not be.

Alarm

The alarm function includes all local or remote indications able to give information on any kind of temporary or extended degradation of the operating conditions of apparatus and systems. Alarms can have different urgency, depending on whether there is the need of immediate intervention or if the apparatus can still operate in an acceptable mode (e.g. thanks to redundancy).

In case of self-recovery after temporary degradation, the alarm may disappear, so it is not generally necessary for this function to be completely immune to low-probability transient phenomena. However, whenever a chronological list (trace) of the alarms is created automatically, this alarm function must not be affected by the electromagnetic phenomena.

Data transmission and telecommunication

Data transmission and telecommunication are auxiliary to other functions. They allow the data acquisition and the remote control of systems installed within an industrial apparatus (the control functions of the process are controlled by local systems).

Interference on data transmission and telecommunications can delay the transfer of command and control, affecting the telecontrol efficiency. Depending on the telecommunication link adopted, electromagnetic phenomena can influence the on-going communication or affect the terminal equipment, producing a bit error rate degradation. Full immunity against electromagnetic phenomena can be obtained only with particular communication technologies, e.g. optical fibre.

Temporary loss of the communication function for a short period is occasionally tolerable, provided that the link is automatically restored within an acceptable time, but the un-corrected receipt of corrupted data is not allowed.

Particular care needs to be taken when the communication link includes a wireless segment. Radio-based communication devices are inherently much more susceptible both to high-level out-of-band transmissions, and to relatively low-level signals, both intentional and unintentional, at or near the operating frequency. It should be assumed that loss of the link will occur on virtually any site under some circumstances, and the safety and operational analyses should be performed under this assumption.

Data acquisition and storage

Data acquisition and storage of relevant parameters from industrial apparatus allows off-line analysis, comparison with reference conditions and computation, etc.

Temporary deviation from precise analogue data acquisition and incorrect time allocation of digital data, due to transient phenomena, are sometimes acceptable due to the possibility of identifying these effects through data validation. No corruption of locally stored data is allowed.

Metering

Metering of parameters of the process gives direct evidence of their values and trend, and is usually carried out by using analogue or digital instruments. These instruments tend to be located on control panels, display panels, or in the proximity of the apparatus being metered.

Temporary deviation of analogue or digital indications as a consequence of a transient disturbance can be accepted, especially where operators can identify the source of the disturbance and hence take it into account; but no degradation outside of normal operational specifications due to continuous electromagnetic phenomena is allowed.

Off-line processing

Off-line processing functions allow simulation of the process, planning of manufacturing, study of models, and analysis of critical working conditions, etc. This function implies the use of data coming from the process or stored data. It does not interact with the on-line process itself.

Temporary degradation of this function due to transient phenomena is often acceptable, on condition that it does not involve the on-line activities and there is no corruption of stored data or processing accuracy.

Monitoring

The process is monitored on displays, showing the entire setting and operating condition of the plants. Information technology equipment with VDU monitors or other devices are generally used to represent the process and its parameters.

Temporary degradation of this function may be acceptable, provided that the consistency of the monitoring with the process condition is resumed. Temporary loss of display with restoration within a given time, e.g. a few seconds, allowing for possible intervention by operators, may also be acceptable. An example would be the distortion of the image on a CRT-type display caused by temporarily high magnetic fields (but note that various Health and Safety legislation applies to the stability of visually displayed images, and the above example must not be taken to imply that EMC considerations can over-ride health and safety issues).

Self-diagnosis

Self-diagnosis capability is increasing in complex electronic systems, and becomes of particular relevance for the credibility of the system itself. Self-diagnostic test cycles are generally assigned low priority in the task sequence.

Temporary loss of the self-diagnostic function can generally be considered as acceptable if it is self recovered within the system working cycle and if it gives rise only to a delay in warning an operator of a system failure condition. Such a loss of function should also not produce spurious alarm conditions which could require attendance at unmanned remote locations.

Some published case studies

(These case studies are extracted from published articles in various sources over the past few years. We gratefully acknowledge the journals concerned.)

C.1 Segaworld

Paul Harris, TRL, Approval, Jan/Feb 97

Being the largest virtual reality theme park in the world Segaworld is crammed full of high speed processing circuitry, interactive displays and motive power actuators. The whole installation is physically large, occupying two floors of the Trocadero building. Sega, the Japanese company building the park, wanted to meet the conformity requirements of the EU EMC Directive, so before the designs for the various rides were finalised they sought help.

TRL has had many dealings with large machines and distributed installations, although our experience did not run to virtual reality theme parks. So when Sega sought our help, one of our staff immediately took advantage of the offer to look at its other virtual reality theme park "Joypolis" at Yokohama. The rides destined for Segaworld were to be European versions of established rides at Joypolis. Thus a detailed review of the Yokohama site was used to plan the EMC conformity project for Sega. We were able not only to discuss the EMC threats with the designers, but also to help incorporate EMC countermeasures at an early stage of the design.

The approach adopted for the Segaworld Trocadero complex was to ensure all games and rides were individually CE marked, providing maximum flexibility for mixing and matching as the theme park evolved. All games machines and consoles produced by Sega were already compliant with the directive, so this left the large rides and virtual reality simulators to be considered.

Rides such as: Aqua Planet; Ghost Hunt; Mad Bazooka; Beast in Darkness; Space Machine; House of Grandish; AS 1. These were too large to be tested against harmonised specifications in laboratory conditions and would thus be tested in-situ after construction and commissioning in the Trocadero building.

Since TRL is a competent body, we were able to define the aspects of the Segaworld complex which would need to be considered to ensure that it met the essential requirements of the directive. We specified what would be required of the TCF to demonstrate that the EMC performance met the requirements.

With the knowledge of the Joypolis complex and the proposed design and layout of the Trocadero, a number of issues were immediately obvious:

- The Segaworld complex is in a building that is in multiple occupation, there would be neighbours on floors above and below it.
- The neighbouring businesses shared the AC power distribution system.
- Some possible immunity problems were safety related.
- Some immunity issues were irrelevant in the context of the games' normal use.

A test plan was devised to take account of these considerations and provide data on each of the rides to support the TCF. Each ride had a test plan and each plan was tailored to the characteristics of the machine technology, the expectation of the user and its location in the Trocadero.

All rides were tested in-situ for conducted interference on the AC power line and for spatial radiation. Limits similar to those used in standards for a residential environment were proposed and eventually achieved for conducted emissions. The measurement techniques involved probing onto heavy current conductors due to the current demand and the permanent nature of such installations.

Spatial electric field radiation was evaluated at the boundary of the Segaworld site. One of the few advantages of in-situ testing is that often the field strength is measured at the genuine point of possible interference and there is no extrapolation or calculation necessary.

The criteria for immunity performance was assessed for each ride and rated in order of priority. The levels of immunity had to be sufficient to ensure:

- Safety.
- That there was no equipment damage.
- That the machines were fit for use.

Since the rides would be used by untrained people, all electrically actuated safety interlocks and countermeasures were fully exercised during the test programmes of each ride. Similarly, we had to make sure that none of the machines had the ability to malfunction and cause damage in the presence of conducted and radiated interference. Because the test environment was inside a conventional building, radiated testing could only be carried out using licensed transmitters (PMR handset and portable cellular phone). Again, this in-situ test was considered a good simulation of the real world operating conditions likely to be encountered. However, radio field susceptibility testing at other than licenced frequencies was simulated with bulk current injection, a technique where interference is coupled into cables judged to be vulnerable.

Some judgement was required in deciding criteria to establish whether the games were fit for use. We took into account the likely expectations of people using the rides. In many cases, we decided that a small on-screen perturbation on an interactive display zapping aliens while hurtling through the streets of an imaginary space city was not significant. However, each phenomena was rated against the level of possible customer dissatisfaction.

The frenetic activity to get Segaworld open on time had to be seen to be believed. TRL's mobile test team worked on each ride as the installations were commissioned by Sega technicians from Japan. We used the test plans we had devised in the way that a specification would be followed in a normal laboratory environment. Where emission levels were found to be exceeding the limits defined by the test plan, remedies were implemented to obtain the required performance. This included using high current power line filters, ferrites and other EMC countermeasures.

Postscript: TRL EMC now employs mobile test engineers with amazing alien zapping, Mad Bazooka scoring and Ghost Hunting capabilities.

C.2   Air Traffic Services

P J Dorey, Assessment Services, Euro-EMC Conference, 7–9 October 1997

Introduction

This paper describes the electromagnetic compatibility (EMC) assessment of large ground based air traffic services (ATS) transmitting equipment. The type of equipment includes radars and landing systems. The Civil Aviation Authority as the UK Notified Body for aeronautical transmitting equipment has a responsibility for ensuring compliance with the EC Type Examination route of the European Directive regarding EMC. As there are no product EMC test standards for ATS transmitting equipment guidance has been required by the CAA, ATS manufacturers and test houses on the EMC requirements and interpretation of existing test standards. This paper is based on a study performed for the CAA Safety Regulation Group who have kindly given permission for its use.

EMC Test Requirements

As there are no specific product standards for ATS transmitting equipment, the following approach has been taken:

- Examine the electromagnetic severity of the environment and select test limits
- Review existing harmonised standards, draft standards and other relevant standards
- List applicable test phenomena and determine whether mandatory or equipment dependent
- Review suitability of test methods for ATS transmitting equipment
- Review necessary test conditions and equipment configuration.

EMC Standards

It is preferable to relate the test requirements to existing generic standards where possible and to adopt harmonised basic standards for individual tests. Existing standards of relevance include:

- Generic emission standards: Light Industrial EN50081-1, Industrial EN50081-2
- Generic immunity standards: Light Industrial EN50082-1, Industrial EN50082-2
- Product emission standards: Information Technology EN55022, Industrial Scientific Medical EN55011
- Basic EMC emission publications series: EN 61000-3-x or IEC1000-3-x
- Basic EMC immunity publications series: EN 61000-4-x or IEC1000-4-x.

EMC Environment

The severity of the electromagnetic environment (EME) in which ATS equipment is installed determines the maximum level of permissible emissions and minimum level of immunity required. To establish levels of EME a series of RF site surveys of airport and radio station sites were performed.

The surveys established that at frequencies below 1GHz, field strengths above 3V/m were often encountered. This is the radiated immunity level specified in the Generic Immunity Standard (Light Industry). This level of protection is therefore inadequate and the more severe environment of 10V/m needs to be adopted as given in the Generic Immunity Standard (Industrial).

Above 1GHz the surveys show that field strengths are highly variable dependent on other transmitters present. Primary radar can produce peak levels up to 100V/m in exposed positions. It is therefore necessary to establish the immunity performance of ATS transmitting equipment above 1GHz. An immunity level of 50V/m is recommended over the frequency range 1GHz to 18GHz.

Examination of power supply arrangements show that ATS transmitting equipment often share common power supplies with other users. Where Uninterruptible Power Supplies (UPS) are used these are often shared with other uncontrolled users therefore degrading the EMC quality of the supply. Many sites do not have a UPS fitted. The Type Examination must therefore be valid for all likely applications of the ATS transmitting equipment i.e. the worst case EMC conditions.

The ATS transmitting equipment usually shares sites with a range of other critical receiving equipment. Limitation of emissions is therefore important and achieved by specifying the lower Class B emission limits in preference to Class A. Class A is applicable where the ATS equipment is only operated in remote or isolated sites.

Applicable EMC Tests

Using the Generic Standards for industrial environments as a guide, the minimum test data must cover radiated and conducted RF emissions and immunity, imported transient immunity and electrostatic discharge. Further equipment dependent tests must be considered. Examples of the rationale required are:

- If the ATS frequency of operation is near or greater than 1GHz then additional testing for emissions and immunity should be considered above 1GHz.
- In the absence of dedicated Uninterruptible Power Supplies (UPS) then tests for power supply harmonics, voltage fluctuation, voltage dips & interrupt are applicable.
- If critical magnetic components (transformer/VDU etc.) are used, consider magnetic tests.
- If external interface signal cables are greater than 3m in length or pass outside the equipment room or building, consider signal line tests i.e. conducted emissions, fast transient/burst, surge, RF conducted immunity.

A range of basic standards and related standards have been selected to address the test requirement.

Suitability of Test Methods

Due to the size, complexity and power consumption of ATS transmitting equipment the test methods given in the standards present difficulties:

Conducted emissions on power lines, EN55022, EN55011: The line impedance stabilisation network (LISN) used must be rated for the line current while maintaining CISPR 16 impedance characteristics. The standards do not currently provide methods or limits for signal lines. In the absence of formal limits, a set of derived current limits are necessary.

Radiated emissions, EN55022, EN55011: The specified method is to use an Open Area Test Site (OATS) which must be able to accommodate the ATS equipment on a turntable and supply the required power. The ATS equipment may be too large or complex to install on an OATS. In this instance an 'on-site' test is a solution. Testing above 1GHz can be performed in accordance with EN55011 in terms of radiated power.

Magnetic emissions. No method or limit is specified in the European standards. If magnetic tests are essential, then the adoption of military standards is recommended, eg. MIL STD 461/462.

Magnetic immunity, EN61000-4-8, EN61000-4-9: These standards provide methods for power frequency 50Hz immunity and induced magnetic field due to lightning. For large equipment a hand held coil can be used to apply the field over the faces and cables of the equipment.

Harmonic current and voltage fluctuations, EN61000-3-2, EN61000-3-3: The scope of the test standards cater for equipment less than 16A/phase. There are no suitable limits available for higher current equipment. Test equipment is designed to supply the line current and as a result most test equipment is limited to 16A. Splitting ATS equipment into groups of sub-equipments allows a method of test.

Electrostatic discharge, IEC1000-4-2: The standard method applies and user/maintenance access to controls and cabinets must be defined to determine the test points.

Immunity RF radiated, ENV50140 or IEC1000-4-3: This test must be performed within a screened enclosure large enough to accommodate the ATS equipment. Ideally the enclosure is anechoically lined and equipped with a turntable to allow testing of each face of the equipment. In the absence of a turntable, the test antenna needs to be moved to each face of the equipment. At microwave frequencies the beamwidth limitation of the test antenna may require multiple testing of each face.

Immunity Fast Transient/Burst, IEC1000-4-4: The test standard requires testing 1m along the cable from the equipment with direct coupling onto power lines. Again a coupling decoupling network (CDN) with adequate current rating is required.

Surge, IEC1000-4-5: Applied to power lines and signal lines that cross the ATS equipment boundary. The standard methods are suitable.

Immunity RF Conducted, ENV50141 or EN61000-4-6: The test applies to cables that cross the ATS equipment boundary. Coupling networks are required for power lines with the current limitations mentioned previously. Signal lines are tested using current injection probes.

Immunity to Voltage Dips, Interrupts, Variations, Fluctuations, EN61000-4-11: The scope of the test standard caters for equipment less than 16A/phase with similar limitations to the harmonic and voltage fluctuation test.

Test Configuration & Conditions

An important part of the manufacturer's type examination application is a definition of the build state of the ATS transmitting equipment including the quantity and type of major items and options which are to be included in the type examination certificate. A boundary must be defined between the ATS transmitting equipment and any other systems which interconnect to it. One area of confusion are items which are bought-in such as modems. Bought-in items supplied as an essential part of the ATS transmitting equipment must satisfy the same Type Examination criteria as the ATS transmitting equipment. CE marked items may not have been tested to the same levels and therefore may require additional tests.

The mode or modes of operation selected must ensure that the transmitter is operating at maximum power and with modulation or coding to produce maximum duty cycle. The antenna would normally be replaced by a force cooled dummy load for laboratory testing. Mounting arrangements of the equipment and cables should be representative of the manufacturer's recommended installation procedures and include earthing and bonding as used in an installation.

The methods of monitoring the ATS equipment during immunity testing must be practical and the performance criteria defined in terms of measurable indications of failure or degradation in performance. Three performance criteria levels are specified, A, B or C. Performance criteria A requires the ATS equipment to continue operating as intended during test. Criteria B allows a pre-defined degradation of performance while maintaining critical functions. Criteria C allows temporary loss of function.

C.3   VDU image problems in a steel rolling mill

Keith Armstrong, Cherry Clough Consultants, EMC Journal, August 1998

A steel rolling-mill drive with drive currents of up to ±8000Amps was designed, constructed, and installed. The supplier promised that it would fully comply with the EMC Directive, and this was part of the contract for the work. The supplier was a major UK drives and heavy power systems manufacturer with wide experience and a good reputation.

During the course of the contract the customer requested information on the suitability of positioning a monitoring room adjacent to the drive cabinets, in particular whether any interference to the computers in this room would occur. The room was to contain a number of computers and screens, including SCADA. The drive supplier said that the proposed position would not suffer from any interference.

When installation of drive system and monitoring room was complete it was clear that the drives were interfering grossly with the images on the monitor screens. They would change colour dramatically, while sliding around on the screen by approximately 50% of the screen's width.

There was a clear correlation between the drive direction (forward/reverse) and the direction of movement of the VDU images, and the amount of image movement could be directly correlated with the drive loading (i.e. motor current). Use of the screens was impossible – not just a question of eye strain, but a complete impossibility of reading the data on the VDUs.

The manufacturer initially claimed that the image problems were not his problem, and continued to maintain that his product met the EMC Directive. His error was in assuming that because the drive met the harmonised emissions standards, it would comply fully with the Directive.

However, compliance with a harmonised standard only gives a legal "presumption of conformity", and since the harmonised standard in question (BS EN 50081-2:1993) contains no tests or limits at all for emissions of DC or low-frequency magnetic fields, it is obvious that an 8000A drive could have problems in meeting the "essential protection requirements" of the EMC Directive.

All suppliers of CE marked apparatus are required to sign a legal declaration that their product meets these "EMC essential protection requirements", which broadly state that products do not cause or suffer unduly from interference in normal use when used as intended. Clearly, the drive could not meet the essential protection requirements when computer monitors were used in proximity, as the manufacturer had indicated they could be.

This is a typical example of a manufacturer not fully considering all the EMC phenomena relevant to his product and the application it was going into, assuming that meeting the harmonised standards was all he had to do, and consequently selling an illegal product. The legality was not the main issue as far as the customer was concerned, but the inability to use the monitoring room was. Luckily (in this instance) the room was only intended to be used part-time – it was not the main control room.

The field strengths in the computer room were measured at over 200µT, and it is a commonly known fact that most CRT-type VDUs start to exhibit visible image changes at around 1µT. A few simple sums along the lines suggested in section 7.1.1.1 of this book would have shown the manufacturer, or the steel mill engineers, that it would not be feasible to site the computer room where they wanted to. It should have been at least 30 metres further away from the drive cabinets and their supply and motor cables, or else used special VDUs which would be immune to the magnetic fields.

After a considerable amount of time (and cost, in terms of delays to other projects) had been spent on the problem, during which the monitoring room was to all intents unusable, a solution was found involving LCD and plasma type displays for the smaller VDUs, these technologies of VDU do not have the susceptibility to magnetic fields that CRT types always have.

Towards the end of the solution LCD screens became available in the size/resolution which suited the SCADA displays. These were still very high cost, though, so a manufacturer of ruggedised CRT monitors was found that included active cancellation of their local magnetic fields. Even so, they were not tough enough for the fields present in the room, although the supplier felt confident that they could engineer a special version which would cope with the actual field strengths.

At the time of writing it is not known whether the final SCADA screens used LCD or ruggedised CRT monitors, but it is known that the drive supplier agreed to pay for the new VDUs. The contingent costs to the steel producer in lost time and consultancy fees had to be written off.

C.4   Large explosion at the Texaco Refinery, Milford Haven, July 1994

Keith Armstrong, Cherry Clough Consultants, EMC Journal, June 1998

At 1:23 pm on Sunday, 24th July 1994 there was an explosion at the Texaco Refinery, Milford Haven. Its force was equivalent to 4 tonnes of high explosive and it started fires that took over two days to put out. Shops in Milford Haven 3km away had their windows blown in. 26 people sustained minor injuries, and the fact that it was Sunday lunchtime and the site was only partially occupied meant it could have been very much worse. Damage to the plant was substantial. Rebuilding costs were estimated at £48 million. There was also a severe loss of production from the plant – enough to significantly affect UK refining capacity. The incident was initiated by an electrical storm between 7:49 and 8:30 am on the Sunday morning which caused a variety of electrical and other disturbances across the whole site.

For more detail on this incident, although not on its relationship to EMC, read IEE

Computing and Control Engineering Journal April 1998 pp 57–60. There is also an HSE report on this incident: "The explosions and fires at Texaco Refinery, Milford Haven, 24th July 1994" HSE Books, May 1997.

Comments: The author has not read the HSE report, but understands from private conversations with HSE experts that the large explosion was caused by the electrical storm giving rise to power surges which tripped out a number of pump motors whilst leaving others running. As there was a great deal of panic and confusion due to the information overload caused by the numerous small fires and equipment outages from the time of the storm, it was not noticed that flammable substances which should have been flared off were accumulating in pipework and vessels. After five hours something ignited the total accumulation, resulting in the large explosion.

The general incidence of surges in the UK's AC power distribution network is quite low, and this often leads people to believe that qualifying the power surge immunity of their products, systems, or installations is not important. This belief is often supported by the observation that neither of the original generic immunity EMC standards included surge testing in their normative sections. But it only takes a single incident such as the above in the lifetime of even a very large plant to make an excellent economic case for a proper preventative strategy. Suitable basic test standards include IEC 61000-4-5 or IEC 61000-4-12 (ring wave), both of which are intended to simulate the indirect effects of electrical storms on power networks.

Engineers are always under pressure to save costs, and the costs of preventative measures are easy to quantify. However, many engineers are uncomfortable with estimating the risks of infrequent and unpredictable events such as thunderstorms so do not effectively communicate the actual risk/cost and safety implications to their managers.

As someone said recently: Doctors kill people in ones, but engineers do it in hundreds. Careers and personal liability are also at stake here too, so it is always best to make an informed cost/risk case and get a written decision from management. There is no shortage of advice and assistance.

C.5   Wide area interference created by large inverter drives

Keith Armstrong, Cherry Clough Consultants, EMC Journal, February 1998, taken from a paper by M J V Wimshurst of Hill Graham Controls, High Wycombe and Allan Ludbrook of Ludbrook and Associates, Ontario, Canada

To cope with increased North Sea oil production, two new pumping stations with 6MW adjustable speed induction motor drives were built and installed in Scotland, one in Netherly and one in Balbeggie. Soon after commissioning the local power utility and the telephone company received a flood of complaints.

Geographically the complaints came from concentrated pockets spread over an area up to 12.5 miles away from the 33kV overhead supply lines feeding the drives. A payphone over 4 miles away from the power line was noisy enough to be almost unusable, whereas just across the street a householder's telephone was relatively unaffected.

Other symptoms included loss of synchronisation on TV sets (rolling pictures) and ringing on the supply to fluorescent lighting circuits. Although the drives had been designed to (and met) the supply industry's G5/3 harmonic limits, the problems turned out to be with higher order harmonics than it covered, up to the 100th in fact (i.e. 5kHz).

The problem became a public relations nightmare for all involved, and culminated in questions being raised at Government level. Remedial EMC work was urgently required and was in fact accomplished, although under extreme difficulties because the cost of any downtime of the oil pumping stations was so high.

C.6   Release of chlorine gas in semiconductor processing plant

Simon Brown, HM Specialist Electrical Inspector, Health and Safety Executive, EMC Journal, February 1999 and April 1999

The HSE recently prosecuted the supplier of an item of equipment which led to a release of chlorine in a semiconductor plant. The equipment was not sufficiently immune to mains transients (and proven to be so by our own labs). We prosecuted under section 6 of the Health and Safety at Work Act because the supplier, though aware of the problem, did not inform the users of the equipment. The defendant entered a plea of guilty to a charge brought under S6 (1) (d) of the Health & Safety at work etc. act , 1974. The magistrates imposed a fine of £5,000 and made an order for the defendants to contribute £7,000 towards HSE's costs of £9,482.

The case concerned a microprocessor based valve control panel used to control the flows of chlorine and nitrogen in a semiconductor plant. There had been a release of chlorine resulting from all of the valves in the control cabinet being set to an open position. Investigation by the HSE found that the unit was susceptible to conducted transients on the mains supply. There were no precautions against electrical interference in the power supply and the microprocessor watchdog was not effective in ensuring a safe state following detection of a fault.

C.7   Traction current interference to safety circuits

Modern Railways, November 1994, p653

Track circuits that detect the presence of a train on the 750V dc electrified lines of the former Southern Region use alternating current at the mains frequency of 50Hz. While the three-phase inverter drives for the traction motors in Networker and Eurostar trains are designed to minimise this frequency in the earth currents which return through the rails, there is the possibility that a drive could malfunction and interfere with the track circuits.

To protect against such a situation, the trains are fitted with an Interference Current Monitoring Unit (ICMU). If it detects a current at 50Hz above a certain threshold for more than 450mS the ICMU shuts the drive down.

Because Eurostar power cars are ... particularly prone to arcing when crossing gaps ... they generate interference at all frequencies, causing the ICMU to operate.

To overcome this problem about 1,600 track circuits in the London area have had to be modified to be either immune to interference or to react more slowly. As a result the Eurostar ICMU can now be set to a two second delay and ... can run ... between London and Dollands Moor without the ICMU tripping.

See also: **The real engineering need for EMC**, *John Whaley, IEE Colloquium on Electromagnetic compatibility in heavy power installations, 23rd February 1999, Middlesbrough, IEE Colloquium Digest no 99/066*

The EU and EEA countries

D.1　The European Union

Established by the Treaty of Rome, 1957.　　* = founder members
= joined 1st January 1995

Austria	#
Belgium	*
Germany	*
Denmark	
Spain	
Finland	#
France	*
Greece	
Ireland	
Italy	*
Luxembourg	*
The Netherlands	*
Portugal	
Sweden	#
United Kingdom	

Applications for membership have been received from several other countries

D.2　The European Economic Area

As above, plus:

Norway
Iceland
Liechtenstein

Glossary

Term	Meaning	See page
AC	Alternating current	
AMN	Artificial Mains Network	239
BCI	Bulk current injection	269
BRC	Bonding Ring Conductor	117
CB	Competent body	
CBN	Common Bonded Network	108
CCTV	Closed Circuit TV	
CDN	Coupling/decoupling network	268
CEN	Comité Européen de Normalisation	
CENELEC	Comité Européen de Normalisation Electrotechnique	
CISPR	Comité Internationale Speciale des Perturbations Radioélectriques	
CM	Common mode	86
CMET	Customer's main earthing terminal	
csa	Cross-sectional area	
DC	Direct current	
DM	Differential mode	86
DRAM	Dynamic Random Access Memory	
EEA	European Economic Area	
EMC	Electromagnetic compatibility	2
ESD	Electrostatic discharge	8
ETSI	European Telecommunications Standards Institute	
EU	European Union	
EUT	Equipment under test	
FCC	Federal Communications Commission (US)	
FE	Functional earth	
FEXT	Far end crosstalk	
GDT	Gas discharge tube	229
HF	High frequency	
HV	High voltage (supply)	
IBN	Insulated bonding network	127
IDC	Insulation displacement connector	
IEC	International Electrotechnical Commission	
I/O	Input/output	

IT	Information Technology	
ITU	International Telecommunications Union	
kHz	kiloHertz (measure of frequency)	
λ (lambda)	Wavelength (metres), = frequency(MHz)/300	
LISN	Line Impedance Stabilising Network	239
LAN	Local Area Network	
LCD	Liquid Crystal Display	
LCL	Longitudinal Conversion Loss	161
LEMP	Lightning electromagnetic pulse	212
LF	Low frequency	
LPS	Lightning protection system	213
LV	Low voltage (supply) (< 1kV)	
LVD	Low Voltage Directive	18
MESH-BN	Mesh Bonding Network	108
MESH-IBN	Mesh Insulated Bonding Network	127
MET	Main Earthing Terminal	
MHz	MegaHertz (measure of frequency)	
MOV	Metal oxide varistor	229
MRI	Magnetic Resonance Imaging	
MSD	Machinery Safety Directive	17
MV	Medium voltage (supply)	
NEXT	Near end crosstalk	
NSA	Normalised Site Attenuation	249
OATS	Open Area Test Site	249
PCB	Printed circuit board	
PEC	Parallel Earth Conductor	188
PEN	Protective Earth Neutral	
PLC	Programmable logic controller	
RCD	Residual current device	
RF	Radio frequency	
RTD	Resistance temperature detector	
SCADA	System control and data acquisition	
SERP	System earth reference point	
SPC	Single Point of Connection	127
SPCW	Single point connection window	129
SPD	Surge protection device	229
SRPP	System Reference Potential Plane	123
STP	Shielded twisted pair	
TCF	Technical Construction File	31
UHF	Ultra high frequency (> 300MHz)	
UTP	Unshielded twisted pair	176
VDU	Visual display unit	
VHF	Very high frequency (30-300MHz)	

Bibliography

Books and company publications

[1]   **Earthing Practice**
       CDA Publication 119, Copper Development Association, Potters Bar, Herts

[2]   **Earthing and Lightning Protection Consultants Handbook**
       W J Furse & Co. Ltd, Publication ref: CHB/4/95, phone 0115 986 3471, fax 0115 986 0538

[3]   **Electronic Systems Protection Handbook**
       W J Furse & Co. Ltd, publication ref: ESPHB1, phone 0115 986 3471, fax 0115 986 0538

[4]   **Electromagnetic Compatibility**
       J J Goedbloed, Prentice-Hall, 1992, ISBN 0-13-249293-8

[5]   **A drive user's guide to installation and EMC**
       IMO Precision Controls Ltd, March 1994

[6]   **The Guide to the EMC Directive**
       C Marshman, 2nd Edition, EPA Press 1995

[7]   **Grounding and Shielding Techniques in Instrumentation**
       R Morrison, 3rd Edition, Wiley, 1986

[8]   **Noise Reduction Techniques in Electronic Systems**
       H W Ott, 2nd Edition, Wiley, 1988

[9]   **Introduction to Electromagnetic Compatibility**
       C R Paul, Wiley, 1992

[10]   **Coupling of External Electromagnetic Fields to Transmission Lines**
       A A Smith, Wiley 1977

[11]   **EMC Practical Installation Guidelines**
       Telemecanique (Group Schneider), reference TEGB1444 March 1996, available for free from Telemecanique +44 (0)1203 416 255, fax: +44 (0)1203 417 517, or from their agents and distributors (Square D in the USA).

[12]   **Cable Shielding for Electromagnetic Compatibility**
       A Tsaliovich, Van Nostrand Reinhold, 1995

[13]   **EMC for Product Designers**
       T Williams, 2nd Edition, BH Newnes 1996

Papers

[14]   **EMC enforcement around Europe**

Approval magazine, Mar/Apr 1997, pp 4–5

[15]   **Long–term survey of the background electromagnetic environment in Switzerland**

J Baumann, G J Behrmann, H Garbe, Federal Office of Environment/EMC Baden, 10th
International Symposium on EMC, Zurich, March 9–11, 1993

[16]   **The technical and economical optimization of interference tests and the analysis of risks
with regard to the design of interference-free apparatus and systems**

S Benda, ABB Automation AB, EMC 93–"Achieving Whole–life EMC: Management, Design
and Compliance", ERA Technology, Heathrow, 18th February 1993, ERA report 93–0013

[17]   **Earthing of screened copper cables between buildings**

I Braithwaite, Telematic UK Ltd, Conference "Earthing Solutions – standards, safety, and good
practice", Solihull 23–24 June 1997, ERA Report 97–0533 pp 5.1.1 to 5.1.12

[18]   **Surge protection devices – installation issues**

I Braithwaite, Telematic UK Ltd, Colloquium on Surges, Transients and EMC, IEE, London,
18th February 1998, IEE Colloquium Digest 1998/197

[19]   **Selling EMC in a large organization**

T E Braxton, AT&T Bell Labs, IEEE 1988 International Symposium on EMC, Seattle, August
1988

[20]   **Protection Concepts in Telecommunication Equipment**

A Bremond, SGS–Thomson Microelectronics AN579/0393, in STM Protection Devices
Databook, 2nd Edition, March 1993

[21]   **Electromagnetic compatibility of building cables**

K Bromley, BRE, IEE Proceedings–A, Vol 141 No 4, July 1994 (special issue on EMC),
pp 263–265

[22]   **Designing Power-line Filters for their Worst-Case Behaviour**

F Broyde, E Clavelier, EXCEM, 9th International Symposium on EMC, Zurich, March 12–14
1991 pp 583–588

[23]   **In the pipeline**

BSI News, May 1996

[24]   **Coupling Reduction in Twisted Wire**

W T Cathey, R M Keith, Martin Marietta, International Symposium on EMC, IEEE, Boulder,
Aug 1981

[25]   **A Review of Transients and their Means of Suppression**

S Cherniak, Motorola Application Note AN–843, 1982

[26]   **Earthing and Bonding for Installations and Buildings**

M Clark, NICEIC, Conference on Earthing and Bonding of Information Technology
Equipment, Inter–Continental Hotel, London, 6th November 95, Proceedings published by
Starform Communications, phone 0181 201 9888, fax 0181 201 8339.

[27]   **European Enforcement of the EMC Legislation**

D Davis, International Product Compliance, May/June 1998, pp 25–28

[28]   **Cable Management Systems**

M Gilmore, The Cabling Partnership, Conference on Earthing and Bonding of Information
Technology Equipment (see [26] for contact)

[29]   **Transients in Low-Voltage Supply Networks**

J J Goedbloed, Philips, IEEE Transactions on Electromagnetic Compatibility, Vol EMC–29,
No 2 May 1987, pp 104–115

[30]   **Characterization of Transient and CW Disturbances induced in Telephone Subscriber
Lines**

J J Goedbloed, W A Pasmooij, Philips Research, IEE 7th International Conference on EMC,
York 28–31st Aug 1990 pp 211–218

[31]   **Investigation of fundamental EMI source mechanisms driving common-mode radiation
from printed circuit boards with attached cables**

D M Hockanson et al, IEEE Transactions on Electromagnetic Compatibility, Vol EMC–38,
Nov 1996 pp 557–566

[32]   **Experiences in making on-site EMC measurements**

H R Hofmann, AT&T, 11th International Symposium on EMC, Zurich, 7–9 March 1995,
pp 35–40

[33]   **How Switches Produce Electrical Noise**

E K Howell, General Electric, IEEE Transactions on Electromagnetic Compatibility, Vol
EMC–21, No 3 August 1979, pp 162–170

[34]   **LAN Networks**

R Hughes, BNR Europe Ltd (Nortel), Conference on Earthing and Bonding of Information
Technology Equipment (see [26] for contact)

[35]   **On the design and use of the average detector in interference measurement**

W Jennings, H Kwan, DTI, 5th International Conference on EMC, York 1986, IERE
Conference Publication 71, pp 311–314

[36]   **Achieving Compatibility in Inter-Unit Wiring**

J W E Jones, Portsmouth Poly, 6th Symposium on EMC, Zurich, March 5–7 1985

[37]   **Reliable protection of electronics against lightning: some practical examples**

P C T van der Laan, A P J van Deursen, IEEE Transactions on Electromagnetic Compatibility,
Vol EMC–40 No 4, November 1998 (special issue on Lightning), pp 513–520; also

Reliable protection of electronics against lightning

P C T van der Laan, A P J van Deursen, 23rd ICLP International Conference on Lightning
Protection, Firenze, Italy, September 23–27, 1996, pp 370–373

[38]   **Design Philosophy for Grounding**

P C T van der Laan, M A van Houten, 5th International Conference on EMC, York, 1986,
IERE Conf Pub 71

[39]   **Average Measurements using a Spectrum Analyzer**

S Linkwitz, Hewlett Packard, EMC Test & Design, May/June 1991, pp 34–38

[40]   **An active approach to harmonic filtering**

P Lynch, ABB Power, IEE Review, May 1999 pp 128–130

[41]   **Some observations on the protection of buildings against the induced effects of lightning**

E Montandon, M Rubenstein, IEEE Transactions on Electromagnetic Compatibility, Vol EMC–40 No 4, November 1998 (special issue on Lightning), pp 505–512

[42]   **A Structured Design Methodology for Control of EMI Characteristics**

U Nilsson, EMC Services AB, IEE 7th International Conference on EMC, York 28–31st Aug 1990

[43]   **Earthing and Bonding of Information Technology Equipment**

R Nye, GPT, Conference on Earthing and Bonding of Information Technology Equipment (see [26] for contact)

[44]   **Ground – A Path for Current Flow**

H W Ott, Bell Labs, International Symposium on EMC, IEEE, San Diego, October 9–11 1979

[45]   **Shielded Flat Cables for EMI and ESD Reduction**

C M Palmgren, 3M, International Symposium on EMC, IEEE, Boulder, Aug 1981

[46]   **Effect of Pigtails on Crosstalk to Braided-Shield Cables**

C R Paul, Univ of Kentucky, IEEE Transactions on Electromagnetic Compatibility, Vol. EMC–22 No 3, Aug 80

[47]   **A Comparison of the Contributions of Common-Mode and Differential-Mode Currents in Radiated Emissions**

C R Paul, Univ of Kentucky, IEEE Transactions on Electromagnetic Compatibility, Vol EMC–31 No 2 May 89

[48]   **Shielding effectiveness of a rectangular enclosure with a rectangular aperture**

M P Robinson et al, University of York, Electronics Letters, 15th August 1996, Vol 32 No 17, pp 1559–1560; also published in more thorough form in

Analytical Formulation for the Shielding Effectiveness of Enclosures with Apertures

M P Robinson et al, University of York, IEEE Transactions on EMC, Vol 40 No 3, August 1998, pp 240–248

[49]   **EMC for Building Designers, Installers and Managers – a practical approach**

D Sawdon, IBM, Conference on Earthing and Bonding of Information Technology Equipment (see [26] for contact)

[50]   **Approximate Simulation of the Shielding Effectiveness of a Rectangular Enclosure with a Grid Wall**

R de Smedt et al, Alcatel Telecom, 1998 IEEE International Symposium on EMC, Denver, August 24–28 1998, pp 1030–1034

[51]   **Attenuation of Electric and Magnetic Fields by Buildings**

A A Smith, IBM, IEEE Transactions on Electromagnetic Compatibility, Vol EMC–20 No 3 Aug 78

[52]   **There is only one earth**

P L Smith, PLS Partnership Ltd, Conference "Earthing Solutions – standards, safety, and good practice", Solihull 23–24 June 1997, ERA Report 97–0533 pp 4.1.1 – 4.1.9

[53]   **Weighting of Interference According to its Effect on Digital Communications Services**

M Stecher, Rohde & Schwarz, 1998 IEEE International Symposium on EMC, Denver, August 24–28 1998, pp 69–73

[54]   **European safety marks**

P Stokes, Approval, May/June 1996

[55]   **Lightning protection by natural components**

M Summers, "Lightning Protection 98: buildings structures and electronic equipment", ERA Technology, Solihull, 6–7 May 1998

[56]   **Shielding Quality of Cables and Connectors: Some Basics for Better Understanding of Test Methods**

B T Szentkuti, Swiss PTT, 1992 IEEE International Symposium on EMC, Anaheim, August 17–21 1992, pp 294–301 (pp 294–333 of this publication is a record of a session on shielded cable testing and contains other useful papers)

[57]   **Characterisation of the Shielding Effectiveness of Loaded Equipment Enclosures**

D W P Thomas et al, University of Nottingham, IEE Conference on EMC, York, July 12–13th 1999

[58]   **Characterisation of the Shielding Effectiveness of Equipment Cabinets containing Apertures**

J D Turner et al, University of Nottingham, EMC'96 ROMA International Symposium on EMC, September 17–20 1996, Rome, pp 574–578

[59]   **Earthing and bonding in communications rooms**

P Wellman et al, Ove Arup, Conference "Earthing Solutions – standards, safety, and good practice", Solihull 23–24 June 1997, ERA Report 97–0533 pp 4.2.1 – 4.2.17

[60]   **How to use component data as a shortcut to EMC compliance**

J Whaley, SGS, Approval, Sep/Oct 1998 pp 24–27, Nov/Dec 1998 pp 23–25

[61]   **The Lightning Threat**

J Whiting, Conference on Earthing and Bonding of Information Technology Equipment (see [26] for contact)

[62]   **Installation Design**

J Whiting, Hawksmere Seminar on Electromagnetic Compatibility Conference 21 May 1996 Café Royal, London, proceedings from Hawksmere on 0171 824 8257 fax 0171 730 4293

[63]   **Earthing arrangements and equipotential bonding for information technology installations**

J Whiting, Conference "Earthing Solutions – standards, safety, and good practice", Solihull 23–24 June 1997, ERA Report 97–0533 pp 1.2.1 – 1.2.6

[64]   **Information technology: revised emissions standard**

T Williams, Elmac Services, Approval, November/December 1998, pp 26–28

Official publications and standards

European standards: available in the UK from

BSI Customer Services
BSI Standards
389 Chiswick High Road
London   W4 4AL
UK
Tel 0181 996 7000; Fax 0181 996 7001

Note: draft standards are included in this listing with their corresponding BSI document number. These should not be regarded in the same light as a published standard.

[65]   **EN 50 065-1:1992 Mains signalling equipment: specification for communication and interference limits and measurements**

[66]   **EN 55 011: Limits and methods of measurement of radio disturbance characteristics of industrial, scientific and medical (ISM) radio-frequency equipment**

[67]   **EN 55 014: Limits and methods of measurement of radio disturbance characteristics of electrical motor-operated and thermal appliances for household and similar purposes, electric tools and electric apparatus**

[68]   **EN 55 022: 1998: information technology equipment – radio disturbance characteristics – Limits and methods of measurement**

[69]   **EN 50 081: Electromagnetic compatibility – generic emission standard**

         Published as part 1:1992 for the residential, commercial and light industrial environment, and part 2: 1993 for the industrial environment

[70]   **EN 50 082: Electromagnetic compatibility – generic immunity standard**

         Published as part 1:1997 for the residential, commercial and light industrial environment, and part 2: 1995 for the industrial environment

[71]   **EN 50 160: 1994: Voltage characteristics of electricity supplied by public distribution systems**

[72]   **EN 50 173: 1995: Information technology – Generic Cabling Systems**

[73]   **prEN 50 174-2: Information Technology – Cabling installation Part 2: Installation planning and practices inside buildings**

         BSI document no 98/637058 DC

[74]   **prEN 50 310: Application of equipotential bonding and earthing at premises with information technology equipment**

         BSI document no 98/637068 DC

[75]   **EN 60 950: 1992: Safety of information technology equipment**

[76]   **ETS 300 127: 1994: Equipment engineering (EE): Radiated emission testing of physically large systems**

         (Draft revision 1.2.1 circulated 12th Nov 98, BSI document no 98/660704 DC)

[77]   **ETS 300 253: 1995 : Earthing and bonding of telecommunication equipment in telecommunication centres**

[78]   **HD 384-3 S2: 1995 : Electrical installations of buildings Part 3 Assessment of general characteristics of installations (IEC 60364-3:1993 modified)**

         Details the various power systems (TN-S, TN-C, TT, and IT).

IEC, CISPR and ISO standards: available from BSI as above or direct from

         IEC Central Office
         1 Rue de Varembé
         1211 Geneva 20
         Switzerland
         Tel (022) 340150; Telex 28872 CEIEC-CH; Telefax (022) 333843

Note: draft standards are included in this listing with their corresponding BSI document number. These should not be regarded in the same light as a published standard.

In 1997 the IEC changed their standards numbering system to add a "6" in front of the original four-figure number; so that e.g. IEC 60096-1 is equivalent to IEC 96-1

[79]   **IEC 50 (161): (BS4727 : Pt 1 : Group 09) Glossary of electrotechnical, power, telecommunication, electronics, lighting and colour terms: Electromagnetic compatibility**

[80]   **IEC 96-1: 1986, Amendment 1 March 1988: Radio frequency cables Part 1: General requirements and measuring methods**

[81]   **IEC 364-4-444:1996 : Electrical Installations of Buildings – Part 4: Protection for safety – Chapter 44: Protection against overvoltages – Section 444: Protection against electromagnetic interference (EMI) in installations of buildings**

[82]   **IEC 364-5-548 Section 548 : Electrical installations of buildings – Part 5: Selection and erection of electrical equipment – Chapter 54: Earthing arrangements and protective conductors – Section 548: Earthing arrangements and equipotential bonding for information technology equipment**

[83]   **IEC 801: (BS6667) Electromagnetic compatibility for industrial-process measurement and control equipment**

IEC 61000: Electromagnetic compatibility

(This listing does not cover all parts of IEC 61000 but only those which are particularly relevant to this book)

[84]   **Draft IEC 61000-1-2: General – Methodology for the achievement of the functional safety of electrical and electronic equipment**

Committee Draft 77/197/CD, November 1997

[85]   **IEC 61000-2-1 (1990): Description of the environment – electromagnetic environment for low-frequency conducted disturbances and signalling in public power supply systems**

Equivalent to BS7484:1991

[86]   **IEC 61000-2-2 (1990): Compatibility levels for low-frequency conducted disturbances and signalling in public low-voltage power supply systems**

Equivalent to DD ENV61000-2-2:1993 with certain CENELEC modifications

[87]   **IEC 61000-2-5 (1995): Classification of electromagnetic environments**

[88]   **IEC 61000-3-2: 1995: EMC – Part 3 Limits: Section 2: Limits for harmonic current emissions (equipment input current ≤ 16A per phase)**

[89]   **IEC 61000-4-2: 1995: EMC – Part 4 – Testing and measurement techniques – Electrostatic discharge immunity test**

[90]   **Draft amendment to IEC 61000-4-2: Definition of accessible parts under normal operating conditions**

IEC 77B/231/CD, January 1998, BSI document no 98/260317 DC

[91]   **IEC 61000-4-3: 1995: EMC – Part 4 – Testing and measurement techniques – Radiated radio frequency electromagnetic field immunity test**

[92]   **IEC 61000-4-4: 1995: EMC – Part 4 – Testing and measurement techniques – Electrical fast transient/burst immunity test**

[93]   **IEC 61000-4-5: 1995: EMC – Part 4 – Testing and measurement techniques – Surge immunity test**

[94]   **IEC 61000-4-6: 1996: EMC – Part 4 – Testing and measurement techniques – Immunity to conducted disturbances induced by radio frequency fields**

[95]   IEC 61000-4-8: 1993: EMC – Part 4 – Testing and measurement techniques – Power
        frequency magnetic field immunity test

[96]   IEC 61000-4-11: 1994: EMC – Part 4 – Testing and measurement techniques – Voltage
        dips, short interruptions and voltage variations immunity test

[97]   IEC 61000-5-2: 1997 : Part 5: Installation and Mitigation Guidelines – Section 2:
        Earthing and cabling

[98]   Draft IEC 61000-5-7: Degrees of protection against electromagnetic disturbances
        provided by enclosures (EM code)
        IEC 77C/54/CD, BSI document no 98/260339 DC

[99]   IEC 61024-1: 1993: Protection of structures against lightning – Part 1: General
        principles

[100]   IEC 61312-1: 1995: Protection against lightning electromagnetic impulse – Part 1:
        General principles

[101]   IEC 61508-1: 1998: Functional safety of electrical/electronic/programmable electronic
        safety-related systems – Part 1: General requirements

[102]   Automation and control apparatus for generating stations and high-voltage sub-stations
        – Electromagnetic Compatibility – Immunity requirements
        IEC New Work Proposal 77/199/NP, October 1997

        CISPR ##: see EN 55 0## above

[103]   CISPR 16-1: 1993 Specification for radio disturbance and immunity measuring
        apparatus and methods: Part 1: Radio disturbance and immunity measuring apparatus

[104]   CISPR 16-2: 1996 Specification for radio disturbance and immunity measuring
        apparatus and methods: Part 2: Methods of measurement of disturbances and immunity

[105]   Draft Amendment to Subclause 4.1 of CISPR 16-2: Emission measurement in the
        presence of ambient signals
        CISPR/A/202/CD, July 1997, BSI document no 97/260651 DC

[106]   CISPR 17: 1981 Methods of measurement of the suppression characteristics of passive
        radio interference filters and suppression components

[107]   Proposed amendment to CISPR 22: Clause 10.7 User installation testing
        CISPR/G/154/CD, BSI document no 98/264577 DC

[108]   ISO 11452-4:1995, Road vehicles – Component test methods for electrical disturbances
        by narrowband radiated electromagnetic energy – Part 4: Bulk current injection (BCI)

[109]   ISO/IEC 11801:1995, Information technology – Generic cabling for customer premises

Other British Standards

[110]   BS5049:1994 (CISPR 18) Radio interference characteristics of overhead power lines and
        high voltage equipment

[111]   BS6651:1992 Code of practice for protection of structures against lightning

[112]   BS6656:1991 Guide to prevention of inadvertent ignition of flammable atmospheres by
        radio-frequency radiation

[113]   BS6657:1991 Guide for prevention of inadvertent initiation of electro-explosive devices
        by radio-frequency radiation

[114]   BS7671:1992 Requirements for electrical installations (The IEE Wiring Regulations)

[115]   BS7697: 1993 (HD 472 S1) Nominal voltages for low voltage public electricity supply
        systems

[116]   DEF STAN 59-41 Part 1 Section 2: Guide to the specification and selection of EMC
        requirements
        Ministry of Defence 29th September 1995

[117]   DEF STAN 59-41 Part 3: Technical Requirements, Test Methods and Limits
        Ministry of Defence 6th October 1995 (Issue 5)

[118]   DEF STAN 59-41 Part 3: Supplement H: Test Method DCS02, Conducted Susceptibility,
        Power, Control and Signal Lines 50kHz–400MHz
        Ministry of Defence 6th October 1995 (Issue 5)

The EMC Directive

[119]   Council Directive of 3rd May 1989 on the approximation of the laws of the Member
        States relating to Electromagnetic Compatibility (89/336/EEC)
        Official Journal of the European Communities No L 139, 23rd May 1989

[120]   Council Directive 92/31/EEC of 28th April 1992 amending Directive 89/336/EEC
        Official Journal of the European Communities No L 126/11, 12th May 1992

[121]   Guidelines on the application of Council Directive 89/336/EEC of 3 May 1989 on the
        approximation of the laws of the Member States relating to Electromagnetic
        Compatibility
        Commission of the European Communities, Brussels, July 1997 (distributed electronically)

[122]   Council Directive 93/68/EEC of 22 July 1993 amending [various Directives]
        Official Journal of the European Communities No L 220/1, 30 August 1993

[123]   Commission communication in the framework of the implementation of the Council
        Directive 89/336/EEC – December 1998: Publication of titles and references of
        harmonized standards under the directive
        Official Journal of the European Communities No 1999/C57/02, 27th February 1999

[124]   The Electromagnetic Compatibility Regulations
        Statutory Instrument 1992 No. 2372, 28th October 1992, HMSO;
        amended by

        The Electromagnetic Compatibility (Amendment) Regulations 1994
        Statutory Instrument 1994 No. 3080, 30th December 1994, HMSO

[125]   Electromagnetic Compatibility: Guidance document on the preparation of a Technical
        Construction File as required by EC Directive 89/336
        Department of Trade & Industry, October 1992

[126]   Minimising the cost of meeting the EC Directive on Electromagnetic Compatibility
        (EMC) 89/336/EEC as amended by EC Directive 92/31/EEC
        Department of Trade & Industry, August 1995

Other official publications and standards

[127]   ITU-T handbook : Earthing of telecommunication installations (Geneva 1976)

[128]   ITU-T Recommendation G.117: 1996 : Transmission aspects of unbalance about earth

[129]   ITU Recommendation K.27 (1996) : Bonding configurations and earthing within a telecommunications building

[130]   ITU-T Recommendation K.31 : Bonding configurations and earthing of telecommunications within a subscriber's building

[131]   ITU Recommendation K.33 : Protection of telecommunication lines against harmful effects from electric power and electrified railway lines

[132]   ITU Recommendation K.35 (1996) : Bonding configurations and earthing at remote electronic sites

[133]   ITU Recommendation K.39 : Risk of damages to telecommunication sites due to lightning discharges (1996)

[134]   The Treatment of Uncertainty in EMC Measurements
        NAMAS Information Sheet NIS 81, NAMAS Executive, NPL, Teddington, May 1994

[135]   Board Statement on Restrictions on Human Exposure to Static and Time Varying Electromagnetic Fields and Radiation
        Documents of the NRPB Volume 4 No 5 1993, National Radiological Protection Board, Chilton 1993

[136]   Review of Occupational Exposure to Optical Radiation and Electric and Magnetic Fields with Regard to the Proposed CEC Physical Agents Directive
        NRPB R265, National Radiological Protection Board, Chilton 1994

[137]   Guidance to the certification of aircraft electrical/electronic systems for operation in the high intensity radiated field (HIRF) environment
        EUROCAE WG33 Subgroups 2 & 3, January 1992

US publications

[138]   FCC Rules Part 15 – Radio Frequency Devices
        FCC, August 1997

[139]   Methods of Measurement of Radio-Noise Emissions from Low-Voltage Electrical and Electronic Equipment in the Range of 9kHz to 40GHz
        ANSI C63.4-1992

[140]   IEEE Recommended Practice on Surge Voltages in Low-Voltage AC Power Circuits
        IEEE C62.41-1991, IEEE, October 1991

[141]   Interface between Data Terminal Equipment and Data Circuit-Terminating Equipment Employing Serial Binary Data Interchange
        ANSI/EIA-232-D-1986, January 1987

[142]   Telecommunications earthing networks
        ANSI/TIA/EIA-607

[143]   Telecommunications cabling systems
        ANSI/TIA/EIA-568A

Index